Sustainability in Fashion: Creativity, Education, and Innovation

Alexandra Cruchinho • José Carlos Neves
Editors

Sustainability in Fashion: Creativity, Education, and Innovation

Proceedings of FL_Fashion & Sustainability International Conference, 2022

Springer

Editors
Alexandra Cruchinho
ECATI
Lusófona University - CICANT
Lisbon, Portugal

José Carlos Neves
Lusófona University - CICANT
Lisbon, Portugal

ISBN 978-3-031-80990-3 ISBN 978-3-031-80991-0 (eBook)
https://doi.org/10.1007/978-3-031-80991-0

This Springer imprint is published by the registered company Springer Nature Switzerland AG
The registered company address is: Gewerbestrasse 11, 6330 Cham, Switzerland

Preface

Sustainability in Fashion: Creativity, Education, and Innovation—Proceedings of FL_Fashion & Sustainability International Conference, 2022, is a book that follows the Fashion & Sustainability—FL International Conference and presents a set of chapters that deal with this topic.

Fashion & Sustainability—FL International Conference is an international conference that seeks to focus on the discussion of Sustainability in the areas of Fashion, Design and Media.

Increasingly, it is important to discuss Sustainability in all its variants, at an economic, environmental and social level, especially when the focus is on areas such as fashion, whose industry is one of the biggest responsible for the environmental damage that has been increasingly occurring over recent times.

If, on one hand, the topic of sustainability takes us to the environmental aspect, it is emerging to discuss this topic from a social perspective where minorities are involved in important processes for communities, where knowledge, values are valued and are included in active life, citizens, which, from the outset, will be able to maintain a much more discreet experience in the environment that surrounds them.

Economic sustainability is crucial to the success of business, brands, industry, small and medium-sized companies and is also an important area to keep in focus in the discussion.

However, Sustainability and Circularity are very rich themes that still herald many discoveries to be revealed. Portugal is one of the countries at the forefront of ITV's investment in sustainability. The projects that emerge whether by stakeholders, by the Academy, or even by the business community are important to share in order to enrich the knowledge of the environment.

The FL_Fashion Sustainability—International Conference aims to be a space for union between Industry, the business community and Academia around a common theme, however, with several very different approaches and with an exchange of very diverse experiences and knowledge.

Regarding the specific theme of this book, the question of sustainability, in fashion, is a subject that still has room for discussion in the search for solutions, with a focus on its practicality. The healthy relationship between fashion and teaching aims to find ideas, study them, investigate them and apply them in everyday dynamics. This book brings together discussions about these synergies and how they can find space in the dynamics of the fashion sector today.

The book has a total of 31 chapters that address Sustainability in Fashion with a focus on Creativity, Education and Innovation.

Lisbon, Portugal

Alexandra Cruchinho
José Carlos Neves

CICANT: Centre for Research in Applied Communication, Culture and New Technologies

This book arises from the strategy of the research center that produces the FL_ Fashion Sustainability—International Conference, the CICANT—Centre for Research in Applied Communication, Culture and New Technologies.

The Centre for Research in Applied Communication, Culture, and New Technologies (CICANT) results from the combined efforts of Lusófona University (Lisbon and Porto) researchers in the field towards the goal of creating an autonomous research unit which will allow for the integration and harnessing of the research activities conducted within the centre. It promotes theoretical and applied research at the cross of media, society, literacies, arts, culture and technologies. CICANT distinguishes itself due to the solid theoretical and applied nature of its research and knowledge transfer activities oriented to teaching and research activities.

CICANT is geared towards innovative and interdisciplinary research in its two main subject areas: Media, Society and Literacies (MSL) and Media Arts, Creative Industries and Technologies (MACIT).

Created in 2004, in the past years, the Centre has been the leader of over 40 national and international projects supported by several funding European programmes, such as Erasmus +, FP7, ISF, H2020 and Media Programme. Also at national level, the Centre has several projects funded by national funds, such as FCT, Calouste Gulbenkian Foundation and the EEA Grants.

https://cicant.ulusofona.pt/

Organization

Organizing Committee

Alexandra Cruchinho – General Chair
José Carlos Neves – Co-Chair

Anna Coutinho
Benilde Reis
Catarina Rito
João Barata
Luís Sanchez
Sara Vaz
Mónica Lopes
Mafalda Soares
Margarida Santos
Ruiana Fraga
Sofia Fernandes

Honor Committee

Manuel José Damásio – UL-ECATI
José Gomes Pinto – UL-ECATI
Fernando Moreira da Silva – FAUL

Scientific Committee

Contents

Sustainability Importance in Co-creation Fashion Brands

Paulo Martins, Rui Miguel, Liliana Pina, and Cláudia Pinheiro

Contextualization

According to Fletcher (2008), the fashion industry creates positive advantages because it generates products that are an asset for the culture as it simultaneously produces wealth and employment. However, many consumers buy clothes addictively, so the environment and the society end up being impacted by the fashion industry (Pero et al., 2020). In part, because the high consumption of materials is due to new trends being launched and because of the changes during the fashion cycles that are supported by the evident insatiability of the consumer's needs. So, it is obvious that the connection between consumption and fashion is not in line with sustainable objectives.

It is also a certainty that fashion design and the fashion industry faced negative environmental and social impacts in the last decades (Fletcher, 2008). In effect, the clothing and textile social impact has been widely examined (Östlund et al., 2020). From production to product consumption, use, and product disposal, sustainability is one of the major problems that the fashion industry faces today (Fletcher, 2008). So, concerning environmental and social issues, this contemporary frame of mind leads fashion brands and designers to be more thoughtful when developing new products (Mizrachi and Tal, 2022).

It is a reality that some of the longest and more complex supply chains with a large number of suppliers and subcontractors produce fashion products and textiles (Fletcher, 2008; Östlund et al., 2020). But the position of the fashion industry has

P. Martins (✉) · L. Pina
Instituto Superior Manuel Teixeira Gomes, Portimão, Portugal
e-mail: paulo.martins@ismat.pt; liliana.pina@ismat.pt

R. Miguel · C. Pinheiro
Universidade da Beira Interior, Covilhã, Portugal
e-mail: rmiguel@ubi.pt; cisp@ubi.pt

A. Cruchinho, J. C. Neves (eds.), *Sustainability in Fashion: Creativity, Education, and Innovation*, https://doi.org/10.1007/978-3-031-80991-0_1

also changed over the last years, as brands and manufacturing companies are starting to recognize their responsibilities economically, environmentally, and socially, and viewing sustainability as an opportunity for their processes of marketing, sourcing, or manufacturing (Pero et al., 2020). According to Pookulangara and Shephard (2013), sustainability is increasingly becoming more important for the fashion industry in what concerns their operations and its workforce, and consequently, their strategy to reach the consumers.

But what is sustainability in its pure state? In order to improve well-being and culture and generate economic value, Pal (2017) defines sustainability as a form of design aimed to reduce environmental impacts based on renewable resources and ecoefficiency. Simultaneously, McDonough and Braungart (2002) mention the importance of reaching new ways to think about product design and production processes. As Granskog and Lee (2020) explain, a brand that promotes sustainability increases the purchasing factor. So, new demands from consumers will change brands, as fully sustainable brands will be on-demand (Granskog et al., 2017). However, even though products are produced in a sustainable way, the way fashion products are consumed might be unsustainable. So, variables that can reduce the product lifespan must be avoided. Instead, designing products that could be repaired and upgraded, as well as improving functionality and product quality, must be considered as a solution to optimize the product life cycle (Karaosman, 2016).

ATP, for example, suggests for the Portuguese Textile and Garment Sectors, till 2030, environmental sustainability, either for products, materials, or services (Pamésa et al., 2021). The main sustainable goals are related to:

- Eco-friendly fashion with the resource of digital technologies and new materials for developing new products.
- Companies' ethics need to be defended and worked on social media.
- New ways of advertising and selling changed due to the Covid-19 crisis and need to become more digital.
- Omnichannel 2.0 with a new relationship between online and offline channels for the demand and offer.
- E-commerce and marketplaces for a broader model of reaching consumers.

Also, design and manufacturing processes that incorporate fair working conditions, nearshore production, small lines, and high quality are related to companies that also tend to be more active in terms of being more sustainable, transparent, and have social responsibility (Pookulangara & Shephard, 2013).

Additionally, consumers are now more conscious about their consumption decisions toward sustainable fashion products (Wang et al., 2022). A McKinsey survey made in Germany, Spain, and the United Kingdom—that collected 1050 responses—found, for example, that 66% of consumers would not, or almost do not, make any purchases at a brand that does not treat fairly their employees or even suppliers (Amed et al., 2021). In fact, a market for ethical products and sustainable products emerged due to the ever-higher consumer awareness about social and environmental impacts of fashion products on the society and environment (Dissanayake et al., 2017). Even though brands act nowadays with more transparency, the complexity of supply chains makes it difficult for consumers to track where and under what conditions the garments are produced. So, it is a requirement that fashion brands are committed to embracing new relationships between products and consumers and have traceability of all the processes that are vital for a better relationship with the brand (Pamésa et al., 2021).

Consumer needs relate mainly to self-actualization; i.e., consumers seek realization of their highest potential (Fletcher & Tham, 2014). They also search for self-fulfillment, personal growth, and emotional experiences (Maslow, 1970; Desmet & Fokkinga, 2020). So, co-creation can provide consumers with all of those, as it is viewed as a path to innovation through consumer participation for the customization of products or services (Voorberg et al., 2017). By integrating consumers' values into the companies' marketing and retail (Maltzahn, 2016) strategies, the consumer can be integrated into the value creation chain and go beyond the consumption of products by gaining an emotional involvement with the brand's philosophy of sustainability. This involvement can also generate more satisfaction and can enable the creation of more meaningful garments for consumers that, in turn, will only be disposed of or replaced later (Mcgrath, 2012; López-Navarro et al., 2014). The possibility of consumers participating in co-creation processes can create a good relationship between the brand and the consumers because it "develops brand trust through positive attitude orientation by enhancing customer repeat purchase desire and brand loyalty" (Chen et al., 2022, p. 17).

Thus, instead of a process of uniformization, through customization and personalization of products, a process of individualization occurs (Carvalhal, 2016). This method allows companies to find better and more sustainable ways to maintain a relationship with consumers and consequently to make a positive impact on the society and environment (López-Navarro et al., 2014). Brands such as Awaytomars and Threadless have been developing co-created products (e.g., clothing and accessories) inside their communities (Martins et al., 2020), but it is not a certainty what consumers desire and expect from co-creation as a slow fashion process and regarding sustainability attributes of the creative and production processes.

So, the main objective of this chapter is to understand what consumers desire and expect from co-creation when confronted with sustainable concerns. Regarding the main objective, a research question can be proposed:

Q1.: What are the co-creation negative factors regarding sustainability that can lead to a loss of interest in the product?

Methodology

The methodology of the chapter is based on two methods that complement each other. Based on a qualitative methodology, the first method focuses on the contextualization of the theme through the reading and analysis of books, scientific articles, theses, and reports in the scope of sustainability and co-creation in fashion design.

The second method of a quantitative nature focuses on the analysis of data from a questionnaire of a sample of 404 valid responses from Portugal. In this chapter, to study the consumers, a positivist approach was applied. A questionnaire with three closed questions, apart from the questions meant to characterize the sample, was created, being based on a literature review and the question of study. To reach a larger number of respondents, the questionnaire was done online. The gathering process was based on non-probabilistic and snowball techniques. This was also the easier way to control the data collection process that was done in Portugal (Sampieri et al., 2013).

Instruments Used for Data Collection

Data collection was done from October 27, 2021, to 4 January, 2022, through email to personal contacts, through social networks in groups dedicated to academic research and also websites and through the University of Beira Interior. The respondents should have already co-created a fashion product or would like to co-create a fashion product in the future and needed to be between 18 years and 64 years. In this sense, the first question asked to the participants was if they had or had not co-created a fashion product. In the event of answering "No," a second question was made for the author to fully assess if the participants really wished to co-create a fashion product. Participants were questioned if they like or would like to participate in the co-creation of fashion products. The respondents could state their opinion about the six topics, through a Likert scale, which were if the rewards for participating interested them, if they would like to feel more satisfied with the products available in the market, if they participate or would participate to express themselves and feel proud of what they are capable of creating, if they participate to gain new knowledge about fashion products and their co-creation, if it is for feeling curious in exploring the co-creation process, or if it is because it increases their social status in the brands' community. Some respondents answered question 2 with low values on all six topics and also responded "No" to question 1, revealing that they also did not intend to participate in the co-creation of fashion products in the future. So, they did not meet the criteria to be eligible for participating in the study. It was the case in 4 answers that were not considered valid. For the questionnaire, a dichotomous scale was used for question 1, and a Likert scale from 1 to 7 was used for questions 2 and 3.

Sociodemographic Characterization of the Sample

Tables 1 and 2 present the sample characterization.

Table 1 Sample characterization

Characterization	Items	(N)	(%)
Gender	Female	321	79
	Male	83	21
Age	18–24	162	40
	25–34	111	28
	35–44	74	18
	45–54	49	12
	55–64	8	2
Level of education	Doctoral	31	8
	Master's	95	23
	Bachelor's	181	45
	High School	91	22
	Middle School	4	1
	Other	2	1
Employment situation or occupation	Employed	192	47
	Self-employed	37	9
	Unemployed	17	4
	Student	156	39
	Retired	2	1
Field of work	Arts and design	95	24
	Textiles and garment industry	78	19
	Administration, business, and services	74	18
	Social and human sciences	31	8
	Science and technology	20	5
	Health and well-being	20	5
	Communication and information	11	3
	Engineering and production	10	2
	Other	65	16

Data Analyses and Discussion

Descriptive statistics were used for the demographic and social data and for question 1, and IBM SPSS Statistics was used for question 2 and for question 3 to create a mean and standard deviation analysis. Considering the question, "What are the co-creation negative factors, regarding sustainability that can lead to a loss of interest in the product?," consumers gave higher importance to knowing that the brand does not have social ethics. Also important was not knowing where or how the product was produced and finally, knowing that the brand and its products are not sustainable (Table 2).

So, considering the data obtained (Table 3), knowing that the brand and its products are not sustainable is also important, and according to Granskog et al. (2017), new demands from consumers will change brands, as fully sustainable brands will be on-demand, and also, Granskog and Lee (2020) consider that a brand that promotes sustainability increases the purchasing factor. So, sustainability's importance is also visible in co-creation and should be taken into high account side by side with social ethics and traceability, as it is referred by Dissanayake et al. (2017), when

Table 2 Cont. Sample characterization

Characterization	Items	(N)	(%)
Net monthly income	Less than 635€	41	10
	636–799€	67	17
	800–999€	38	9
	1000–1499€	68	17
	1500–1999€	29	7
	More than 2000€	26	6
	I do not have a monthly income	135	34
Residence	City	266	66
	Town	79	19
	Village	59	15
County	Açores	8	2
	Aveiro	18	4
	Braga	27	7
	Bragança	6	1
	Castelo Branco	116	29
	Coimbra	14	3
	Évora	6	1
	Faro	9	2
	Guarda	47	12
	Leiria	13	3
	Lisboa	59	15
	Madeira	2	1
	Portalegre	2	1
	Porto	30	8
	Santarém	6	1
	Setúbal	14	3
	Viana do Castelo	4	1
	Vila Real	4	1
	Viseu	19	5

Table 3 Mean and standard deviation analysis

	N	Mean	Std. deviation
Knowing that the brand and its products are not sustainable	404	54,554	152,904
Not knowing where or how the product was produced	404	56,139	139,635
Knowing that the brand does not have social ethics	404	61,658	125,020

saying that a market for ethical products and sustainable products emerged due to the ever higher consumer awareness about social and environmental impacts of fashion products on the society and environment. And also that traceability of all the processes are vital to a better relationship with the brand as indicated by Pamésa et al. (2021). So, interest in co-creation will also be dependent on its transparency and communication regarding the brands' level of sustainability and on social ethics and product traceability.

Conclusion

Co-creation success is largely based on the consumers' interest and participation in the process, but as discussed, consumers understand that co-creation brands success will also be dependent on their transparency and communication regarding the brands' level of sustainability and of their social ethics and product traceability. These are seen as negative factors for the brand when not considered.

Future studies should explore the possibility of building a framework that includes the findings of this chapter among other important aspects that can define and improve co-creation fashion brands so that this could be applied in the development of brands or improvement of the methods on already established fashion brands.

Endnotes
The authors would like to express their gratitude the support given by the research unit Fiber Materials and Environmental Technologies (FibEnTech) on the extent of the project reference UIDB/00195/2020, funded by the Fundação para a Ciência e a Tecnologia (FCT) and to the Research Unit in Business Sciences (NECE) at the University of Beira Interior.

Bibliography

Amed, I. et al. (2021) *The state of fashion 2021*. McKinsey & Company. Available at: https://www.mckinsey.com/industries/retail/our-insights/state-of-fashion

Carvalhal, A. (2016). *Moda com propósito* (1st ed.). Paralela.

Chen, L., et al. (2022). Research on influencing mechanism of fashion brand image value creation based on consumer value co-creation and experiential value perception theory. *Sustainability, 14*(13), 7524. https://doi.org/10.3390/su14137524

Desmet, P., & Fokkinga, S. (2020). Beyond maslow's pyramid: Introducing a typology of thirteen fundamental needs for human-centered design. *Multimodal Technologies and Interaction, 4*(3), 1–22. https://doi.org/10.3390/mti4030038

Dissanayake, D. G. K., Perera, S., & Wanniarachchi, T. (2017). Sustainable and ethical manufacturing: A case study from handloom industry. *Textiles and Clothing Sustainability, 3*(1). https://doi.org/10.1186/s40689-016-0024-3

Fletcher, K. (2008). *Sustainable fashion and textiles: Design journeys*. Earthscan.

Fletcher, K., & Tham, M. (2014). *Routledge handbook of sustainability and fashion*. https://doi.org/10.4324/9780203519943

Granskog, A., & Lee, L. (2020). *The fashion industry can reduce emissions across the entire value chain*. Available at: https://www.mckinsey.com/business-functions/sustainability/our-insights/sustainability-blog/this-surprising-change-can-help-the-auto-industry-tackle-emissions-goals

Granskog, A., Lobis, M., & Magnus, K.-H. (2017) *The future of sustainable fashion*. McKinsey & Company. Available at: https://www.mckinsey.com/industries/retail/our-insights/the-future-of-sustainable-fashion

Karaosman, H. (2016). *Sustainability in fashion. The Factbook*.

López-Navarro, M., Lozano-Gómez, C., & Mcgrath, A. (2014). Co-creation experiences as the basis for value creation in the sustainable fashion industry. In *Customer experience management: Enhancing experience and value through service management* (pp. 133–152).

Martins, P. et al. (2020). Co-creation fashion brands: A case study. *Global fashion conference*. Lyon. Available at: http://gfc-conference.eu/wp-content/uploads/2020/11/MARTINS-ET-AL_Co-creation-fashion-brands-a-case-study.pdf

Maslow, A. H. (1970). *Motivation and personality*. Harper and Row.

McDonough, W., & Braungart, M. (2002). *Cradle to Cradle: Remaking the way we make things*. North Point Press.

Mcgrath, A. (2012). Fashioning sustainability: How the clothes we wear can support environmental and human well-being. In *Sustainable fashion* (pp. 1–24).

Mizrachi, M. P., & Tal, A. (2022). Regulation for promoting sustainable, fair and circular fashion.

Östlund, Å. et al. (2020). *Investor brief: Sustainability in textiles and fashion*.

Pal, R. (2017). Sustainable design and business models in textile and fashion industry. In *Textile science and clothing technology* (pp. 109–138). https://doi.org/10.1007/978-981-10-2639-3_6

Pamésa, C. et al. (2021). *Visão Prospetiva e Estratégias ITV. Contributo para um Plano Estratégico para o Setor Têxtil e Vestuário Português até 2030*.

Pero, M., Arrigo, E., & Fionda-Douglas, A. (2020). Sustainability in fashion brands. *Sustainability, 12*(14), 10–13. https://doi.org/10.3390/su12145843

Pookulangara, S., & Shephard, A. (2013). Slow fashion movement: Understanding consumer perceptions-an exploratory study. *Journal of Retailing and Consumer Services, 20*(2), 200–206. https://doi.org/10.1016/j.jretconser.2012.12.002

Sampieri, R., Collado, C., & Lucio, M. (2013). *Metodologia de Pesquisa*. Quinta Edi. Porto Alegre: Penso.

von Maltzahn, C. F. (2016). *Co-creating individuals: A roadmap to value creation in fashion retailing*. Available at: https://citeseerx.ist.psu.edu/viewdoc/download?doi=10.1.1.527.3521&rep=rep1&type=pdf

Voorberg, W. H., Bekkers, V. J. J. M., & Tummers, L. G. (2017). A systematic review of co-creation and co- production: Embarking on the social innovation journey. *Public Management Review, 17*(9), 1333–1357. https://doi.org/10.1080/14719037.2014.930505

Wang, L., et al. (2022). Preferred product attributes for sustainable outdoor apparel: A conjoint analysis approach. *Sustainable Production and Consumption, 29*, 657–671. https://doi.org/10.1016/j.spc.2021.11.011

Clothing Transformation Platform: A Vehicle for Upcycling

Bárbara da Silva Barrigana Bule

Introduction

Concern about the circular economy, especially sustainability, is in vogue. Consumers and organizations are more concerned about the environment, which includes the expenditure and destruction of natural resources, especially those that are scarce, difficult to renew, and even finite. Addressing this concern, upcycling emerges as a proposal for the reformulation of garments created and sold by brands, avoiding waste and saving the environment.

According to Vezzoli et al. (2022), the product life cycle encompasses pre-production, with the purchase of raw materials; production, with the joining of components and respective finishing; distribution, with packaging, transport, and storage; usage, with the use of that product and its maintenance; and disposal, where it goes through different paths until reaching the end of its useful life (p. 134).

Waste with clothing and fabrics needs to stop transmitting an image that is still considered negative and instead reach consumers as new raw materials (Marques et al., 2019). The fashion industry has been calling for a constantly changing sustainable development, which calls for creativity on the part of designers and brands (Minh & Ngan, 2021).

Online platforms, considered digital marketplaces, unite buyers and sellers who meet for transactions. With the Covid-19 pandemic, new opportunities have emerged in the digital environment for the fashion industry.

Therefore, it is proposed to combine upcycling with online platforms, facilitating the delivery service of clothing that no longer uses the brands directly, through the digital medium.

B. da Silva Barrigana Bule (✉)
Universidade da Beira Interior/IADE, Covilhã, Portugal

A. Cruchinho, J. C. Neves (eds.), *Sustainability in Fashion: Creativity, Education, and Innovation*, https://doi.org/10.1007/978-3-031-80991-0_2

Upcycling and the Circular Economy

The fashion industry has a high impact on the environment, so the circular economy becomes essential. It provides sustainable production through cleaner technologies and promotes the recycling of existing ones, which can be done through upcycling (Marques et al., 2019). According to the circular economy, natural resources extracted from the environment must be continuously in circulation, avoiding the creation of new waste (Fernandes et al., 2018).

To avoid environmental degradation, it is possible to follow three principles: reduce, reuse, and recycle. These lead to the field of reverse logistics, where discarded products are collected. Subsequently, a sorting is carried out where they are classified. Finally, they arrive at the processing stage, where they undergo activities such as washing, repair, and others that allow restoring and assigning a new use (Paras & Curteza, 2018).

Subsequently, according to Marques et al. (2019), William McDonough and Michael Braungart present a new R, rethink or redesign, which leads to ecological design, connected with sustainability. In 2015, the European Commission launched a "Circular Economy Package," with the hierarchy of multi-Rs: rethink, redesign, repair, re-manufacture, redistribution, and recovery (Fernandes et al., 2018; Fernandes, 2020).

Luxury brands have begun to present ethical and sustainable practices and products. The LVMH group created the LIFE 360 program, which focuses on the complete product life cycle, obtaining durable and repairable goods (Park et al., 2022). The Kering group, a direct competitor, had already implemented a circular approach and argues that upcycling consists of transforming used materials into high-quality products (Kering, 2022).

Designers are also starting to take initiatives in the luxury context. Duran Lantik, a Dutch designer who was shortlisted for the LVMH Award 2019 competition, created his autumn/winter 2021 collection by reinventing clothes from brands such as Louis Vuitton, Balenciaga, Balmain, Prada, and Gucci (L'officiel, 2022).

According to some studies, luxury and sustainability are compatible, since luxury itself conveys high quality, durability, and scarcity, favoring moderate consumption (Park et al., 2022).

Clothing reform still encounters barriers as we only verify its existence in small confections and some laundries, in addition to being done in a home environment (Kim et al., 2020).

For the collection of products to be processed, it is important that the points are close to the consumers. There is a preference for doing it close to home or in convenience stores and gas stations (Paras & Curteza, 2018).

Despite the growing concern of luxury consumers with ethical values, communication, persuasion, and education efforts are needed so that consumption, especially sustainable luxury, becomes a trend (Park et al. 2022).

The Online Platforms

The platform concept is not new. It started with the Persian bazaar and the Roman markets, where physically buyers and sellers carried out transactions. Currently, these platforms have moved to shopping centers, where transactions take place in physical stores (Montenegro, 2021).

The platforms stimulate the global economy, not only with new products and services but also with the stimulation of innovation and even improve the economy in different sectors. They relate several companies and activities, present a less formal and hierarchical structure when compared to companies, and are more united when compared to traditional markets (Kretschmer et al., 2020).

A platform combines features found in traditional markets and organizations. However, the investment is not in production, but in infrastructure and tools that promote a "networked market or community" (Montenegro, 2021). They offer valuable resources like information about what customers want and feedback. Something positive are the incentives. While organizations always pay the same low salary directly, platforms, although the compensation is indirect, are more attractive and occur in transactions between participants. Every involved party creates value among themselves and collects part of the value (Kretschmer et al., 2020).

In order for platforms to work and be successful, there must be coordination, cooperation, and involvement between the various participants, who do not have the same goals and interests (Kretschmer et al., 2020).

The Covid-19 pandemic has driven an acceleration of technology adoption, responding to consumer needs. Thus, brands must also follow these changes to the same extent that consumers do (de Villiers, 2020).

Despite the pandemic having negatively impacted the fashion industry, companies have brought their activities to technological platforms, becoming a solution for the present and the future. Brands begin to explore new technologies for their marketing, production, and business development solutions (Minh & Ngan, 2021).

Case Studies: LIFE 360: LVMH and the Circular Approach of the Kering Group

LIFE 360, from the LVMH group, appears in May 2021 as a sustainability plan (Cernansky, 2022). It aims to establish a link between the group and nature in the creation of products, contributing to biodiversity and the fight against climate change. Several objectives have been established, and all brands in the group must contribute by creating their own action plans (LVMH, 2021).

To achieve the goals, they point out four priority strategies (Fig. 1), which promote a "bold new vision of luxury" (LVMH, 2021). One of the goals is to earn 25% profits from circular services, instead of creating new products, by 2030 (Cernansky, 2022).

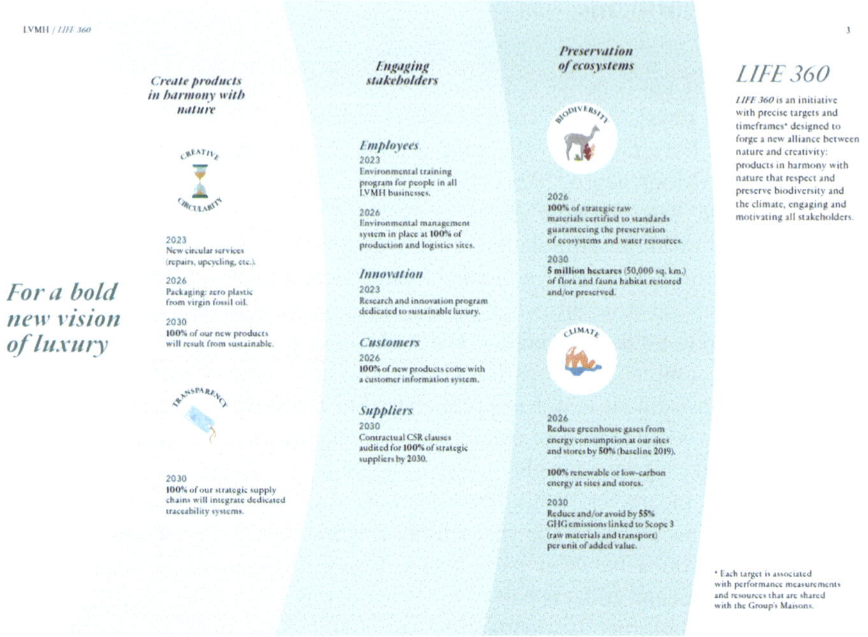

Fig. 1 Goals to be achieved in the LIFE 360 sustainability program. (LVMH, 2021)

According to Alexandre Capelli, deputy director of the group's environmental development, regenerative agriculture is being adopted, which should be inserted throughout the product cycle, applying regenerative materials and processes (Cernansky, 2022).

Upcycling is part of the goals to be achieved by 2023 with regard to products created in harmony with nature, specifically in the "new circular services" (LVMH, 2021). Capelli gives as an example Nona Source, a platform for selling deadstock materials, which makes these circular services possible. It also mentions the possibility of repair, as is the case with the Louis Vuitton, Berluti, and Loewe brands, which offer this service to increase the useful life of the products. They offer sophisticated restoration, upcycling, new use of precious raw materials, and use of alternative materials to support the circular economy (Cernansky, 2022).

An example of a transformation already carried out was the LV sneakers by designer Virgil Abloh, in which the oldest version of the same footwear was dismantled (Cernansky, 2022).

According to the group, it is only possible to achieve the proposed objectives if all those involved are willing to change. Environmental training is being given to the different agents, according to their roles and responsibilities. In addition, suppliers must be equally aligned (LVMH, 2021).

The LVMH group has been adopting sustainability more slowly when compared to the rest of the sector. Brands such as H&M, the Inditex group, and the luxury group Kering were already developing sustainability plans. The latter "rethought" its brands to make them sustainable through initiatives such as the gradual removal of fur. The LVMH group, for example, is known for the use of furs, especially exotic ones, which makes a change necessary (Cernansky, 2022).

The Kering group states that sustainability is its motto, which includes the concepts of "upcycling, recycling and regeneration," and the main objective is to contribute with "new and sustainable" procedures through "collaboration and innovation" (Kering, 2022).

The Kering group has been committed to upcycling since 2018. It has partnered with cooperatives and NGOs, such as La Réserve des Arts, with the aim of collecting materials considered waste, such as leather and other textiles. In addition, it started supporting Tissons la Solidarité in 2019, with the aim of providing professional reintegration training to seamstresses, with the focus on upcycling. In addition, it partnered with ECONYL® with the aim of generating and supplying nylon made from plastic waste (Kering, 2022).

Gucci, belonging to this group, was highlighted as one of the first luxury brands to adhere to a more sustainable future; proof of this is the launch of the Off The Grid collection, in which it made use of organic or recycled materials, highlighting the ECONYL®. Windbreaks, high-top sneakers, and bags were highlighted (Farra, 2020; Kering, 2022). In addition, the packaging is made from recycled paper, with a bag made from recycled nylon powder. For Alessandro Michele, creative director at Gucci, this collection is "a symbol for the new era" (Farra, 2020).

Online Platform as a Vehicle for Upcycling

Covid-19 brought new uncertainties, attacking the economy and technology (The National Intelligence Council, 2021, p. 11). In technology, it is expected that in the next two decades its impact will accelerate and the human being will face new challenges. Companies gain an advantage by using technology, as long as they have the right resources and long-term global reach (The National Intelligence Council, 2021, p. 54).

Countries seek to recover their economies, and the circular economy has become important for this. High turnover products must be designed to avoid waste and pollution. Products with a high-quality design emerge, both for long-lasting and for packaging (Ellen Macarthur Foundation, 2021, pp. 7–8).

The B2C platform, provisionally named UpCycliMe, not yet implemented, is based on Vestiaire Collective and ASOS, enabling garment transformation. According to Fernandes (2020), the creation of this type of business model contributes to the perception that a disposable product can become durable.

The Vestiaire Collective is considered the largest resale platform for luxury products (Duffy, 2021). It emerged in Paris in 2009 and aims to promote a sustainable future. It promotes a circular fashion, selling "pre-loved and unique" pieces. You can find thousands of luxury items, clothing, and accessories, with up to 70% off when compared to the real sale price. In addition, all articles are authenticated by experts (Vestiaire Collective, 2022).

According to this platform, the fashion industry contributes to excess consumption and production, in addition to bringing problems in climate change and work ethics (Vestiaire Collective, 2022). However, the circular economy has been mentioned as an important contribution to the evolution and growth of second-hand luxury sales, something provided by the younger people who defend sustainability (Duffy, 2021).

To sell items at Vestiaire Collective (2022), the interested party must fill in a form with the item they wish to discard and insert the respective photographs. At that time, they will receive support to increase product visibility. Subsequently, when the product is sold, it must be sent to the address of the platform, free of charge, so that the quality and authenticity are confirmed and, thus, proceed to the address of the buyer. At that time, the seller will receive a payment of up to 80% of the sale value.

As a buyer, you can find sellers from around the world as well as authentic luxury items. It is possible to filter products by novelty, designers, men, women, children, "we love," vintage, bags, watches and jewelry, and express shipping and direct shipping (Vestiaire Collective, 2022).

ASOS is an online platform based in England, which offers products from more than 850 brands, in addition to its own brand of clothing and accessories (ASOS, 2022).

According to ASOS (2022), it should value: the democracy of fashion, offering various sizes at the same price and body positivity, defending a healthy image, without stereotypes. It features zero tolerance for the transaction of counterfeit goods and advocates for animal welfare, demanding that materials be a by-product of the meat industry so that they do not suffer "in the name of fashion."

On the ASOS platform it is possible to find the ASOS Marketplace, where small or vintage brands can apply. Brands are committed to focusing on the customer, taking the time to serve them. They will pay the platform a 20% commission on sales, must have a PayPal account, be 100% genuine, respond within 24 h, send and return within two business days, and have a rating above 90% (ASOS, 2022).

One of the methodologies of the circular economy is to create collections with discarded materials. There is talk of upcycling with the aim of recycling, in addition to using biological and biodegradable fibers (Fernandes, 2020). Therefore, UpCycliMe should allow the consumer to obtain a new collection if it chooses to modify several pieces of the same brand.

In common with the analyzed platforms, UpCycliMe's objective will be to promote the circular economy, offering unique pieces. It will reach brands and buyers from all over the world, guaranteeing the authenticity of the piece even if it is transformed. In addition, it must defend ideals such as fashion democracy and sustainability. Finally, all purchase and sale processing must meet deadlines, similar to those previously indicated by the ASOS Marketplace.

Unlike the analyzed platforms that offer products, UpCycliMe will have a set of upcycling services according to the type of clothing and respective changes. When choosing the service(s), the consumer fills in a form, where they attach the photograph(s) of the clothes and explain details that they would like to change (Fig. 2). This request will go for approval, and the brand will have to respond within 24 h for the customer to accept it or not. When proceeding with their purchase, the consumer will have to make the payment and send the clothes to the address indicated by the platform, already with the prepaid stamp. Upon receipt, the brand must meet the transformation deadline and subsequently send the new product(s) to the customer's address, closing the purchase process (Fig. 3).

To create this platform, resources are needed, such as the purchase of the domain, which can be purchased, for example, on the NameSilo website internationally, and hosting that can also be purchased on this or another more intuitive website such as Wix. According to Wix (2022), it is possible to build a platform with creative freedom, advanced SEO, and optimization for the mobile version. They offer a variety of editing tools and provide a professional email account, statistical analysis, contact management, member login, chat for website visitors, connectivity to social networks, different languages on the platform, and an SSL certificate to protect the website, specifically personal data and payments. In addition to PayPal payments, they accept credit cards, and brands are notified when a purchase occurs, either by email or via the Wix app. It is also possible to hire professionals to create online platforms, indicated by Wix.

In order to cover the gap pointed out by Paras and Curteza (2018), the new platform should facilitate the collection of products. To this end, it is important to establish a partnership with a distributor that collects the part from the consumer's address, sends it to the brand, and, later sends the product to the customer, with these costs already included in the service. According to Kargo (2021), an intelligent delivery management platform, the objective is to provide a fast delivery service at low cost. It calculates transport costs in real time and monitors deliveries, optimizing the entire process.

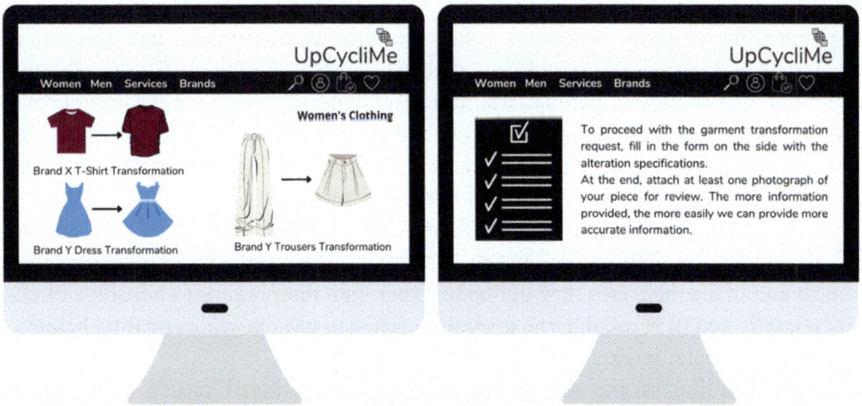

Fig. 2 Mockups of the new platform. (Own authorship, based on the ASOS Marketplace, 2022)

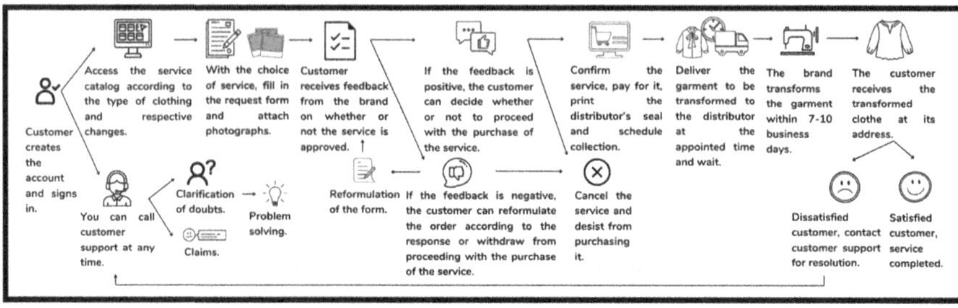

Fig. 3 Platform functioning model. (Own authorship, based on the ASOS Marketplace, 2022)

Finally, good communication, persuasion, and instruction of this type of upcycling service to consumers are essential, as already indicated by Park et al. (2022), as well as strong partnership links with the brands involved.

For the platform to be functional, all the resources indicated above must be used, as well as an initial investment in the design of the infrastructure and tools (Montenegro, 2021) and continuous investment in the platform itself, in the distribution network and in the communication of services. Available and the upcycling itself.

Conclusion

Through the State of the Art, it was possible to understand how upcycling and online platforms can connect in order to drive brands toward a better future. It is important that brands and designers adapt to new technologies and drive creativity to create change.

The fashion industry continues to have a high impact on the environment. Conversely, brands are ever more conscious of their part in it, so upcycling, despite encountering barriers, has been a focus, even for luxury brands. However, it is important that clothing collection points are close to consumers, and consumers need to be educated and persuaded to do so.

Online platforms, compared to traditional markets, are present in the digital environment with less formal and more united structures. With Covid-19, technologies saw their adoption increase, and brands felt the need to keep up with this acceleration and thus brought their activities to these platforms.

With the case studies, it was possible to perceive that luxury brands are starting to adapt to the new reality that supports sustainability, notable in the collections they launch and in the measures they define together with their brands to boost the circular economy and in particular, the upcycling. This, in turn, provides quality, being in line with the characteristics defended by luxury.

The promotion of the search for unique and pre-loved pieces by Vestiaire Collective combined with the platform's operating model, as well as the guidelines defended by ASOS, namely democratization in terms of prices, body positivity, the originality of the pieces and the well-being being animal, were key points to take into account for the suggestion of the new UpCycliMe platform. However, it had to be adapted, as those analyzed sell products and not services.

As resources, examples of websites for purchasing domains, hosting, and distribution were indicated, with the focus being on Namesilo, Wix, and Kargo, respectively, all of which the author has already used.

With the study of methodologies, the creation of the online platform as a clothing transformation service was considered justified, viable, and relevant, connecting customers to brands. The entire proposal was based on future trends in technology as well as the circular economy. This platform is equally important to solve the problem of difficult access to upcycling services by customers, as well as the difficulty of delivering their clothes.

Bibliography

ASOS. (2022). *ASOS Marketplace terms & conditions (Boutiques).* Available at: https://marketplace.asos.com/help/terms-and-conditions?ctaref=mktp%7Cfooter%7Ctsandcs. Accessed 18 July 2022.

Cernansky, R. (2022). Inside LVMH's biggest sustainability plan yet. *Na: Vogue.* Available at: https://www.voguebusiness.com/sustainability/inside-lvmhs-biggest-sustainability-plan-yet. Accessed 09 May 2022.

de Villiers, R. (2020). Accelerated technology adoption by consumers during the COVID-19 pandemic. *Journal of Textile Science & Fashion Technology, 6*(5). https://doi.org/10.33552/jtsft.2020.06.000647

Duffy, K. (2021). *Luxury resale: A circular strategy by Vestiaire collective.* Bloomsbury Fashion Business Cases.

Ellen Macarthur Foundation. (2021). *Universal circular economy policy goals: Enabling a large-scale transition.* Ellen Macarthur Foundation.

Farra, E. (2020). *"A Symbol for the New Era" – Gucci Off the Grid Is Alessandro Michele's First Step Toward Circularity.* Na: Vogue. Available at: https://www.vogue.com/slideshow/gucci-off-the-grid-alessandro-michele-circularity-sustainability. Accessed 07 June 2022.

Fernandes, S. (2020). *Creative collaborative circular economics, exponential system design for sustainability in fashion design (tese de doutoramento).* Universidade do Minho.

Fernandes, S., Lucas, J., Madeira, M. J., Cruchinho, A., & Honório, I. D. (2018). Circular and collaborative economies as a propulsion of environmental sustainability in the new business models in fashion. In *International conference on innovation, engineering and entrepreneurship* (pp. 925–932). Available at: https://doi.org/10.1007/978-3-319-91334-6_127. Accessed 16 May 2022.

Kargo. (2021). *Plataforma #1 na gestão inteligente de entregas.* Available at: https://www.kargo.pt/. Accessed 18 July 2022.

Kering. (2022). *Sustainability. Upcycling, recycling and renegeration.* Available at: https://www.kering.com/en/sustainability/innovating-for-tomorrow/upcycling-and-recycling/. Accessed 11 May 2022.

Kim, B., Son, J., Song, S., Kubono, S., Noguchi, R., Sato, M., Yang, W., & Yeoun, M.-H. (2020). Proposal of service platform for reforming used clothes. *Design Works, 3*(1), 38–47. https://doi.org/10.15187/dw.2020.05.3.1.38

Kretschmer, T., Leiponen, A., Schilling, M., & Vasudeva, G. (2020). Platform ecosystems as meta-organizations: Implications for platform strategies. *Strategic Management Journal., 43*, 405. https://doi.org/10.1002/smj.3250

L'officiel. (2022). *Meet Duran Lantink, the designer reimagining luxury pieces.* Available at: https://www.lofficielusa.com/fashion/duran-lantink-fashion-designer-upcycle-louis-vuitton-gucci-balmain. Accessed 10 May 2022.

LVMH. (2021). *Life 360 An alliance of nature and creativity.* Available at: https://r.lvmh-static.com/uploads/2021/05/life_360_en_externe_def.pdf. Accessed 09 May 2022.

Marques, A. D., Moreira, B., Cunha, J., & Moreira, S. (2019). From waste to fashion – A fashion upcycling contest. *Procedia CIRP, 84*, 1063–1068. https://doi.org/10.1016/j.procir.2019.04.217

Minh, N. T., & Ngan, H. N. (2021). Digital fashion – An optimal solution for fashion industry during Covid-19 pandemic. *1st van lang international conference on heritage and technology conference proceeding*. https://doi.org/10.1063/5.0066478.

Montenegro, D. (2021). A new definition of platform market dominance for the platform economy. *SSRN Electronic Journal*. https://doi.org/10.2139/ssrn.3893910.

Paras, M. K., & Curteza, A. (2018). Revisiting upcycling phenomena: A concept in clothing industry. *Research Journal of Textile and Apparel, 22*(1), 46–58. https://doi.org/10.1108/rjta-03-2017-0011

Park, J., Eom, H. J., & Spence, C. (2022). The effect of perceived scarcity on strengthening the attitude–behavior relation for sustainable luxury products. *Journal of Product & Brand Management, 31*(3), 469–483. https://doi.org/10.1108/jpbm-09-2020-3091

The National Intelligence Council. (2021). *Global trends 2040: A more contested world.* Global Trends.

Vestiaire Collective. (2022). *Pre-owned fashion shop.* Available at: https://www.vestiairecollective.com/journal/welcome-to-vestiaire-collective/. Accessed 15 Oct 2022.

Vezzoli, C., Conti, G., Macrì, L., & Motta, M. (2022). *Designing sustainable clothing systems: The design for environmentally sustainable textile clothes and its product-service systems.* FranAngeli.

Wix. (2022). *Wix resources: Discover all the resources to create and manage your website online.* Available at: https://pt.wix.com/features/main. Accessed 18 July 2022.

Costumes in Early Years of Portuguese Television: Broadcasting Visual Messages

João A. B. Barata and Carolina M. Fadigas

Introduction

Costumes are imperative elements in visual communication for both cinema and media. They are visual messages and carry, as a constructed design, the power to foster knowledge, increase culture, and enlarge the information in a public domain. Costumes can be a message to be decoded in a semiotic fashion.

Often, for a better understanding of these elements, it is necessary to investigate past narratives, connecting costume designs' visual data and their values in terms of meanings and association. The following research collects information from communication, design theory, and historical, political, and cultural events, aiming to understand the narratives encapsulated in two examples from the early years of television in Portugal.

Costumes = Messages

According to (Costa, 2002), in "The costume as an essential element for the narrative," costumes should be highlighted for its great relevance in communicating in cinema and audio-visual pieces. The author states garments are responsible to help the understanding of historical timelines and locations of the narratives, as well as the mood, shapes, and personal traits. As a result, costumes and costume design is an act of nonverbal communication.

J. A. B. Barata (✉) · C. M. Fadigas
Universidade Lusófona-Centro Universitário de Lisboa – CICANT; DESIGN-ID, Lisbon, Portugal
e-mail: joao.barata@ulusofona.pt

A. Cruchinho, J. C. Neves (eds.), *Sustainability in Fashion: Creativity, Education, and Innovation*, https://doi.org/10.1007/978-3-031-80991-0_3

Costumes can be analyzed both in cinema and in audiovisuals as part of the narrative through the materiality of clothing and its communicative capacity. These signs may be depicted in garments, accessories, and the overall styling of a given character. When applying the communication process, one can assume three components: the sender, the message, and its audience (the receiver). The gap between the source of information and the receiver is bridged by a message, the sign between the two ends of the communication pipeline (Barata, 2012). Understanding a costume design is subjective; once to communicate between parts is a process by which one makes sense (codes and decodes) of a design. It is co-dependent on the artifacts known to both the sender of a message/design and the one to whom it was sent (Watzlawick, 1991; Watzlawick et al., 2002).

To communicate is to make something common, and storytelling with costumes implies using symbology common to the audience, aiming to reduce the gap for misleading interpretations; semiotics, quality, time, and space of signs are fundamental to ensure the transmission of visual messages. In cinema, for example, the receiver knows the character and his/her surroundings; that is, the messages sent by the costumes must be organic and unforced. Character's psychology is understood and perceived over a certain exposure period. In cinema, there is time for the costume to evolve along with the character in its various psychological stages during the narrative. In the audiovisual sector, specifically television, the receiver's interpretation timespan is shorter than in the cinema, and so costume designing usually applies more direct approaches and less complex designs/messages. Whereas cinema is set up for a closed and focused audience, watching television is a daily routine. Watching television does not require a focused attention to the screen and coexists with other simultaneous activities, and that is why media costumes are built on repeating visual concepts, signs, and stereotypes. In cinema and media, costumes' understandings are to be natural and organic once they are part of the contexts surrounding both the sender and the viewer. Costumes' visual messages are semiotic, collaborative, and culturally rooted. (Wigley, 1998; Barthes, 1999; Miranda, 2003).

A sign system is fundamental to fully understand a certain costume, it is part of a given context (the narrative) (Telles, 2009). In this research, historical, cultural, and political circumstances are determinant to describe past narratives, connecting costume designs and their meanings.

Historical Contexts

Portuguese television started broadcasting over 60 years ago, and the new technology changed the society. Combining sound and image, television, just like cinema, conquered the national public with the little *magic box* as a new form of entertainment and information. This audiovisual spectrum, as a new element of communication, increasingly brought closer the highly divided country between cities and countryside. When television emerged during the Salazar dictatorship, it allowed to educate how to behave, including how to dress properly in accordance with the rules imposed by the regime, and explored new social habits.

Television gradually entered Portuguese homes, as stated in Tables 1 and 2 (Barreto, 1996). According to the presented data, one may argue that during the 1960 s and mid-1970 s, watching television was mainly a social event, a gathering in cafes and other establishments. In 1960, there were only 5.2 televisions per 1000 inhabitants, and in 1975, data shows 77.6 sets per 1000 inhabitants. Analyzing the data, watching television was increasingly becoming a familiar and continuous act for Portuguese families. Public demand has led to augmenting the annual number of broadcasting hours (Table 2), starting with 1285 h in 1960 and registering 27,433 h in 1994. It is noticeable that in almost four decades, television went from being practically mysterious to a component of Portuguese culture.

Table 1 Number of TV sets from 1960 to 1990 in Portugal

Year	Number of registered television sets	Number of TV sets per thousand inhabitants
1960	46,372	5.2
1961	68,174	7.6
1962	89,642	9.9
1963	119,381	13.2
1964	151,464	16.8
1965	180,095	20.1
1966	213,775	24.0
1967	271,025	30.6
1968	305,878	34.7
1969	347,399	39.9
1970	387,512	44.7
1971	473,363	54.9
1972	542,846	62.9
1973	608,527	70.5
1974	674,686	76.0
1975	722,315	77.6
1976	914,354	97.2
1977	1,137,100	119.6
1978	1,174,815	122.3
1979	1,149,700	118.4
1980	1,131,900	140.7
1981	1,460,902	147.8
1982	1,500,519	151.0
1983	1,522,597	152.7
1984	1,566,519	156.5
1985	1,605,058	160.3
1986	1,618,391	161.7
1987	1,618,313	162.1
1988	1,649,222	165.7
1989	1,671,484	168.5
1990	1,701,000	172.3

Source: Barreto (1996, p. 146)

Table 2 Emission hours on TV from 1960 to 1994 in Portugal

Year	Emission hours	Profit (contos)	Costs (contos)
1960	1285	–	–
1961	1825	–	–
1962	1903	–	–
1963	2011	–	–
1964	2309	–	–
1965	2359	–	–
1966	2930	–	–
1967	2996	–	–
1968	3126	–	–
1969	4178	–	–
1970	4435	–	–
1971	5053	–	–
1972	5946	–	–
1973	7167	–	–
1974	7385	–	–
1975	6912	–	–
1976	7179	–	–
1977	10,223	822,692	892,401
1978	12,065	1,136,491	1,269,529
1979	12,653	1,390,824	1,754,063
1980	12,804	2,486,819	2,462,897
1981	10,918	3,430,366	3,272,635
1982	12,962	4,984,895	5,027,790
1983	13,649	5,395,540	6,411,176
1984	12,649	7,431,697	8,117,638
1985	12,904	8,970,483	11,702,813
1986	12,604	13,680,085	12,463,112
1987	14,741	17,074,198	15,628,913
1988	17,679	22,079,515	19,561,714
1989	18,596	24,951,187	22,939,690
1990	19,798	28,242,405	27,880,854
1991	21,102	37,506,853	37,006,096
1992	22,791	42,242,515	46,351,848
1993	26,818	39,753,223	47,635,832
1994	27,433	28,255,710	47,813,440

Source: Barreto (1996, p. 146)

António Barreto and Joana Pontes (2007) explain how television changed minds in Portugal. The major reason is its connectivity and its role in fading the differences between the countryside and big cities. "In most of the territory, there was no access to the mail or the telephone. There were few roads"[1] (Barreto & Pontes, 2007, 03:34 min). This was the reality of communication in Portugal until the 1960s. There was very little access to information, radios were expensive, and most people did not know how to read or write. All existing information was analyzed and censored (by the International State Defense Police—PIDE). The documentary "We and Television" (2007) also shows the power of television as a social phenomenon in its early beginning because "men started going out with women to watch television"[2] (Barreto & Pontes, 2007, 06:35 min). Having a television at home was very expensive, so families and friends would gather in cafes to watch broadcasts. These events gave women a little more freedom: "TV showed different behaviors, women smoking and drinking. Other ways to dress and styling hair. In real life, women continue to cook, but with a more modern appearance"[3] (Barreto & Pontes, 2007, 09:40 min).

From a fashion perspective, it is relevant to state that until television appeared and demonstrated this new image, only a few people in Portugal had access to what was happening in the rest of the world. Valter Carlos Cardim (2014) states that even fashion articles were victims of censorship. Only those who lived in bigger cities were lucky enough to see the new trends. The access to the good fashion was restricted to wealthy people, usually living in Lisbon or Porto (Soares & Cantista, 2016). Cardim (2014) also added that: "Despite the situation, Chiado continues to be in the 1960s, the greatest meeting point and reference point for fashion lovers"[4] (Cardim, 2014, p. 41).

Information, Media, and Costumes

Televised information integrated the Portuguese dialect in a broad manner. Meanings could be broadcasted in different ways; costume design can be analyzed as an interface for communication and democratization of information.

RTP Archive has a video with several compressed television fragments entitled "1st Experimental Transmission from RTP" (RTP-Memória, 1956). The archive shows the backstage of the first experimental transmission on September 4, 1956, at Feira Popular in Palhavã, Lisbon. It is noticeable two people, one male and one

[1] Free Translation from: "Na maioria do território, não se tinha acesso ao correio nem ao telefone. Havia poucas estradas" (Barreto & Pontes, 2007, 03:34 min).

[2] Free Translation from: "os homens começaram a sair com as mulheres para ir ver televisão" (Barreto & Pontes, 2007, 06:35 min).

[3] Free Translation from: "a televisão mostrou diferentes comportamentos, mulheres a fumarem e a beberem. Outras maneiras de vestir e pentear. Na vida real, as mulheres continuam a cozinhar, mas com uma aparência mais moderna" (Barreto & Pontes, 2007, 09:40 min).

[4] Free Translation from: "o Chiado, desprezando a situação, continuou a ser na década de 1960, o maior ponto de encontro e de referência para os amantes de moda" (Cardim, 2014, p. 41).

female, getting ready to start broadcasting. Their costumes can be analyzed as they were vital once visual communication was highlighted in this new media, unlike radio before it. The lady is well dressed in a celebratory fashion, respectfully adorned with a pearl necklace. The gentleman, in a tailored suit as expected at the time. The significance of these costumes demonstrates a rich and educated point of view on a governmental spectrum. The flyer (Fig. 1) that promoted this televised emancipation shows a traditional family illustrated on the cover (RTP, 1956). The beautiful, submissive wife, the hardworking husband, and the happy little children, precisely how António de Oliveira Salazar would like to paint a typical Portuguese family. The flyer's cover shows a small photograph on the left panel, demonstrating Portuguese traditionalism, with typical costumes, usually used in folk music dance. One must pay attention to the similarity of fitting between the first Portuguese television broadcast costumes and the imagery on the announcing flyer. This showcases the power of dictatorial forces in the visual communication process.

Fig. 1 Flyer promoting RTP's first emission. (Source: RTP, 1956)

Two Media(Tic) Costumes: Selling Aesthetics and Freedom

Considering the contexts widely described in the previous sections, two costumes were selected to be analyzed in terms of their messages and meanings in each moment in the early years of Portuguese television. Both were chosen considering the iconic and cultural importance that the women who wore them had in the Portuguese culture. In the following cases, costumes showed new aesthetics and became a symbol of hope during dictatorship.

Chef Maria de Lourdes Modesto's innovative culinary program in the Portuguese context, during the late 1950s and 1960s, devoted a space to females in media. As mentioned earlier, the craft of cooking was still present in women's routine, but for the first time, they interconnected with what they saw in the *magic box*. Recently deceased, Maria de Lourdes was the star of the first television cooking program in Portugal, which debuted in 1958, becoming an icon of style and a celebrity herself. *Flama* magazine, traditionally used to writing about stars and singers, placed her on the cover of the May 29, 1968, edition. The apron she wore on television became a bestseller in Portugal. Many stores had replicas to sell, and many seamstresses tried to recreate it, as she herself said in the documentary "We and Television" (Barreto & Pontes, 2007). Her popularity surpassed borders, and on March 4, 1987, the New York Times considered her to be "Portugal's Julia Child" (Agência Lusa, 2022), about the equivalence of her feminine relevance in the kitchen and in television programs dedicated to the thematic. The power of television as a money-making machine is not new, and, in this case, the chef's power portrait in a costume influenced shopping aesthetics and behaviors. Giving women the opportunity to be influenced to dress more modernly, hoping to encapsulate Maria de Lourdes Modesto's status.

Following on the costume's ability to transport a constructive message, another meaning is evident when shown on television. In 1969, Portugal was still living on a dictatorial regime; however, it took a few minutes from a song in a song contest to give the spectators a glimpse of freedom. Icon Simone de Oliveira sang "Desfolhada" in Eurovision and carried a powerful message into Portuguese homes. From a semiotic point of view, the green dress she wore can be associated with the same green as the national flag from Portugal. Simone's dress, as sender, on the Eurovision stage, can be interpreted as "hope" to the viewers in Portugal. Green, according to the meanings in the Portuguese flag, represents the future, the hope, and the progress (Museu da Presidência da Republica, n.d.). In addition, large, typical Portuguese gold earrings in the shape of Viana's Hearts were added to the costume. This accessory can further ensure a connection to the Portuguese people, as a receiver, engaged on an emotional level. Because, as in the previous case, there is a personal connection and recognition in what one is watching on television. Simone de Oliveira (RTP2, 2021) recalls how people went crazy seeing her return to Portugal, stopping the train from the frontier to Lisbon, taking her and her team 10 h to cross the country. The occurrence of this costume on national television and on an international stage did not arouse the desire of the receiver to acquire a similar costume; however, it transported the receiver to a dream place, forgetting dictatorship.

Conclusion

In Portugal, television appeared during a dictatorship. Despite its first broadcast in 1958, the shy initial growth is visible in Tables 1 and 2, associating a natural post-revolution boom after 1974. Between 1960 and 1970 there is a considerable change in habits influenced by television. The analyzed documentary from RTP reveals that the constraint of not having the possibility of owning a television at home led to meetings in public spaces of both men and women. Not only were women able to gain more social visibility, but they also saw new behaviors and clothing aesthetics (new fashions). This innovation also took advantage of censorship to collaborate with the uniform appearance of the clothes of the Salazar's "good family" ideal, being visible right from the first experimental television broadcast and even in the advertising flyer itself. Meanwhile, the *magic box* also allowed to influence the wishes and hopes of its viewers. Whether it is the desire to wear clothes like the idol of Portuguese cuisine, Maria de Lourdes Modesto, or be thinking about the hope of having a better country with the internationalized Simone de Oliveira's green dress.

This study was only possible by embracing external contemporary contexts and points of view. Enlarging the knowledge of one's culture and dialect can be connected to costume design and its meanings.

References

Agência Lusa. (2022). Morreu Maria de Lourdes Modesto, referência da cozinha portuguesa. *dnoticias.pt*.

Barata, J. A. B. (2012). Design de Moda e/é Comunicação: O desenvolvimento de um objecto mutável. Universidade da Beira Interior

Barreto, A. (1996) *A Situação Social em Portugal 1960–1995*. Instituto de Ciências Sociais-Universidade de Lisboa.

Barreto, A., & Pontes, J. (2007). Nós e a Televisão. Portugal: RTP 1. Available at: https://arquivos.rtp.pt/conteudos/nos-e-a-televisao/

Barthes, R. (1999) *Sistema da Moda*. Lisboa: Edições 70.

Cardim, V. (2014). *A moda em Portugal: 1960 a 1999*. Edições IADE.

Costa, F. (2002). O figurino como elemento essencial da narrativa. *Sessões do Imaginário*, 8(Agosto).

Miranda, J. B. de (2003). O design como problema. *Autoria e Produção em Televisão Interactiva* (pp. 294–312).

Museu da Presidência da Republica. (n.d.). *Bandeira Nacional*. Available at: https://www.museu.presidencia.pt/pt/conhecer/simbolos-nacionais/bandeira-nacional/

RTP. (1956). Folheto de apresentação da RTP. Available at: https://cdn-images.rtp.pt//icm/arquivo/docs/d3/d39fad04a851f431fe13a31171b5a26c_5ee889d2e005733027e53d663dacbb8f.pdf

RTP2. (2021). Seis Décadas de Televisão – Simone de Oliveira. Portugal: Minidocs. Available at: https://www.rtp.pt/play/p9099/e563800/seis-decadas-de-televisao-minidocs

RTP-Memória. (1956). 1ª Emissão Experimental da RTP. Portugal. Available at: https://arquivos.rtp.pt/conteudos/1a-emissao-experimental-da-rtp/

Soares, P. C., & Cantista, I. (2016). Portuguese fashion in the 20th century. In I. Cantista (Ed.), *Fashion spaces – geographical, physical, and virtual* (pp. 43–60). Grupo Almedina.

Telles, L. (2009). Elementos da Comunicação e suas formas de Planejamento. *Anuário da Produção Acadêmica Docente, III*(5), 149–162. Available at: http://repositorio.pgsskroton.com.br/bitstream/123456789/1355/1/Artigo12.pdf.

Watzlawick, P. (1991). *A Realidade é Real?* Relógio D'Água.

Watzlawick, P., Beavin, J., & Jackson, D. D. (2002). *Pragmática da comunicação humana: Um estudo dos padrões, patologias e paradoxos da interacção.* Edited by Editora Cultrix. São Paulo.

Wigley, M. (1998). Whatever happened to total design? *Harvard Design Magazine, 5*, 25–30.

Handicrafts, Fashion, and Sustainability: Possible Connections: A Case Study of the Meroca Sustainable Fashion Brand

Juliana Bezerra Gomes de Pinho Pessoa, Germana Maria Fontenelle Bezerra, and Maria Rozimeire Bezerra Augusto

Introduction

Reflecting on "Fashion Sustainability" leads us to think about the textile/fashion/clothing chain, from the choice of raw materials to their disposal. Our proposal aims to show how it is possible to create a brand with a sustainable purpose, using waste from the food industry, manual work, and exclusive, high-quality design. These are some of the attributes of the Ceará accessories brand Meroca moda sustentável.

The brand was born out of a course completion project, in which a Fashion Design student proposed a reflection on the "Kardume de artesanato" group, which is part of the Association of Fish Processors and Producers in the municipality of Jaguaribara in the state of Ceará, Brazil. The symbolic and cultural aspects that handicrafts bring to the municipality were identified. Another aspect analyzed was how design interferes with the craft activity and how it can contribute to adding value to the local product. "Observes that the aspects that were collected in the research direct us to new paths, showing us that there are several aspects that can be addressed in the Crafts theme, characterizing one of them as the relationship of exchange of experiences between the designer x artisan, or the artisan designer" (Augusto et al., p. 1).

We will present how the fish skin is transformed into leather, its processes, dyeing, and the development of the brand's products. The brand started being built

J. B. G. de P. Pessoa (✉)
Universidade do Minho, Braga, Portugal

G. M. F. Bezerra
Universidade Federal do Ceará, Fortaleza, Brazil

M. R. B. Augusto
Meroca – Moda Sustentável, Petrópolis, Rio de Janerio, Brazil

A. Cruchinho, J. C. Neves (eds.), *Sustainability in Fashion: Creativity, Education, and Innovation*, https://doi.org/10.1007/978-3-031-80991-0_4

when the owner lived in the city of São Paulo, and the factory is currently located in the municipality of Eusébio, in the metropolitan region of Fortaleza in Ceará, Brazil.

We can consider the brand's products to be handicrafts with a sustainable design difference.

> The Brazilian economic and business scenario has become increasingly complex and multifaceted, characterized by changes and advances in which we can see that creative and purposeful businesses and entrepreneurs have been consolidating. Fashion, in this context, is one of the areas that has shown such changes, evidenced by some brands that are concerned with issues such as sustainability and purpose, concepts that are increasingly sought after by consumers. (Barbiero et al., 2022, p. 39)

We have noticed this trend in the state of Ceará with the encouragement of the creative economy. Many graduates from fashion design courses are starting their own brands.

Ceará's handicrafts are known for their beauty and creativity and are some of the most diverse in the country. The state of Ceará is a benchmark in Brazil's handicraft market. The variations in types further enrich this business hub, which has been generating income for Ceará's economy. Lace, embroidery, crochet, straw, leather, and other handmade items are all produced. According to Banco do Nordeste (2002), Ceará is among the four states that produce the most handicrafts made from leather, accounting for thirty-one point one percent (31.1%) of the twenty-eight occurrences of the typology. For Barroso (2002), it is necessary to identify and promote products with a strong regional influence to highlight the roots, history, and trajectory of a people.

> Ceará is one of the main handicraft production centers among the states of the northeast region, and the places with the highest concentrations of handicraft products in Ceará are the central market, the pine nut market, the seafront fair and the Ceará Handicraft Center (CEART), which promote and enrich Ceará's handicrafts, generating income and cultural identity for the state. The diversity of products is rich and varied, and the fair held on the promenade of Avenida Beira Mar offers tourists all kinds of choices. (Augusto et al., p. 3)

Tilapia leather handicrafts are still in their infancy among the various types found in our state, and the Meroca sustainable fashion brand meets the objective of placing an ecological and handmade design product on the local, national, and even international fashion market, where each piece is unique, with an intuitive creative process and a poetic narrative.

Tilapia Leather

The Skin

The production of fish fillets generates a large volume of waste, including the skin, which is discarded, causing environmental damage at the processing sites. Tilapia skins have been used in the health sector to treat wounds and burns, and in the fashion sector for garments made with eco-leather.

Making use of waste from fishing and fish processing is a way of promoting creative and productive autonomy in a sustainable way among the community involved. In addition, the use of this waste can serve as a raw material for obtaining various products with high commercial value and generating income, especially for communities that live off fishing (Vidal-Campell et al., 2020).

According to De Souza et al., 2021, when the fish are slaughtered, the skin is removed from the fish and they are selected according to previously defined categories based on their weight. They are then placed in ice and water at a temperature of zero degrees Celsius, followed by the filleting process. The skins are then measured and sent for tanning and processing into leather.

> (...), the skin was removed with the aid of special pliers, starting with the traction of the skin removal, the dorsal region and the cut around the head in a caudal and ventral direction. This resulted in whole skins and, consequently, hides with a larger utilization area, as described by Franco et al., which can be seen in

Fig. 1. (De Souza et al., 2021, p. 4)

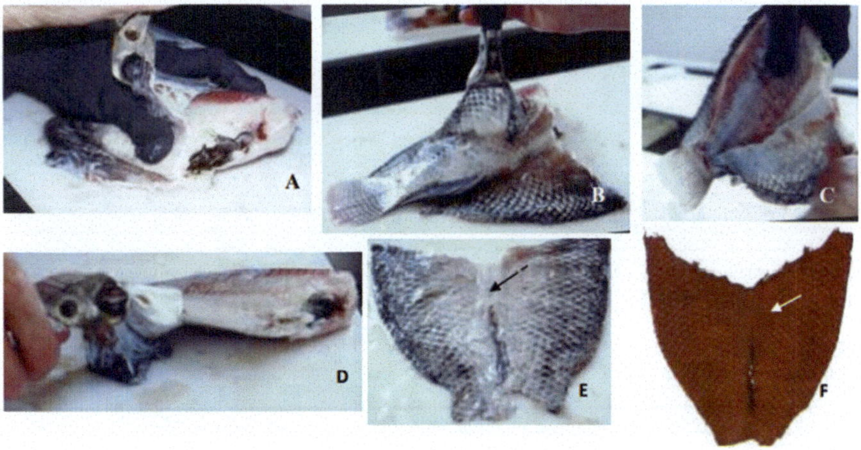

Fig. 1 Procedure for removing tilapia skin: (**a**, **b**, **c**, **d**, **e**)—using pliers to keep the skin on both sides together (joined by the ventral region, arrow). (White arrow = showing region where the lamellae are smaller). (Source: Franco et al.)

From the images, we can see that the skins are very small in size and are very suitable for use in the production of fashion accessories or even for small details in garments. Designers have used them to make garments using the patchwork technique to join the leather. There is also an industrial process for producing tilapia leather as a blanket.

De Souza et al., 2021, carried out tests on the "Resistance of tilapia leather as a function of slaughter weight classification and direction of removal of the test body when tanned with chromium salts" and reached the following conclusions: the study was carried out with Nile tilapia, the same species used by the brand under study.

According to the aforementioned study:

Nile tilapia have a hide with a standardized distribution of lamellae protecting the insertion of the scales (Figs. 1f and 2). The lamellae are distributed over the entire area of the skin, showing a pattern in the size and length of the lamellae, except in the ventral region, where they are smaller (Fig. 1f, arrow, and Fig. 2). Figure 5 shows details of the flower design, where it can be seen that the lamellae have a more defined size (position where the scale was) as well as length (tissue located in the anterior portion of the scale) (De Souza et al., 2021, p. 8).

As a result of the research, we can conclude that larger skins are more resistant, so for the purpose of fashion products, these skins are more suitable. For the brand under study, as it deals with the type of tanning of the skin and the dyeing process, it is of fundamental importance that these processes are sustainable and measure their impact on the environment.

As leather is marketed by weight, it is of fundamental importance that the fashion designer has knowledge of the characteristics of resistance, elongation, and durability, as this will produce leather that makes the most of it.

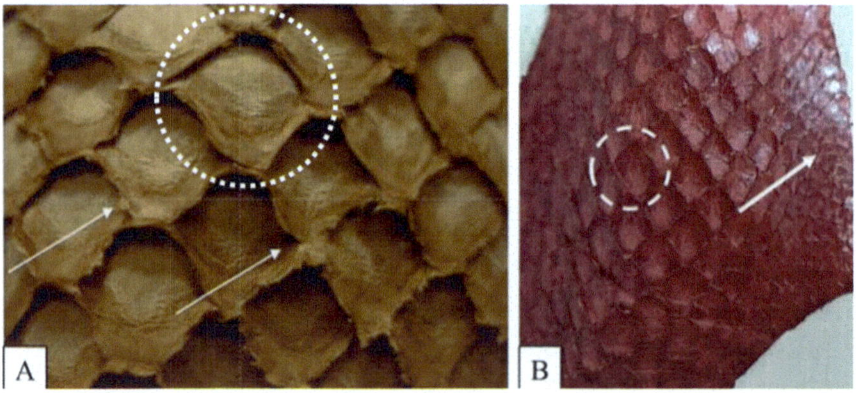

Fig. 2 Drawing of tilapia leather flower. (**a**) showing the opening (dotted circle) and length (arrow) of the protective lamellae and insertion of the scales. (**b**) showing lamellae (dotted circle) and arrow indicating small lamellae in the ventral region. (Source: De Souza et al., 2021, p. 9)

Also according to the aforementioned authors, when analyzing the direction of the leather (transverse or longitudinal) in relation to the length of the fish, it was found that the leather has more resistance to traction and progressive tearing in the transverse direction compared to the longitudinal direction. The amount of force load and the maximum force applied in the tests were greater (p. 10).

Tilapia leather has a very distinctive appearance characteristic of the fish species, giving fashion products a peculiar aesthetic and adding value to the product's design. This appearance is given by the fish's scales, which in turn are protected by lamellae, forming "mosaics on the surface of the leather" (Fig. 3).

This leather flower design is a characteristic of the species, i.e., it has a specific design pattern per skin, taking into account the size of the scales inserted into the fish's skin. The design differs from that of other fish because it is larger, giving it a different texture from other leathers. "100 people were interviewed, 63% of whom considered Nile tilapia leather to be more beautiful and attractive compared to mirror carp (33%) and only 4% liked both types of leather" (De Souza et al., 2021, p. 14).

Tensile strength is important in the development of fashion products using Tilapia leather, especially those that require tensile strength, such as the manufacture of fashion accessories, more specifically belts. In the case of Nile tilapia leather, the species used by the brand, the leather is more resistant transversally than longitudinally, due to the collagen fibers that are superimposed in thick layers on the fish's skin. "Therefore, the way in which the collagen fibers are arranged in the dermis

Fig. 3 Tilapia leather bracelet. Source: Instagram of the brand under study, @meroca.modasustentavel

provides a bond between them, and consequently the leather is more resistant." "Nile tilapia skins (from 500 to 800 g) can be used to make clothing, as the average resistance value was within the reference values for mammals" (De Souza et al., 2021, p. 13/14).

The tilapia leather used by the brand under study is made from Nile tilapia skin, so it is essential to know the characteristics of the raw material. Belts need to be more resistant to traction, so the designer should make the pieces in the direction of the leather's greatest resistance, which is transversal.

Leather

Given the characteristics presented, we can infer that Nile tilapia leather (and other tilapia species) can be used in the clothing industry.

Cultivation

The state of Ceará has fish farming projects to promote fish, including Nile tilapia, which are used in food, health, and the fashion industry.

The leather used by the brand is currently 100% bought from the artisan community in the town of Guabaribara—Ceará—Brazil, a town that was moved to build the state's largest dam, called Castanhão, in an attempt to solve the state's water needs. The group where the leather is purchased is called "Grupo Kardume."

Faced with such an adverse situation, Jaguaribarenses felt the need to reaffirm the bonds of belonging to a land, to unite and recreate symbols that would give meaning to the struggle to remain in a place, which can be understood when we attribute values and meanings to a space (Augusto et al., p. 4).

The artisans are all housewives and organize their time between leather production and household chores. They also produce their own handicrafts using tilapia leather. According to one of the artisans, working outside the home with handicrafts, although it is not the basis of her family's livelihood, is something that is very pleasurable and has opened up new paths for her and given her a new way of behaving in society.

The leather that is now used in the production process used to be thrown away, polluting the environment. Today, this leather goes through a tanning and dyeing process, which is used to shape the pieces. The design applied to the products is inspired by elements from the region, or even fish farming, so the cultural aspect of the municipality is preserved in the pieces produced, creating a cultural identity for the group and the region.

Tanning and Dyeing

The Association of Fish Processors and Producers of Jaguaribara and Lages (APLAGES) treats the fish to make food and donates the skin they don't use to the Kardume group, which reuses tilapia skin that used to be thrown away to turn it into leather and later into fashion products.

After the skins are removed from the fish, they are placed in a tank with water and ice so that they don't lose their characteristics and deteriorate. The skins go to the shed where the "Kardume" group meets to tan the skin and dye the tilapia leather.

There are various methodologies used for tanning, but we will only report on the tanning that is considered to be the most ecological and sustainable, which is carried out by the group that uses vegetable products that are abundant in the region.

This process does away with the use of chromium, a heavy metal that causes high levels of contamination to the ecosystem. Vegetable tannin, an extract from the bark of certain trees, is used as a tanning agent. In this way, there is a significant reduction in pollution (Lopes, 2011, p. 14).

The first stage of tanning the skin is done in a very artisanal way: it is scraped with a spoon to remove all the excess fat and scales from its surface, so that the skin is more malleable.

In the second stage, the hides are placed inside the Fulão (Fig. 4), which is the machine that does the tanning and dyeing process of the leather, using lime and neutral detergent to completely remove impurities and fat. The process lasts around 24 h, and it is not advisable to spend less than 7 h, as the cleaner the hide, the more elongated it becomes when it is removed from the machine.

In the third stage, after 24 h, an industrial powder called acacia tannin is added to the Fulão, **which** can be replaced by cashew bark or angico bark, common plants in the backlands of Ceará (Fig. 5).

Fig. 4 The Soot. (Source: Augusto et al.)

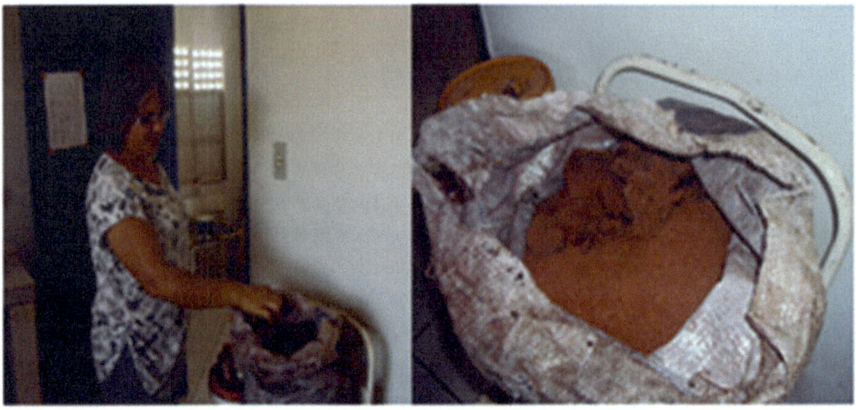

Fig. 5 Vegetable tannin powder. (Source: Augusto et al.)

Fig. 6 Removing the skin from the Fulão, Leather on the wooden board and dyed leather. (Source: Augusto et al.)

The fourth stage is dyeing by adding aniline and dye according to the desired colors, after which the leather pieces are laid out on wooden planks side by side, fastened and left to dry (Fig. 6).

After drying, the leather pieces are cut according to their intended purpose. The process can be done by hand or with a cutting machine. They are marketed according to the weight and thickness of the leather (Fig. 6).

Methodology

Initially, a bibliographical survey was carried out to find out about the studies already carried out on tilapia leather and its processing.

The methodology used was a case study of the accessories brand "Meroca Moda Sustentável," through research into articles on the subject, analysis of the brand's website, and informal conversations with the brand's owner. This is a qualitative study.

One of the authors made an on-site visit to the Kardume group to learn more about the tanning and dyeing processes and other important aspects of the research. The technique used was to interview the producers of tílapia leather and to film some of the processes that were being carried out.

The Brand

The Meroca sustainable fashion brand was created in January 2017 by designer, artist, and craftswoman Rozi Bezerra.

On the occasion of her graduation from the Fashion Design course in 2013, she addressed the topic in her final paper: Design and handicrafts made with Tilapia fish leather, seen with an immense possibility of transforming work into an authorial, sustainable, and poetic brand. The designer faced the challenging world of female entrepreneurship and idealized the brand, imprinting in each creation the aspects of her childhood in the interior of Ceará and her connection with elemental nature, her taste for manual work and was notable in the College period, as she always aroused interest in handicrafts.

With this poetic narrative, the designer, who also writes poems and intuitive short stories, creates each bag with unique designs that are not repeated, emphasizing her authorial and intuitive process.

In line with this process, we can mention the author André Carvalhal, who states in his book: Fashion with purpose: "Fashion is an extension of our body. It can be a way of seeing and showing life. It helps us to discover 'who we are,' and also to reveal our identity, a way of floating in the world, of being new every day and at the same time unique" (Carvalhal, p. 85).

Our production is based on *slow fashion*, everything on demand and to order, meticulous work where every detail adds up and makes the tilapia leather patchwork come together. The poetic narrative directs us to the details that will be created, and each one imitates our essence, imprinting the way designer Rozi Bezerra sees the world around her.

Our product catalog includes handbags and accessories—maxi necklaces, maxi earrings, belts, bracelets, and shoes, all made from sustainable and ecological waste, produced with minimal environmental impact, using an ecologically correct tanning and dyeing process. It is also interesting to highlight a passage by Carvalhal that speaks of this change that must take place in the world of fashion, because for the author "we need to look at ourselves and the planet in a different way" (Carvalhal, p. 47).

The brand's purpose is to provide ecological and artisanal luxury fashion that can align our values with the way we consume, through a unique and poetic experience of wearing sustainable accessories.

Concept

The narrative of Meroca sustainable fashion is artistic, poetic, and free. Our inspiration is intuitive and comes first and foremost from the closeness and connection with the elementals, especially branches, leaves, and roots. "To co-create a new world, we need to accept freedom" (Carvalhal, p. 365).

It is a concept of meeting and bringing together human life, the human gaze in contrast to the scraps of tilapia leather and its inimitable design and aesthetics. It is also the coming together of memories of the child the designer once was, of the desire to transform and resignify childhood memories, creating textures and transforming them into perception and emotions, placing herself authentically in the world, transforming and inspiring other people too.

As a human being and as a designer at the head of the Meroca sustainable fashion brand, Rozi Bezerra makes a point of expressing her total freedom when it comes to creating. She also says that the process has no rules, that her intuition is the master of every detail, that it's a true encounter with her divine self.

In the book Fashion with Purpose there is a very beautiful definition of who we are, in which the author mentions when he was in the holistic space for the realization of Being in Piracanga, the speech is by the founder of this place Amélia Clark, "we are different pieces of a great spark of light, a great energy, which many call God, the meeting of many selves." We can say that the concept of the brand is pure poetry and art, through the reuse of tilapia fish skin. So, to paraphrase Carvalhal, the purpose of the brand is "before being recognized, you have to recognize yourself."

Technique Applied to the Creative Process

In addition to the fundamental topics covered in the book Fashion with Purpose, its author points out that "we are part of this living organism called planet earth, that our connection with nature has been lost, we have forgotten that it is also our life."

Awareness of the processes used in each brand plays a fundamental role in the sustainable fashion chain, and the *upcycling* technique is a concept that has been applied in the textile waste reuse market.

The *upcycling* technique we use to reuse tilapia fish leather waste starts with the process of collecting the skin by artisan groups.

At this point, the artisans collect and transform the skin that came from the garbage into genuine, sustainable, and ecologically correct leather. Following the idea of this concept that preserves the shape and original design of the raw material, without taking away any aspect of its look, be it functional or aesthetic, the *upcycling* technique can be used with a variety of materials and has a significant effect on the sustainable fashion chain.

Fig. 7 Meroca sustainable fashion products. (Source: Instagram @mmeroca.modasustentavel)

At the Meroca sustainable fashion brand, every piece, every scrap is a possibility, and this is undoubtedly one of the brand's great differentials, the designer's peculiar and transformative vision. With the leftovers from the bags, necklaces, bracelets and many unique products full of personality are created. Some of the brand's products are shown below (Fig. 7).

This technique and concept add up, because leather is produced in the form of scraps, small pieces as possible, and you need to know how to look at it and have this perception of how to make the most of this waste.

Social Responsibility

The brand is a partner of artisanal groups that have chosen to reuse the waste from tilapia fish leather, which would otherwise be discarded by the fishing industry.

These are artisan cooperatives that work together to produce and process textiles from this waste. Production is local, with a focus on their town, which also cultivates and creates fish farming businesses. In most cases, the artisans who process the leather are the wives of the fish farmers, creating a link and a collaborative chain around their own locality.

In addition to generating family income, this activity makes a significant contribution to the social development of these craft groups. They add value to local handicrafts and transmit the culture of their municipality.

Our partnership is with the "Kardume de Artesanato" group from the city of Jaguaribara-Ceará, 230 km from the state capital. The group is supported by the municipal government and takes part in events to promote their work with leather.

It is a work of great innovation and artisanal technology, in which all the effort and training of the artisans is put, who are also great women and human beings.

Finally, it would be good to quote Carvalhal (p. 367) who says "(...) this is the moment to remember that we are all born free, with the power to choose, ideas and characteristics of our own. Without any information and with everything new to discover, original and authentic."

Conclusion

Exacerbated consumption of industrialized products is having a huge negative impact on the environment and society in general. The textile-clothing-fashion chain generates a lot of waste, and we need to look for alternative solutions for more sustainable production.

The contribution of this research is to show academics and designers how they can undertake to meet all the pillars of sustainability—economic, social, and environmental—with a slow fashion product, in a creative, natural, and innovative way, as each product is unique, personalized, and full of meaning.

The economic pillar is seen in the livelihood of this group of women who tan the skins, improving the quality of life of the fishing community, as well as through the work of the brand's designer, which reflects on the social scene since these women are valued for their work with Tilapia leather. On the environmental front, through the reuse of Tilapia skin, a waste product that was damaging the environment is being transformed into a fashion product, using methods that do not harm the environment and, consequently, human beings.

The fashion consumption cycle is increasingly on the move, and more and more thought has been put into developing more affordable products with a more contemporary design. As a result, the consumer trend of the moment has become more exploited and has caused a desire to buy and a much greater production of textile waste.

As a result of this accelerated fashion consumption, the designer has become more and more visible within the company, taking part in all the product development processes, from aesthetics to functionality and the choice of materials and raw materials, giving consumers a great desire to consume in the face of so many options on the market.

Therefore, in order to be considered sustainable design or eco-design or eco-friendly, product design has to be attentive to all the phases of the product's life cycle and take into account all the environmental, social, and economic factors of sustainability, as it encompasses all these aspects as well as social equality, cultural diversity, respect, the possibility of shared experience, collective work, the ease of resources, and the ability to design for the future.

Therefore, when thinking about Meroca sustainable fashion, we can say that the choice of applicable materials needed to make the accessories is concerned with the impacts produced, collaborating to reduce them and working to raise people's awareness toward a more peaceful economy.

References

Banco Do Nordeste. (2002). Ações para o desenvolvimento do artesanato do nordeste/Banco do Nordeste. 2ª Edição- Fortaleza: Banco do Nordeste.

Barbiero, L. L., Schemes, C., & Bohnenberger, M. C. (2022). Fashion with purpose: A case study of the somos numa brand. *Modapalavra e-periódico, 15*(36), 35–60.

Barroso, E. (2002). Design, cultural identity and handicrafts course. Fortaleza: SEBRAE/FIEC – modules 1 and 2.

De Macedo Quaresma, D. M., & De Moura, H. T. (2016). Design for the broad sustainability of product-service systems: A case study of a leather fashion accessory design company. *Studies in Design, 24*(2).

De Souza, M. L. R., et al. (2021). Resistance of tilapia leather as a function of slaughter weight classification and direction of removal of the specimen when tanned with chromium salts. *Pesquisa, Sociedade e Desenvolvimento, 10*(8), 33310817236.

Lopes, P. D. (2011). MODA, D. E. O COURO DE PEIXE COMO Forma DE materializar O conceito DE sustentabilidade Ao design DE moda.uniritter, Porto Alegre.

Oliveira, G. G. et al. (2019). Leather resistance of NILO TILAPIA (Oreochromis Niloticus) subjected to different types of acids in the skin stage.

Schwarz, K. K., et al. (2017). Methodologies for the transformation of sole, sea bass, Paru and tilapia skins into leather. *Pubvet, 12*, 150.

Vidal-Campello, J. M. A., et al. (2020). Use of tilappia waste as an alternative for local development in the community of açude Saco i, Serra Talhada-PE. *Revista Extensão & Cidadania, 8*(14), 45–59.

What Would Incentivize and Discourage Parents from Buying Secondhand Products for Their Kids?

Ana S. C. Pires, Susana C. Silva, and Fábio S. Sandes

Introduction

The secondhand market is booming, driven by the rise in sustainable consumption among younger consumers (Sandes & Leandro, 2019; Silva et al., 2021). With its shorter lifecycle, children's clothing fits well into this trend due to the tradition of passing down items as "hand-me-downs" (Damme & Vermoesen, 2009; Gregson & Crewe, 1998). Gregson & Crewe (1998) discuss parents' sense of obligation to pass on goods in good condition. Mintel (2012) found that three in ten parents opt for secondhand baby items. Despite extensive research on the secondhand products (SHP) market, more attention should be given to the motivators and barriers influencing consumer behavior in children's clothing markets. This gap is significant as the children's clothing market has unique characteristics that influence motivations and barriers, such as rapid product turnover (Piacentini, 2010), multi-person involvement in the buying process (Chaudhary et al., 2018), and influence from peers/friends (Grant & Stephen, 2005), and parents. Prior studies on secondhand childrenswear (SHC) have mainly focused on sustainability aspects (Waight, 2013, 2014), including social class influence (Waight, 2015) and perceptions of "hand-me-down" products (Waight, 2017).

This study delves into SHC by investigating how past shopping experiences influence parents' subsequent experiences with SHC. We specifically explore parents' motivations and barriers to consuming SHC and assess whether these factors

A. S. C. Pires · S. C. Silva
Católica Porto Business School, Porto, Portugal

F. S. Sandes (✉)
Universidade Lusófona-Centro Universitário de Lisboa, Lisbon, Portugal

ESDES Business School, UCLy - Université Catholique de Lyon, Annecy, France
e-mail: fsandes@univ-catholyon.fr

A. Cruchinho, J. C. Neves (eds.), *Sustainability in Fashion: Creativity, Education, and Innovation*, https://doi.org/10.1007/978-3-031-80991-0_5

change after engaging in SHC. A survey involving 265 parents was conducted, with 169 (63.8%) having prior experience purchasing SHC. Our findings reveal that parents do not face social embarrassment when buying SHC, and the tested barriers—risk and hygiene—were insignificant, suggesting that barriers to SHP consumption may not be readily apparent. Economic motivation emerges as the primary driver for purchasing SHP, with sustainability also playing a significant role in the SHC market. These findings offer crucial insights for managers aiming to optimize SHC offerings by leveraging economic and sustainability arguments in their strategies.

Literature Review

Guiot and Roux (2010) defined secondhand shopping as buying previously owned goods, irrespective of usage frequency (Machado et al., 2019). This market offers distinct advantages over traditional retail stores (Guiot & Roux, 2010). The secondhand childrenswear (SHC) market is notable due to the short usage lifespan of children's clothing, necessitating frequent purchases. Anecdotal evidence suggests a growing trend in the SHC market. For example, a 2021 survey by "Myname Tags," a British company based in Portugal, showed that 32% of Portuguese parents purchase used items for their children, with shoes (28%) and pants/skirts (27%) being the most popular (Correia, 2021). The main reasons cited were cost savings (34%) and sustainability concerns (19%). Humana Portugal (2021), a non-profit, reported a 78% increase in SHC sales since 2016, reaching 61,768 items in 2020, impacted by the COVID-19 pandemic (Correia, 2021).

Motivations for buying secondhand products (SHP) include economic, sustainable, and hedonic factors like finding unique garments and desiring distinctive items (Machado et al., 2019). Economic motivation is primary, with parents seeking cost savings by buying lower-quality functional products (Guiot & Roux, 2010; Baqall & Abdulkhaleq, 2018; Silva et al., 2021). However, concerns about information gaps, like missing original packaging or age and safety details, can deter purchases (Baqall & Abdulkhaleq, 2018; Sandes & Leandro, 2019; Jenny, 2020).

Studies indicate that these motivations and barriers also apply to the SHC market. Yan et al. found that young parents prefer secondhand clothing for uniqueness and fashion, with concerns about clothing condition and hygiene as barriers. Gajjar and Shah (2021) showed younger consumers are likely SHC buyers, citing hygiene and clothing conditions as top barriers. Walters and Mariani (2020) identified economic motivations, sustainability, and distancing from the consumption system as primary drivers. Silva et al. (2022) discussed the positive perceptions of thrift store shopping due to product uniqueness, economic motivations, sustainability, and the thrill of the hunt.

Concerns about social embarrassment and potential judgment also affect parents' attitudes toward SHC (Sandes & Leandro, 2019; Harris et al., 2016; Laitala & Klepp, 2018; Silva et al., 2021). Parents may view their children's clothing as a reflection of their parenting abilities. Income influences secondhand children's clothing purchases, with Norum and Norton (2017) finding a negative correlation between income and the likelihood of purchasing secondhand items. Millennials engage more in acquiring secondhand clothing, potentially leading to social embarrassment. Finally, contamination concerns from previously owned items are another barrier (Horne & Maddrell, 2002; Sandes & Leandro, 2019).

Methodology

We surveyed 167 participants, 154 of whom were parents, which is therefore valid for this study. Thirteen respondents who did not have children were excluded from the sample. The survey was distributed to a non-probability sample using a snowball technique via the online personal emails of parents from two school institutes and published on some Facebook groups of SHP and mothers' groups.

We surveyed parents who may or may have yet to purchase SHP. Our samples are consistent only with parents (respondents without kids were excluded). We collected responses from previously validated scales for all the five constructs (social embarrassment, environmental sustainability, economic motivation, risk, and hygiene) discussed in the literature review.

Results

The final sample comprised 109 women and 45 men ($N = 154$), with more than 53% of the population over 41 years of age. Most participants attended university, almost 79% worked for others, and 16% owned a business. In addition, 95 parents had already bought SHC. We tested the reliability of all scales and found that all scales used were reliable, with Cronbach's Alpha equal to or higher than 0.7. We ran a linear correlation for models of respondents with and without previous experience in buying SHC. We present the results of the regression models in the following section.

The Model with Consumers Without Previous Experience in SHP

In 59 parents without previous SHP purchasing experience, this overall model presented a correlation coefficient of R2 = 0.605, indicating a good prediction level. The overall model explains 60.5% of the variation in parents' intention to purchase SHC. Applying a 5% statistical significance criterion, only two constructs remained statistically significant: environmental sustainability ($\beta = 0.274$; $t = 2.425$; $p = 0.019$)

and economic motivation ($\beta = 0.814$; $t = 4.895$; $p < 0.001$). The other three constructs were not statistically significant: hygiene ($\beta = 0.152$; $t = 0.962$; $p = 0.34\ 1$), social embarrassment ($\beta = 0.087$; $t = 0.518$; $p = 0.606$), and risk ($\beta = 0.019$; $t = 0.147$).

The Model with Consumers with Previous Experience in SHP

With a sample of 95 participants who had already bought SHP, the correlation coefficient of R2 = 0.38 indicates a good prediction level. The overall model with experienced consumers explains 38% of the variance. Only two constructs were statistically significant, related to the motivations to buy SHP: environmental sustainability ($\beta = 0.514$; $t = 5.517$; $p < 0{,}001$) and economic motivation ($\beta = 0.283$; $t = 2.165$; $p = 0.033$). The other three constructs, related to the barriers to buying SHP, were not statistically significant: hygiene ($\beta = 0.05$; $t = 0.396$; $p = 0.693$), social embarrassment ($\beta = -0.127$; $t = -0.952$; $p = 0.344$), and risk ($\beta = -0.051$; $t = -0.523$; $p = 0.602$).

Discussion

The notable discovery in this study is the statistically significant results we found for the constructs, especially when comparing consumers with and without experience in SHC products. These results imply that experience with SHC is a relevant factor in the childrenswear secondhand products market. Surprisingly, social embarrassment does not significantly deter the intention to purchase SHC, regardless of prior experience. This contradicts the concerns raised by Laitala and Klepp (2018) about potential social judgment. It is also inconsistent with Lo et al.'s (2019) findings, where the fear of being judged or associated with lower-income individuals played a role. Silva et al. (2021) also concluded that the negative association of social embarrassment, especially among consumers without experience, signifies a higher level of social embarrassment for inexperienced consumers. One possible explanation for this discrepancy could be the widespread social acceptance of purchasing used clothing for children, driven by the understanding that children outgrow clothes quickly, which are used for a short period.

Both participants with and without experience in purchasing SHC showed a motivation for environmental sustainability, aligning with the conclusions of Sandes and Leandro (2019) and the growing consumer interest in sustainable consumption practices (Vehmas et al., 2018). This finding also resonates with Waight's (2013) study, emphasizing the desire of individuals to be environmentally conscious parents.

The results indicate that competitive pricing is not a crucial factor influencing consumers' decisions to buy secondhand garments for their children. Retailers and marketers should instead offer attractive pricing strategies to entice potential customers.

Although it was anticipated that both experienced and inexperienced parents would be highly motivated to purchase SHC for economic reasons, our results showed that economic motivation was not significant for either group. This contrasts with previous studies that emphasized economic motivation as the main driver for engaging in SHP, such as the work of Gopalakrishnan and Mathews (2018), who interviewed thrift shop store managers emphasizing consumers' perception of value for money when buying SHP. One possible explanation for this finding could be that the childrenswear market is more accustomed to and comfortable with used products. The experience of buying and donating used clothes within families may influence the importance of economic motivation for consumers in this market.

For marketers, investing in first-time buyers could yield long-term benefits, as the experience may effectively reduce significant barriers to their future buying behavior. Overall, the positive influence of experience with SHC on purchase intention indicates that individuals who have prior experience buying secondhand clothes for their children are more likely to have a positive attitude toward such purchases. This underscores the importance of creating a positive customer experience to encourage repeat purchases and foster a loyal customer base.

Limitation and Future Studies

While our study offers valuable insights, it is essential to recognize its limitations. We focused specifically on the children's clothing market, which has unique traits, making it more open to the SHC market. Therefore, we must be cautious when applying our findings to other SHP markets. Future research should explore how the motivators we identified can boost the intention to buy SHC products in different market settings. For example, delving into the influence of brands and how prestigious brand perception affects consumers' intention to buy secondhand products could shed more light on consumer behavior.

Moreover, investigating how managers can effectively utilize the sustainability aspect of the SHC market to positively influence consumer intentions is another promising area for future research. Lastly, understanding what motivates consumers to become producers and sellers of SHC products would provide a comprehensive view of the dynamics of the SHC market and the factors that influence consumer behavior. Addressing these research gaps will deepen our understanding of consumer behavior in the realm of secondhand clothing and aid in developing effective marketing strategies.

References

Baqall, Y. F., & Abdulkhaleq, S. (2018). Analysis of factors influencing the decisions over purchasing secondhand products. *International Journal of Social Sciences & Educational Studies, 4*(4).

Chaudhary, M., Ghouse, S., & Durrah, O. (2018). Young Arab consumers: An analysis of family buying process in Oman. *Young Consumers, 1*(19), 1–18. https://doi.org/10.1108/yc-07-2017-00720

Correia, R. (2021, July 4). *Maioria dos portugueses doa roupa que já não serve.* Revista Sustentável. https://www.revistasustentavel.pt/consumo-etico/maioria-dos-portugueses-doa-roupa-que-ja-nao-serve/

Damme, I. V., & Vermoesen, R. (2009). Secondhand consumption as a way of life: Public auctions in the surroundings of Alost in the late eighteenth century. *Continuity and Change, 24*(2), 275–305.

Gajjar, C., & Shah, P. (2021). Secondhand shopping: Understanding consumer behavior toward pre-owned clothing in India. *WRIPUB.* https://doi.org/10.46830/wripn.20.00035

Gopalakrishnan, S., & Matthews, D. (2018). Collaborative consumption: A business model analysis of secondhand fashion. *Journal of Fashion Marketing and Management.*

Grant, I. J., Stephen, G. R (2005). Buying behaviour of "tweenage" girls and key societal communicating factors influencing their purchasing of fashion clothing. *Journal of Fashion Marketing and Management: An International Journal, 9*(4), 450–467.

Gregson, N., & Crewe, L. (1998). Dusting down Secondhand rose: Gendered identities and the world of secondhand goods in the space of the car boot sale. *Gender, Place & Culture, 5*(1), 77–100.

Guiot, D., & Roux, D. (2010). A secondhand shoppers' motivation scale: Antecedents, consequences, and implications for retailers. *Journal of Retailing, 86*(4), 355–371. https://doi.org/10.1016/j.jretai.2010.08.002

Harris, F., Roby, H., & Dibb, S. (2016). Sustainable clothing: Challenges, barriers, and interventions for encouraging more sustainable consumer behavior. *International Journal of Consumer Studies, 40*(3), 309–318.

Horne, S., & Maddrell, S. (2002). *Charity shops; retailing, consumption, and society*. Routledge.

Jenny. (2020, December 29). What are the Pros and Cons of buying secondhand? *The Air Ambulance Service.* https://theairambulanceservice.org.uk/what-are-the-pros-and-cons-of-buying-second-hand/

Laitala, K., & Klepp, I. G. (2018). Motivations for and against secondhand clothing acquisition. *Clothing Cultures, 5*(2), 247–262.

Machado, M. A., et al. (2019). Secondhand fashion market: Consumer role in circular economy. *Journal of Fashion Marketing and Management, 23*(3).

Mintel. (2012). *Baby and nursery equipment January 2012.* Mintel Group.

Norum, P., & Norton, M. (2017). Factors affecting consumer acquisition of Secondhand clothing in the US. *JFMM, 2*(21), 206–218. https://doi.org/10.1108/jfmm-10-2016-0090

Piacentini, M. (2010). Children and fashion. In *Understanding children as consumers* (pp. 202–217).

Sandes, F. S., & Leandro, J. (2019). Exploring the motivations and barriers for secondhand product consumption. *Global Fashion Management Conference, 2019*, 292–296.

Silva, S., Santos, A., Duarte, P., & Vlaciç, B. (2021). The role of social embarrassment, sustainability, familiarity, and hygiene perception in secondhand clothing purchase experience. *International Journal of Retail & Distribution Management.* https://doi.org/10.1108/IJRDM-09-2020-0356

Silva, S. C., Duarte, P., Sandes, F. S., & Almeida, C. A. (2022). The hunt for treasures, bargains, and individuality in pre-loved luxury. *International Journal of Retail & Distribution Management, 50*(11), 1321–1336.

Vehmas, K., Raudaskoski, A., Heikkilä, P., Harlin, A., & Mensonen, A. (2018). Consumer attitudes and communication in a circular fashion. *Journal of Fashion Marketing and Management: An International Journal, 22*(3), 286–300.

Waight, E. (2013). Eco babies: Reducing a parent's ecological footprint with secondhand consumer goods. *International Journal of Green Economics, 7*(2), 197–211. https://doi.org/10.1504/IJGE.2013.057444

Waight, E. (2014). Secondhand consumption among middle-class mothers in the UK: Thrift, distinction, and risk. *Families, Relationships and Societies, 3*(1), 159–162. https://doi.org/10.1332/204674313X13805551473134

Waight, E. (2015). Buying for baby: How middle-class mothers negotiate risk with secondhand goods. In *Intimacies, critical consumption, and diverse economies* (pp. 197–215). https://doi.org/10.1057/9781137429087_10

Waight, E. (2017). Hand-me-down childrenswear and the middle-class economy of nearly new sales. In *Sharing economies in times of crisis: Practices, politics and possibilities*. Routledge.

Walters, G., & Mariani, M. (2020). Understanding the intention to buy secondhand clothing on sharing economy platforms: The influence of sustainability, distance from the consumption system, and economic motivations. *Psychol Mark, 5*(37), 724–739. https://doi.org/10.1002/mar.21334

Santa Clara Trapologia: The Case of a Traditional Textile Technology

Alexandra Cruchinho, Benilde Reis, and Catarina Rito

Introduction

Trapologia is the motto for this research project. The initial approach is to conduct an exploratory study of the subject to guide the research towards the development of a larger project.

The interest in the study stems from an initial motivation associated with the experiences of the researchers. Their childhood memories of using rag blankets in different contexts, mainly to decorate the interiors of their grandparents' houses and at festivals and pilgrimages, are examples of where *Trapologia* products can be found, both at the time and today.

Here, we differentiate between woven *Trapologia* and sewn *Trapologia.* Thus, woven *Trapologia* is worked on a frame loom, and small pieces of fabric are interwoven in a web-like thread, making up a blanket or rug (Fig. 1). Sewn trappings, which have a more extended history, are the result of joining different pieces of fabric together to create a new surface that allows patterns and shapes to be developed for different products that can be used in everyday life (Fig. 2).

This study aims to better understand the approach to *trapologia* taken by artisans in the Santa Clara—Almodôvar region. This approach is pertinent because it allows us to understand if there are any divergent points in terms of technique, function and aesthetics compared to other existing *trapologia* records in our country, particularly in the inland centre/north regions.

The methodology developed is a mixed methodology, non-interventional and interventional. In the first phase of the study, a search and review of the existing

A. Cruchinho (✉) · B. Reis · C. Rito
Universidade Lusófona – CICANT, Lisbon, Portugal
e-mail: alexandra.cruchinho@ulusofona.pt; benilde.reis@ulusofona.pt; catarina.rito@ulusofona.pt

A. Cruchinho, J. C. Neves (eds.), *Sustainability in Fashion: Creativity, Education, and Innovation*, https://doi.org/10.1007/978-3-031-80991-0_6

Fig. 1 Hunting coat, about between 1620–30, India. Museum no. IS.18–1847. (© Victoria and Albert Museum, London. Available on https://www.vam.ac.uk/articles/an-introduction-to-quilting-and-patchwork)

literature on this subject were carried out, and in the next stage, a practical study was carried out with an interview with the region's main artisan working with woven *trapologia*, as well as wool weaving, in Santa Clara, Mr Carlos Rosa.

The Importance of Tradition in a People's Cultural Heritage

The aim of the older generations, particularly artisans, is not to let traditions die so that the younger generation can learn about them and eventually perpetuate them. However, reality shows that the desire to learn traditional production techniques—processes that date back to other times, some passed down over centuries between mothers, daughters and granddaughters (Fig. 3)—has been lost over the last few decades.

Growing concern on the part of the competent authorities—municipalities, heritage protection institutions and academia—is leading to a growing commitment to recovering a heritage that is not to be extinguished. In this way, a survey of the different traditional techniques that exist in Portugal is being carried out in various projects, trying not only to encourage interest and learning on the part of the younger generations, together with the generations that know this knowledge, but also to document and record all the technical processes for future memory.

Fig. 2 Photo by Olga Ferina. Available on https://unsplash.com/pt-br/fotografias/F-byi_8ekJc

Fig. 3 Children working on a *trapologia* quilt, c. 1880

The dissemination of ancestral knowledge, mastered by people of very advanced age, is the primary concern of organisations such as local authorities, public and private associations, parish councils and researchers. Passing on knowledge between generations could be an essential solution to the issue of keeping this knowledge alive. Other ways of perpetuating this knowledge could be to add innovation to these craft techniques. Thus, developing projects and products resulting from reinterpreting these techniques and traditions could also be a possible solution.

Trapologia Historical Context

Trapologia is a traditional technique that can be found in different regions of the country, and there are differences in approach and inspiration, so it's interesting to understand what influences these differences. The earliest record of the origin of the *trapologia* technique dates back to Ancient Egypt, 3400 B.C., on a garment found in a Pharaonic tomb. The Roman army (Roman Empire—24 BC to 475 AD) had waistcoats or Pteruges in their wardrobe made of leather or strips of plain or quilted fabric, which formed a skirt or sleeves to protect the limbs while maintaining mobility and ventilation. The pteruges had small appliqués and metal plates to provide additional protection. However, this clothing was not waterproof; when wet, it took a long time to dry. To avoid this, soldiers used to wear a leather garment over it. *Trapologia* and Patchwork are (both) craft techniques linked to the textile sector. Their history takes us back to the beginnings of textiles, evolving from the simple interweaving of threads to a laminar structure obtained through the orthogonal crossing of two parallel thread systems, the warp and the weft (Victor et al., 1996). "Quilting [*trapologia*] can be considered a precursor of the popular recycling phenomena in vogue today" (Ziegler et al., 1995, p. 87). *Trapologia* and Patchwork have been essential economic crafts since ancient times and were integrated into the domestic and social life of the time. "The way of making and using the Quilt [*trapologia*] manifested itself in various countries in Asia and Europe. (...) Buddhists travelled from India to China via the Silk Road. (...) Portuguese colonists started trading posts in India; among the goods they ordered, they brought to the West via Goa—cotton, linen and silk quilts with silk embroidery made by talented artisans in Bengal and Cambay, India" (Stapel et al., 2000, p. 48–61). *Trapologia* arose from man's need to dress himself; however, nowadays, its reasons are certainly different: in an age when human beings are more environmentally conscious, recycling makes sense in reusing old fabrics or rags. Traditional *trapologia* uses handlooms (Figs. 4 and 5), knitting needles and even one's fingers, the fundamental value of recycling old rags being the raison d'être of this art (Fig. 6).

It is in this context that traditional tools and art objects are made (Alencastre, 2011). In contemporary *trapologia*, the work is essentially done using sewing machines, according to a methodological plan and by purchasing the fabrics on the market. Because of this, contemporary *trapologia* lacks greater economic strength than traditional *trapologia*, in which only old fabrics are reused, usually leftovers

Fig. 4 Manual wooden loom and bed quilt. Available in the author's collection

from sewing work (Baldini, 2006). This is also known as patchwork sewing: "it is an art of domestic economy that consists of using the waste of coloured fabrics to make bags, blankets, rugs and, more recently, decorative panels" (Victor, et al., 1996, p. 197).

The *Trapologia* of Santa Clara: Almodôvar

The research was carried out by interviewing local citizens to see if we could still find some people working in this craft.

Mr Carlos Rosa is the craftsman constantly mentioned throughout the conversations because he is an expert in weaving and still manages to make a living from this craft. He is also a great enthusiast and defender of this ancient traditional technology. Mr Carlos Rosa has worked in this craft for 37 years and began using used school looms (Fig. 7). He has two medium, two large and one small loom.

A few women still work in *trapologia* but for purely utilitarian purposes.

He currently weaves wool and cotton (rag), but he uses rag suppliers in the north due to a lack of raw materials in the region.

Fig. 5 Manual wooden loom and bed quilt. Available in the author's collection

In methodological terms, Mr Carlos doesn't mention having any specific methodology. He follows a series of processes and stages leading to the final product. The products follow the same lines; they don't add great experience or innovation. What differentiates them may be the colours and finishes.

He also says that products that were once used for day-to-day activities are now used for decoration. An example is saddlebags, traditionally used to transport cereals and vegetables and applied to the backs of donkeys or horses.

Today, saddlebags are used to store magazines and souvenirs.

The saddlebags can be made in woven or sewn *trapologia*, as can the blankets, traditionally used on beds and sofas and at fairs and pilgrimages. Nowadays, according to the craftsman, they are bought in significant quantities by Nordic countries to decorate the walls of houses.

The woven products are made on four-frame looms and bound in twill or taffeta. Mr Carlos makes blankets and rugs of very different sizes to order. The colours are those traditionally used, but orders determine the colours of the products.

Each piece, measuring 2 metres by 0.5 metres, takes around 8 h to make on the loom, and the finishing depends on the order and purpose of the use of each product.

Fig. 6 Trapology quilt made from old scraps/rags. Available in the author's collection

Conclusion

There is a traditional technique typical of the region. It is possible to interpret traditional made-in products and create products with creativity/updated fashion.

This research raises interest in the recovery of a region's cultural heritage in an international context. Involving fashion students in a future project can ensure that future generations are aware of important issues such as a country's identity.

The type of products made does not differ much from those found in other regions of the country, but the finishes can be different. Products such as saddlebags are also distinctive in this region.

The use of sewn trappings is even more distinctive than in other regions, not because of the technique but because of the shape and aesthetics of the product.

In conclusion, it is essential to emphasise the importance of carrying out a research project on this subject for this region of Santa Clara, as it is a study with high technical and scientific potential and, above all, will be an important driving force for preserving a region's history, traditions, customs and knowledge.

Fig. 7 Craftsman Mr Carlos Rosa. Available in the author's collection

References

Alencastre, H. (2011). Ser Artesão, um estado de graça ou uma provação? [autor do livro] Gabriel Cunha. Artesanato- Questões de Comercialização. Coimbra: CEARTE, 2011.

Baldini, M. (2006). A invenção da Moda- as teorias, os estilistas, a história. Lisboa: Ediçoes 70, Lda., 2006. ISBN: 972-44-1260-1.

Frost, P. (2000). *Miller's collecting textiles*. Octopus Publishing.

Group Ltd. (2000). ISBN: 1-84000-203-4.

Johnson, J. (2013). *Emporia State of University*. www.emporia.edu. Citação: 25 de Julho de 2013. http://www.emporia.edu/cgps/tales/quilte~1.html

Stalp, M. C. (2001). http://getd.galib.uga.edu/public/stalp_marybeth_c_200112_phd/stalp_marybeth_c_2 00112_phd.pdf

Stapel, Koning e Vibeke, Hanne. (2000). *Silk Quilts- From the silk road to the quilter's studio*. NTC- Contemporany publishing Grou. Inc.

Victor, Isabel e Gonçalves, Luis Jorge. (1996). *Bordados e outras artes de agulha*. IEFP. Artesanato de Lisboa e Val do Tejo. Coimbra : IEFP, 1996.

Ziegler, Kathleen e Greco, Nick. (1995). *Fabric Sculpture- the step-by-step guide and show case*. s.l.: Dimensional Illustrators, Inc. , 1995. ISBN: 1-56496-133-8. Interview with Craftsman Mr Carlos Rosa—November of 2022.

Changes in Clothing Use and Identity Building During the COVID-19 Pandemic

Mariana Anacleto and Irene Maldini

Introduction

Over two years after the outbreak of the Covid-19 pandemic, many parts of the world experienced a gradual, yet uncertain, return to public life and space as the pandemic entered an endemic phase. The ever-pressing question of "how can we (re)build a better future?" was then faced with two years' worth of experience in adaptation and resilience. In regard to clothing consumption levels, and their associated environmental impact (EEA, 2019), the pandemic seemed to offer opportunities to rethink practices driving overproduction and overconsumption. The experience of extended isolation caused an acute disruption of habits, and since consumption follows habits (Sheth, 2020), this rendered the pandemic a crucial time to study clothing consumption and its implications for the future (Sheth, 2020, p. 283). Clothing has a high potential for emotional attachment (Mellander & McIntyre, 2021) and is susceptible to individual creativity, making it a good setting to study consumer resilience in the context of the pandemic (Liu & Lang, 2021). The sudden disruption in representation practices (how we show ourselves to the world, a significant function of clothing) and consumption (understood as referring to purchase, use and disposal-related activities) provided a unique setting to understand what reflections were triggered in regard to clothing. This chapter therefore aims to capture how the Covid-19 pandemic influenced clothing behaviour; emphasizing the topic of use, yet touching on buying and disposing practices to anchor the observations made. It also aims to explore reflections on appearance and views on

M. Anacleto (✉)
Vrije Universiteit Amsterdam, Amsterdam, The Netherlands

I. Maldini
Oslo Metropolitan University, Oslo, Norway

A. Cruchinho, J. C. Neves (eds.), *Sustainability in Fashion: Creativity, Education, and Innovation*, https://doi.org/10.1007/978-3-031-80991-0_7

the role of dressing in a confined reality and to identify the contextual factors and personal values behind these changes.

Being part of the mapping phase of the *Solidarity in Fashion* research project, which took the temporary standstill of the Covid-19 pandemic on the fashion system as an opportunity to draw lessons for long-term system change (ArtEZ, 2021), this study also explores the potential of the practices observed for sustainable clothing consumption. The shift from overproduction and overconsumption to "survival mode", coupled with the increased visibility of the problems of this globalised industry and a growing desire for more personal connection raises the question: did the Covid-19 restrictions provide an opportunity for a reset of values towards sustainable behaviour?

To our knowledge, three relevant studies on the topic of clothing use during the pandemic were published before this research was conducted, in April 2021. The first one, published in April 2020, is a qualitative study conducted in Italy about changes in habits and meanings of body and dress, gathered from messages and posts from the author's network in March 2020 (Winterhalter, 2020, p. 158). It captured the following state and views of dressing at the time:

1. A preference for "less beautiful and more comfortable clothing", partly as a stress coping mechanism.
2. A lower importance given to the representational role of clothing, upheld only by remote work.
3. Changes in the construction of identity: "the ways we represent ourselves have been reduced or brought back to their true meaning, to their essence. Today, to build our identity, we need less" (Winterhalter, 2020). This behaviour is what Åsa Callmer and others call "sufficiency-related practices" (Callmer, 2019), a topic highlighted by previous studies as key for achieving sustainable clothing use, particularly in affluent societies (Mellander & McIntyre, 2021). Winterhalter refers to Oyserman et al. (2012) to discuss the role of identity-building in clothing, who argue that the *self* juggles both stability and fluidity in the construction of identity. *Stability* because it feels like an anchor that guides the choices we make every day, thus making the self an important motivational and self-regulatory tool; and *fluidity* because context influences these choices, meaning that identities are also dynamically constructed in the moment (Oyserman et al., 2012). The authors also argue that a distant perspective on the self is an opportunity for reflection (Wakslak et al., 2008), which is relevant in the context of prolonged disruption to routine.

The second study is a life-writing article from September 2020, where Zalewska (2020) wrote about her lockdown identity crisis over formal wear, the author's previous day-to-day dressing. In it, the loss of clothing identity was approached as "a death of a certain part of self (formal, official, busy)" (Zalewska, 2020, p. 177), here depicting the strong link between clothing and identity. Zalewska reported a transformation that resulted from dissecting her judgment towards her lockdown dressing, explaining that "by dropping the business identity and integrating effortlessness and idleness, dressing after lockdown is focused more on my own perception (…) and less on visual perception by others" (Zalewska, 2020, p. 188).

The third study, published in April 2021, is a quantitative analysis of all tweets worldwide in English regarding clothing consumption and buying behaviour, between January and September 2020 (Liu & Lang, 2021). This study found evidence of pent-up demand for clothing, yet also a rediscovery of garments forgotten in the closet, and a drastic change in style that favours comfort. Moreover, their findings provide evidence that consumption change is a fundamental stress coping mechanism, which, according to Lazarus and Folkman (1984), is any readjustment to restore a balance that individuals previously had, and clothing behaviour, is particularly susceptible to it (Liu & Lang, 2021).

After we carried out this study, another study exploring sustainable behavioural shifts in clothing consumption during the pandemic was published by Iran et al. (2022). This quantitative and qualitative study, carried out in six countries, detected a shift towards sustainable clothing consumption in a considerable portion of the respondents, with an average of 29% across the countries.

All the above studies establish that clothing use did change because of pandemic restrictions, so the goal of this study was to compare which habits changed or remained stable a year into the pandemic, adding to literature on resilience.

Method

To delve into the phenomena discussed above, semi-structured ethnographical interviews were carried out in April 2021. They included an exercise inspired by two wardrobe study methods: Skjold's "Biographical wardrobe method" whereby clustering garments in piles reveals the participant's "formulas" for dressing (Skjold, 2017); and Klepp and Skuland's "Auditing a section of the wardrobe", which analyses the objects/user's stories quantitatively and qualitatively, thus revealing novel patterns (Klepp & Skuland, 2017).

The interviews were carried out online due to the circumstances and in order to promote geographical variety, allowing a wider perspective on the topic. They included 20 people living in Western countries and the group was varied in terms of gender (12 men, one of which was trans, and 8 women), age (18–72 years old) and household composition. Twelve participants resided in Europe, 4 in South America and 4 in North America; however, nationalities differed because 8 people had emigrated. Most respondents lived in cities, and income varied greatly. The interview duration varied between 30 and 50 minutes, and most participants were working or studying from home.

Participants were recruited via word of mouth and through an open call on social media, promoting demographic variety (gender and age), and variety in their interest in clothing.

During the wardrobe exercise, participants were asked to make three piles of clothing: The first, made of two to five pieces they had worn the most during the

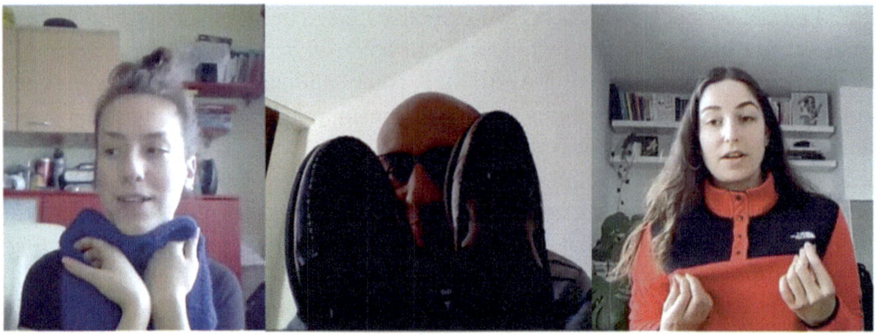

Fig. 1 Screenshot of participants and garments from Exercise 1, April 2021

pandemic. The second pile, of one to three pieces they had worn the least, but missed wearing. The third and last pile, of one to three pieces worn the least, but not missed.

The first pile aimed to capture a global snapshot of lockdown dressing (Fig. 1). Participants were asked about the material, functional, aesthetic and emotional characteristics of the pieces, as well as the time and type of purchase (bought new, second-hand, gifted). These topics were chosen to capture a detailed account of the garments in order to cross-check any behavioural pattern with the pandemic conditions. The second and third piles focused on unused pieces (missed and not missed) to further bring to the surface thoughts and feelings that may fall outside of daily routine.

The interview revolved around three key themes—patterns of use (choice and maintenance), identification (self-expression through appearance), and attachment to clothing—and concluded with a semi-structured conversation. The data collected was interpreted with a thematic analysis, considering the participant's background and demographics. To our knowledge, this is the first study of clothing practices during the pandemic period that includes wardrobe interviews, leading to a more detailed understanding not only of the values behind clothing use, but also about the characteristics of the garments involved, and their relationships.

Findings

On the topic of *use*, only two of the 16 participants working from home were comfortable with, and regularly did stay in, pyjamas for the entire day, debunking the pyjama-ridden narrative often described in the media. Most participants dressed first thing in the morning, and this practice was described as:

- "putting on your 'coffee clothes' to separate your bedroom from your work life". Getting an energising boost, a feeling of productivity from dressing.
- "It seems like it wouldn't, but dressing differently (from pyjamas), more as for work, seems to motivate a little bit more".

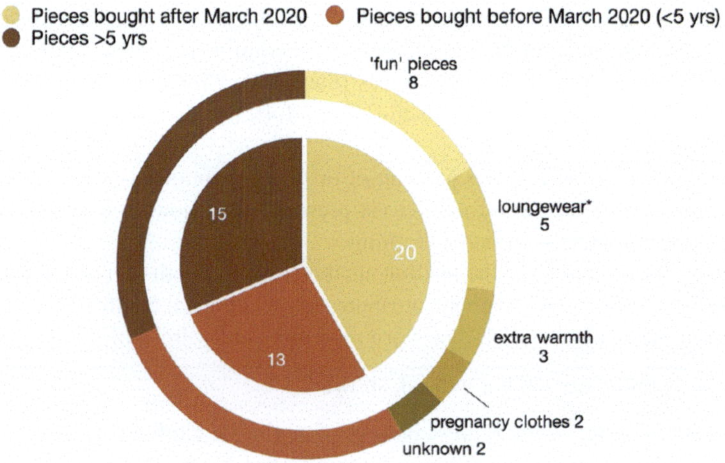

Fig. 2 Analysis of pile 1 by clothing age (<1.5, <5, >5 years old) and motivation for the purchase of pieces bought during the pandemic. * Pieces where aesthetics and comfort were of equal value

- "(keeping a dressing routine) keeps us feeling alive, feeling we're in our normal work routine, feeling well". A reported impression of normalcy and of preserving an internal image, and by extension, maintained one's identity.

Moreover, a significant reduction of the active section of the wardrobe was observed: participants were asked how much of their wardrobe was worn since the start of the pandemic. Their guesses ranged between 5 and 50%, with most guesses at 30%.

On the topic of *disposing*, despite the increase of wardrobe cleanouts (WRAP, 2020), only five participants disposed of clothing in this period, two of which were motivated by a house move. Two reported doing a big clean-out—one motivated by solidarity, the other by reorganisation.

On the topic of *buying*, some participants experienced their altered lockdown dressing with ease, whereas others experienced a certain anxiety from it.[1] Furthermore, the majority of those that felt relaxed in their lockdown dressing bought between zero and two garments (eight participants) since the start of the pandemic, whereas those who experienced boredom from their clothes bought upwards of four pieces (six participants).

Regarding the wardrobe exercise, pile 1—the clothing most worn—showed the following results:

- A preference for newness (see Fig. 2): the majority—33 pieces out of the 56 selected—were classified as new (being less than 5 years old), and of those, 20 had been bought during the pandemic. The main motivations for their purchase were aesthetics and comfort.

[1] There was one exception from the six—one participant didn't buy any garments.

- Participants reported a newfound value for comfort as well as a realisation of how fashionableness often directly conflicted comfort in pre-pandemic times. "if someone is comfortable, that gives (them) other freedoms too". This quote hints to the psychological dimension of comfort, and how it relates to physical comfort.
- The majority of participants gave an effortful description of how the garments currently chosen are "comfortable, yet presentable", showing a reconsideration of the appropriateness of home clothing.
- Comfort was found to be the leading attribute for use, both in material attributes (physical comfort) and aesthetic attributes (psychological comfort). Overall, the attributes of clothes worn most frequently are shown in Fig. 3.

Pile 2, the clothing missed but unused, showed:

- A strong longing for the representational function of clothing: feeling beautiful or confident, standing out through clothing expression. This aspect was more prevalent than missing activities associated with certain garments, such as going out or sports.
- The garments in this pile were older than in pile 1, confirming a preference for newness in the clothing used.

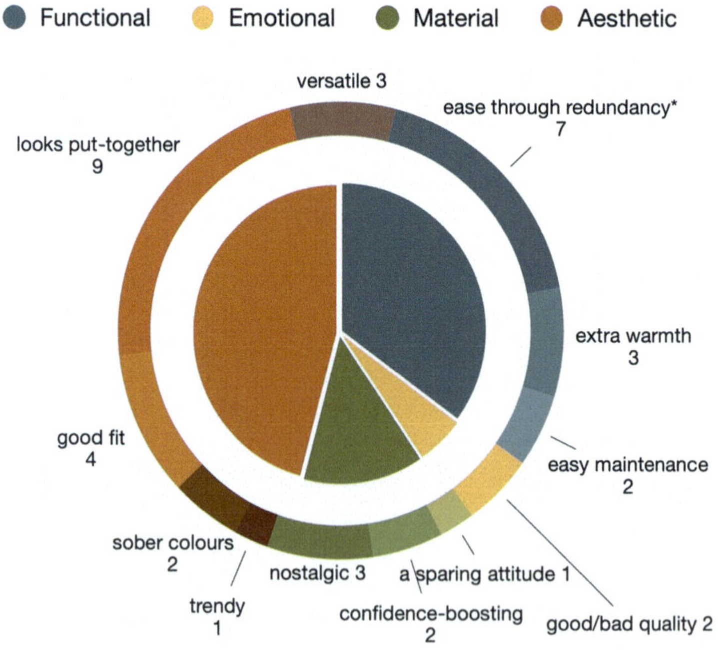

Fig. 3 Analysis of clothing attributes from pile 1, grouped in themes. *A type of clothing that forgoes paring with other pieces. Having multiple of these relieves the wearer of the task of composing a presentable outfit

And pile 3, the clothing not used nor missed, showed:

- For the majority of participants, the disinterest in the pieces selected here showed no discernible link to confinement. Instead, it stemmed from incompatibilities such as discomfort from wearing, or no longer identifying with the garment. This finding suggests that disposal was therefore motivated by external reasons, which is in line with a survey by WRAP that found "donating to charity"—an external driver—to be the leading cause for the increase in clothing disposal in the United Kingdom (WRAP 2020).
- Some participants commented they would discard the pieces from this pile after the interview.

To conclude the exercise findings, there has been a progression from "pure comfort" (Winterhalter, 2020) to "beautiful comfort", including times outside work, thus demonstrating that a concern for representation has re-entered the motivation of dressing choice a year into prolonged confinement. Naturally, this reverted the attitude of "needing less to build identity" observed in 2020 (Winterhalter, 2020)—in this study, some participants explored representation through buying new garments, while others found it in their existing wardrobe.

Finally, on the topics of *identity* and *personal values*:

- At least half of the participants touched on the question "do we dress for ourselves, or do we dress for other people", which explores the connection between identification and representation. None of the interview questions touched on this topic, so the fact that they brought it up shows that disruption in daily clothing practices led to reflection with potential for behavioural change in the long term.
- A few participants reported readjusting their values in light of the devastating impacts of this global crisis. "People started to give importance to certain areas that they didn't really value, for example, the value of life, because of covid. People started to be levelled (and challenged) in the same way, and they can't hide in their clothes to feel different, because that's not going to protect them".
- In another take on this sentiment, "I think during the pandemic people are internalizing who they are, what they really like and what they value", this participant spoke of a genuine way of dressing and how it has helped these realisations come into practice, echoing Zalewska's experience.
- One participant expressed that the lack of pressure to dress or put on make-up allowed for quality time to reconnect to her values and needs. "Now I understand myself on a different level, and because of that, I know how to treat myself better". This realisation was attributed to a "lack of interaction with humans", showing how a break from (physical) social interaction allowed this participant to explore her well-being.

Discussion and Conclusion

To answer how the Covid-19 pandemic, and prolonged confinement in specific, influenced clothing use, we focus on three key findings: the value of comfort, the influence of participants' predisposition to fashion on their consumption and well-being, and the effects of introspection on appearance.

The observed shift from "pure comfort" (Winterhalter, 2020) to "beautiful comfort" (the commonly reported "comfortable, yet presentable" description of garments in use), shows an exploration of what feels right to wear in a scenario that is dually home and remote work. This behaviour shows a redefining of clothing identity within the constraints of lockdown measures. Even though aesthetics regained influence in dressing choice throughout the year, comfort remained the defining attribute of clothing use, which may present an environmental advantage. Previous wardrobe studies suggest that big portions of the wardrobe are not in active use (see e.g. Maldini et al., 2017), and increasing clothing utilisation is seen as a solution to clothing overconsumption (see e.g. Ellen McArthur, 2017). A growing awareness and importance given to comfort by users (and by association, by companies) can to some extent revert the tendency of clothing being designed to be sold rather than to be used. A decline in the purchases that are motivated by representation only (how we show ourselves to the world) could lead to a reduction in demand, and ultimately production.

As for the second finding, participants that experienced some anxiety from a disrupted dressing routine, and bought four pieces or more, were found to be of young age (<26 years) and have a passion for fashion. Clothing expression was an activity that played a central role in the construction of their identities and was therefore difficult to let go of. Conversely, participants who did not experience this anxiety showed less, or a more nuanced, interest in fashion, and bought between zero and two garments. The nuance described by these participants was a mild interest in clothing but with confidence in their dressing choices, with some calling it "expression", others not. Therefore, this study has identified having a "passion for fashion" to be a factor of lower resilience in clothing use, consumption, and overall well-being under prolonged confinement. Another implication of this finding is that predisposition to fashion was found to have a stronger influence on consumption behaviour than routine, or the contextual disruption of it. Importantly, this suggests that only a change in personal values, not external restrictions, can bring about a meaningful reduction in clothing consumption for these individuals.

Lastly, the findings on appearance show two sides. On the one hand, the routine of dressing gave participants a feeling of normalcy and motivation ("it helps to start the day"), which contributed to their well-being. On the other hand, the perception of their appearance was challenged, with participants reporting "a lack of pressure" to dress up, a "freedom" from following trends, and a sudden unimportance of standing out through clothing due to the challenges of the pandemic. Taking the theoretical perspective of Oyserman et al. on the *stability* and *fluidity* facets of identity building, we can see the motivational qualities of the self at play in the former (the guiding anchor to our daily choices); and the dynamic, adaptive qualities in the latter, responding to the prompts brought by isolation ("do we dress for ourselves or for other people?"). It is clear that a meaningful change happened not in clothing itself, but in the wider picture of tending to one's appearance to fulfil a social need (representation). This can be attributed to how reflecting on the self from a more distant perspective, according to Wakslak et al., brings attention to one's broader goals and values (Wakslak et al., 2008). Notably, since identity building is strongly connected to clothing, changes in dressing brought about reflection on the mental constructs of the self—e.g. self-esteem, desires and sense of fulfilment. This realisation prompted an introspection and readjustment of how identities were built before the pandemic ("people are internalizing who they are, what they really like and what they value"), revealing which dressing practices were no longer serving participants. This resulted in an increased confidence, and therefore resilience to change in representation practices. As seen in the second finding, such a resilience may reduce clothing consumption in the long term, indicating this is a positive finding of a change of values towards sustainable consumption.

Limitations and Recommendations

Studying the drivers that may impact clothing consumption in different cultures was beyond the scope of this study. Future studies may consider this factor for a more comprehensive analysis of global clothing consumption. This study suggests implications for a Western mindset, the common trait of the study participants, but the geographical variety within it is not discussed due to the small size of the sample.

We have identified comfort as a characteristic of clothing with potential for environmental strategies for use intensity. Designing and prototyping with a greater emphasis on comfort could therefore be an added value for designers aiming to run a more responsible design practice. Additionally, the transformation of identities during the pandemic is already yielding a growing body of research and news media across many fields (moral, religious, sexual, gender identity, etc.). Considering the significant changes in well-being and consumption behaviour observed in this study, studying the long-term effects of these identity shifts post-pandemic is a compelling topic for future research.

Finally, even in prolonged confinement, people long for newness, seen in how the majority of garments most worn during this time had been bought during the pandemic. To study how this tendency can co-exist with sustainable consumption, future research on how to cultivate a feeling of newness without new products would be vital. Alternatively, solutions to reduce clothing demand should act at a value level to question our need for newness altogether.

Bibliography

ArtEZ. (2021). *Solidarity in fashion*. Research Projects. https://www.artez.nl/en/research/projects/solidarity-in-fashion. Accessed 26 Nov 2022.

Callmer, Å. (2019) *Making sense of sufficiency: Entries, practices and politics*. PhD. KTH Royal Institute of Technology.

EEA. (2019, November 19). *Textiles in Europe's circular economy*. https://www.eea.europa.eu/publications/textiles-in-europes-circular-economy. Accessed 05 Dec 2022.

Ellen MacArthur Foundation. (2017). *A new textiles economy: Redesigning fashion's future* http://www.ellenmacarthurfoundation.org/publications

Iran, S., et al. (2022). When mortality knocks: Pandemic-inspired attitude shifts towards sustainable clothing consumption in six countries. *International Journal of Sustainable Fashion & Textiles, 1*, 9–39.

Klepp, I. G., & Skuland, S. E. (2017). Method 2. Wardrobe studies, wardrobe sections studies. In I. G. Klepp & K. Fletcher (Eds.), *Opening up the Wardobe: A methods book* (pp. 22–24). Novus Forlag.

Liu, C. X. S., & Lang, C. (2021). Clothing consumption during the COVID-19 pandemic: Evidence from mining tweets. *Clothing and Textiles Research Journal, 1*(17).

Maldini, I., Duncker, L., Bregman, L., Duscha, L., Piltz, G., Grevinga, T., Cunningham, G., Tap, R., Balgooi, F., & Vooges, M. (2017). *Measuring the Dutch clothing mountains: Data for sustainability-oriented studies and actions in the apparel sector*.

Mellander, E. P., & McIntyre, M. (2021). Fashionable detachments: Wardrobes, bodies and the desire to let go. *Consumption Markets & Culture, 24*(4), 343–356.

Oyserman, D., Elmore, K., & Smith, G. (2012). Self, self-concept, and identity. In M. R. Leary & J. P. Tangney (Eds.), *Handbook of self and identity* (pp. 69–105). The Guildford Press.

Skjold, E. (2017). Method 11. Biographical wardrobe method. In I. G. Klepp & K. Fletcher (Eds.), *Opening up the Wardobe: A methods book* (pp. 47–48). Novus Forlag.

Wakslak, C., Nussbaum, S., Liberman, N., & Trope, Y. (2008). Representations of the self in the near and distant future. *Journal of Personality and Social Psychology, 95*, 757–773.

Winterhalter, C. (2020). Il corpo, il vestito, il Covid-19. In A. Guigoni & R. Ferrari (Eds.), *Pandemia 2020. La Vita Quotidiana in Italia Con Il Covid-19* (Vol. 1, pp. 157–167). M&J Printing House.

WRAP. (2020, June 23). *67 million items of clothing could be discarded from UK homes post-lockdown*. Love Your Clothes. https://wrap.org.uk/media-centre/press-releases/67-million-items-clothing-could-be-discarded-uk-homes-post-lockdown. Accessed 05 Dec 2022.

Zalewska, A. (2020). Loss and revival of visual identity: Inner work with clothing. *Clothing Cultures, 7*(2), 177–189.

Social and Gender Identity in the 'New Fashion' World: 'Food and Beverage' Made in Italy Television Commercials Year 2022

Federica Maria Marrella

Introduction

Food, besides being pure energy for the body, is actually also an *imaginary*, an *ideology*, a *culture* within which it plays a prominent role. Food is a form of language, like fashion, in its broadest sense. In Roland Barthes, 1961 quotation, the semiologist points out a parallelism between food and fashion. He states:

> From the moment a need is taken over by the norms of consumption production [in short, from the moment it moves to the rank of institution], it is no longer possible in it to dissociate the function from the sign of the actual function... which is true of clothing, and is equally true of food.[1]

Fashion and food are the two clearest and most defined places of social language and culture. Barthes talks about food as an object of creativity and social consciousness, just as a dress or as clothing. He does so in the same essay from which the above passage is taken, in which he describes the value of sugar in the culture of the United States of America.[2] But above all, I would like to focus on what he writes in *Mythologies* (1957), in which he illustrates the role of wine, milk, steak, French fries and ornamental cooking. Three short essays that light up around like a bright lantern: so ironic, so cynical, so exquisite. What does Roland Barthes say about wine? Wine for the French is an institution. Wine is an act; it is a gesture, and it is even related to existence, to a person's way of life. Furthermore, wine allows irrational acts. It justifies drunkenness, because it is actually an entity for the French of complete integration. Wine, therefore, has a profound sense and is profoundly

[1] Barthes R., 1961, trad. it. Marrone G., pp. 33–34, 2001, in Ivi, pp. 14–15

[2] Barthes R., 1961

F. M. Marrella (✉)
IULM University, Milan, Italy

A. Cruchinho, J. C. Neves (eds.), *Sustainability in Fashion: Creativity, Education, and Innovation*, https://doi.org/10.1007/978-3-031-80991-0_8

artistic, dramatic and theatrical, even in gesture, and, no less, in the semiotic construction of social classes.

Wine is 'socialised', he states, because it is based not only on a *moral* but also on a *scenario*. It is the centre of the 'ceremonials' of the French. It is the pivot around which anything can happen. And, if we notice well, the words that the semiologist repeats are *moral, national, scenario, social, mythology and function*. Everything that lives in a reality far beyond wine itself, but which makes it better understood.

And what is the opposite of wine? Is it water? No, it's milk.

Milk is the real anti-wine. 'It is opposed to fire', he states, 'because of all its molecular density, because of the creamy, and therefore sopitive, nature of its surface. Wine is mutilating, surgical, it transmutes and gives birth. Milk is cosmetic, it connects, covers, restores. Moreover, its purity, associated with childish innocence, is a pledge of strength [...] calm, white, lucid, in all equal to the real'.[3]

Wine is blood, milk is placenta.

Steak also participates in the same mythology as wine.[4] Meat, for the way it is cooked and put together with other ingredients, becomes an oneiric, prenatal, placental image. This sounds like the description of a Salvador Dali painting. It is no coincidence that, to explain his work *The Persistence of Memory* (1931), the painter stated that it was inspired by a dream he had one night after eating too much 'Camembert' for dinner the night before. Food, thus, becomes the place of the social but also of the dream, of the most hidden and unpredictable desire, like the unconscious.

Another word repeated by the scholar is 'national' or 'nationalised'. Food becomes Homeland for the French, sometimes more Homeland than Society. But at the same time, it describes, in the way it is 'processed' and cooked, the social environment in which it is served and savoured and also advertised.

In fact, another essay in *Mythologies* is crucial for understanding the *communication* of food, namely the one on ornamental food. Here, Barthes describes the food displayed in the photographs of the *Elle* newspaper.[5] He traces different images displayed by the newspaper, characterised by exaggeratedly processed and covered, packaged, embellished, ornamented food: glazed, coloured, shiny, baroque food, but at the same time, as if hidden, and magically, among other things, it is only photographed from above. This photographic viewpoint accentuates the fact that that sweet is an object, close but elusive, whose consumption can only be accomplished through the eyes, through sight. So, the way the image of food is made, produced, displayed, photographed, narrated, is also a *sign*, a clue to the society to which it is offered. Just, if we think about it, like the image of fashion. And like the image of advertising.

[3] Therein, p. 69

[4] Therein, pp. 71–73

[5] Therein, pp. 125–126

Food and *Beverage* could, therefore, be the new *Fashion*: an aesthetic and cultural quest that goes beyond the purely functional, more so perhaps than any other social language.

Research Questions

The main objective of this study is to investigate the representation of social identity in the most played Made in Italy Food and Beverage commercials on television in April 2022. The analysed channels are: Rai 1, Rai 2, Rai 3, Rete 4, Canale 5, Italia 1, La7. The time examined is the evening time, from 19.00 to 23.00. The choice was made to focus on adult, youth and family target brands.

The most played commercials of food brand Made in Italy are:

Riso Flora, Nutella Be Ready, Magnum Michelangelo 'The Universal Pleasure,' Spaghetti Barilla Pasta al Bronzo, Pesto Barilla, Pan Bauletto Mulino Bianco, Insalata Bonduelle, Gelato NUII, Gocciole Pavesi, Biscotti Gran Cereale Frutta, Pesto Tigullio, Galletto Vallespluga and Sabelli Mozzarelle e Ricotte.

The most played commercials of beverage brand Made in Italy are:

Birra Peroni, Birra Moretti, Birra Ichnusa, Acqua Vitasnella, Acqua Levissima, Estathé and Crodino.

This resulted in a **total** of **20 spo**ts to be analysed: **13** of food and **7** of beverage.

Methodology

An interesting point from which to approach the analysis of the commercials is the Semiotic System table expressed by B. Martin and R. Felizitas, in *Key terms* in Semiotic[6] (Table 1):

However, it is worth highlighting the fact that advertising in contemporary society is increasingly imposing itself not only as an autonomous language but also as a true *discursive form*, as 'the configuration that a given set of social practices,

Table 1 Semiotic system table

Sign	Commercial film
Signified	Visual text (all visual elements in the commercial film and their presentation style), verbal text (external sound, slogan, speech, music and sound effects), written text (slogan, logo)
Signifier	The connotations that took part beside signified in the commercial movie
Analysis	Assessment of the relationship between the signified and signifier

[6] Martin B., Felizitas R., (2006), in Balci V., Özgen, C. 2017, p. 25

interest systems, social groups, ethical convictions and cultural orientations takes on at a given time and in a given society'.[7] From a marginal role in social dynamics, advertising is increasingly becoming an 'advertising form'.[8] That is, it has developed its own discourse, its own way of representing social characters and their relationships. Therefore, in addition to the table above, the contribution of Jean-Marie Floch's work on the consumption value system is fundamental in this perspective. The square of consumption values is proposed by Floch in an essay dedicated to various car advertising campaigns, with a particular focus on that of Citroën.[9] In this essay, the semiologist creates the classification of value types, dividing consumption values into four groups, structured in a theoretical square. The famous square that the scholar calls 'of the axiologies of consumption behaviour' is as follows[10]:

The categories identified by Floch seldom appear divided in advertising campaigns; often campaigns focus on a mixture of the four types of values (Fig. 1).

Speaking of 'values', another necessary point to illustrate, before concluding this presentation of the research methodology, is the subject of brand identity. Mauro Ferraresi is the one who recently has deeply investigated the combination of brand identity and image.[11] The brand performs the functions of *differentiating* and *enhancing*; as it creates a difference between itself and the context, it builds its own identity with own values.[12] This process is possible thanks to the *image system*.

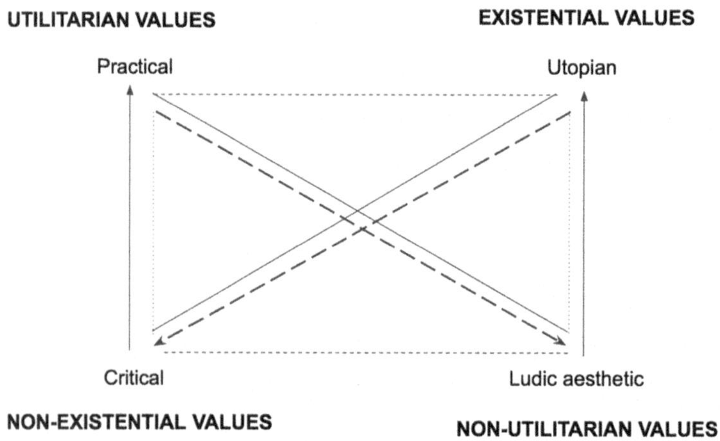

Fig. 1 Types of values

[7] Semprini A., 2003, p. 101, in Bianchi C., 2005, p. 77

[8] Codeluppi V., 1996, in Ibidem

[9] Floch J. M., 1990, cap. 6: 'J'aime, j'aime, j'aime,' Car advertising and the consumer value system

[10] Floch J. M., 1990, in Pozzato M. P., 2013, pp. 152–153

[11] Ferraresi M., 2003, in Bianchi C., 2005, p. 99

[12] Ferraresi M., 2003

Ferraresi[13] points out that many of the typical characteristics of the brand world are common to what Umberto Eco defines as the 'possible world of narration':

> cultural construction of the narrated universe, the high level of cohesion (i.e. internal connections within the text at surface level) and coherence (i.e. stability and continuity of meaning), the suppression of disbelief necessary to understand them, the production of moods, sensations and synaesthesia.[14]

But it is also true that what is represented—which is often narrated as a great revolution in form and content—is nothing more than a tiny variation within the same theme of values. Furthermore, the types of possible narratives that the brand discourse proposes vary between the 'verisimilar' and the 'improbable', and only rarely do new narratives arise. Advertising communication must propose values compatible with those the consumer is willing to accept; advertising must first of all be reassuring, maternal and parental, citing Jean Baudrillard's articulate thought.[15] In fact, Volli states: 'brand discourses are weak, confused, usually careful not to create divisions in the target audience (and therefore not to be too incisive)'.[16] It is as if everything is always fiction, a narrative that wants to seem real but is always far from the truth. René Magritte's 1928–1929 work 'The Betrayal of Images' or 'Ceci n'est pas une pipe', a masterpiece of Surrealism and the psychoanalytic link between objects and the human being's vision, between objects and their subjective meaning, thus comes to mind once again. Dealing with advertising thus seems like a game between intentions and ideas, between active and passive—or unconscious and unpredictable—meaning, between subject and representation of the subject. That *Cut* that Lucio Fontana, years later, would make on a canvas seems to mean: 'go beyond representation, this is only fiction'. The spatial concept is nothing other than understanding 'La Trahison des Images'.

Furthermore, conscious of having to deal with continuous betrayals of images, with iconological memories, with improbable and sometimes grotesque narratives, the aforementioned methodological approaches were used in the analysis of the chosen spots, in which Erving Goffman's ritualisations and hyper-ritualisations and the representations and archetypes found and investigated in my previous research work were also analysed.[17]

The Interpretative Framework

> Every era has the revival of antiquity that it deserves.[18]

---

[13] Ferraresi M., 2003, pp. 102–103, in Ibidem

[14] Bianchi C., 2005, p. 100

[15] Baudrillard J., 2003, p. 213

[16] Volli U., 2003, p. 91, in Bianchi C., 2005, p. 101

[17] Marrella F. M., 2019

[18] E. H. Gombrich, 2003, p. 206

The results of the research can be summarised as follows:

Observing Tables 2 and 3, we can summarise as follows:

- In general, there is an almost continuous presence of playful-aesthetic enhancement, both in food and beverage: in a total of 18 out of 20 spots, except in the Mulino Bianco Bauletto Bread and Ichnusa Beer spots.
- Secondly, there is practical valorisation, in 10 commercials but, interestingly, only in 2 out of 7 beverage commercials, and instead in a much higher percentage in food: in 8 out of 13 commercials.
- Utopian valorisation can be found in 8 spots: 4 out of 13 (Food), 4 out of 7 (Beverage).
- Critical valorisation is completely absent.

Food and beverages, in Italy, are not the place where the Italian consumer wants to save money. The virtually absent critical appreciation seems to affirm just this, as it is equally affirmed that on food the Italian demands quality (practical appreciation is very present).

Both food and drink are an invitation to joy, fun, beauty, pleasure and sociability (Ludic-Aesthetic Valuation). Through this world of *Food* and *Beverage,* they also want to address utopian and chimerical themes: union, peace, humanity, fight against prejudice, love and happiness.

As far as 'fashion bodies', archetypes, stereotypes, ritualisations and hyper-ritualisations are concerned, we can summarise as follows:

- The *female 'touch-taste'*: it is almost omnipresent. It is the woman who tastes, especially in Food. In Beverage, tasting is more masculine, except in water commercials—Levissima and Vitasnella.

Table 2 Results of the research

Spot Food	Practical V.	Utopian V.	Critical V.	Ludic—Aesthetic V.
Riso Flora	X			X
Nutella B Ready				X
Magnum Michelangelo				X
Spaghetti Barilla		X		X
Pesto Barilla	X	X		X
Pan Bauletto Mulino Bianco		X		
Insalata Bonduelle	X			X
Gelato NUII	X			X
Gocciole Pavesi Extra Dark				X
Biscotti Gran Cereale Frutta	X			X
Pesto Tigullio	X			X
Galletto Vallespluga	X			X
Sabelli Mozzarelle e Ricotte	X	X		X

Table 3 Results of the research

Spot Beverage	Practical V.	Utopian V.	Critical V.	Ludic— Aesthetic V.
Birra Peroni		X		X
Birra Ichnusa	X	X		
Birra Moretti		X		X
Acqua Vitasnella	X			X
Crodino		X		X
Estathé				X
Acqua Levissima				X

- *Female ecstasy*: the moment of female ecstasy, often represented with close-ups and sighs, occurs immediately after the product has been tasted. Female sensuality is linked to the pleasure of taste.
- The *female 'touch'*: the female sensual touch, narrated by Erving Goffman, is present here mainly through the touch of the product with the mouth, and thus through the sense of taste. However, in some commercials, we have the touch of the woman caressing her face, or her hair, an attitude reminiscent of sensuality.
- The *authorised female estrangement*: we encountered it in two commercials, and it indicates a somewhat distracted, scatterbrained, immersed-in-her-own-thoughts female representation. Elements that do not make the spectator experience her as a reassuring and credible character. Or rather, she is such in the tasting, in the taste of the product, not in the practical enhancement of it. A situation that is also related to and expressed in the next point.
- The *male voice—functional hierarchy*: the male voice-over is present in 14 out of 20 commercials. The female voice-over can be found only in 2 out of 20 spots. In the other commercials, there is no voice-over. This detail is extremely interesting. It suggests that the voice—which represents not the aesthetic and sensual image, but knowledge, reassurance and accompaniment to understanding the product—is almost always male. This detail ties in with Goffman's ritualisation of the functional hierarchy. When it comes to the explanation of a product, it is always the man who is the protagonist, and the woman stands by to listen to his reading of instructions. This happened in the representation of the advertisements analysed by Goffman. Here it happens with hearing. The female spectator—the main target of these products—feels more reassured by a male voice, which is warm and welcoming but, at the same time, evokes security, professionalism, knowledge and rigour.
- *The traditional body (often Nordic)* vs *the multicultural body*: an element that is repeated in several commercials is the presence of bodies that I would call 'natural' and also 'Nordic': women not wearing make-up, simple, dressed in light colours, blue eyes, light hair. We could find the multicultural body in just two spots, and in just one spot, the Italian Mediterranean body. The traditional idea of 'wellness' is linked to this iconography: if you eat and drink healthy, you are blond, tall and blue-eyed. We found the multicultural body in two commercials that clearly target young people: Barilla's 'Le Roommates' and Birra Moretti.

We find it also in the Crodino ad, but it is a multicultural body mainly related to the message of internationality and world unity, not multicultural in an event-sharing place.

- The *'sacred rite'*: food, the sharing of it, white as the colour of purity, the breaking of bread or the cutting of mozzarella cheese, are elements that recall the Christian rite of the sacredness of sharing food: an idea present in several commercials.
- The *fable and romance*: the fable of happiness is an undisputed protagonist of these narratives. Love, dreams and pleasure are the most repeated words in the claims of these commercials.
- *The colours of Italy:* the value of 'Made in Italy' is not only told through words—Pesto Tigullio and Sabelli commercials, for example—but also through colour, as in the advertisement analysed by Roland Barthes of 'Panzani'. Here, green, white and red are the colours that praise the Homeland: Italy as a place of quality raw materials and as a place and thought of national pride.
- *Food and beverage, the two genders*: it seems, from the representations analysed, that food has a more feminine valorisation, while beverages, except for water, have a more masculine one. As if food and drink were actually the visual representation of two genders: female and male. This makes us think that the target audience of the representation of food is mainly women, while that of drinks is mainly men. With the exception of water, that is associated to well-being and health, and to purity, elements that in our narratives seem to be purely feminine, while fun and goliardia are considered to be mainly masculine.
- *Sexual and gender identity*: in the analysed advertisements, we have only found homosexual representation once, in Barilla's 'The Roommates', and in a fairly stereotyped and superficial way. The world of food and drink in Italy is only male and female, and heterosexual; anything else is non-existent.
- *The family*: once again, the family represented is generally traditional: mother, father, grandparents, children, grandchildren: a basic heterosexual family, or a mother with daughters/children. Single mothers and single fathers are never shown. Single grandparents with grandchildren. Adult women and men alone with parents. Everything remains very close to the perfect ideal of a large, loving family and also the family of friends. The idea of group, of number, is present. There is little loneliness and little representation of families other than the 'traditional' and 'ideal' one.
- *The pandemic*: it cannot be overlooked that these commercials were made and also broadcast in the years of a pandemic. *Smart working*, for example, has a specific value in these narratives. But above all, the continuous reiteration of ideals such as love, dreams, feelings, happiness, sweetness, companionship, pleasure, friendship, hugs and sharing underlines how much these are thoughts structured around their lack and the tragedy that occurred in the years 2020–2022. Food and drink so, even more, meant in the language of these advertisements sharing, sacredness, closeness and affection. 'Staying human', the ironic phrase of the Gorilla in the Crodino spot, actually seems to be the *leitmotif* of the majority of the commercials taken into consideration.

Research Limits and Possible Next Steps

The analysis of each image, and of the moving image, with quotations from Erwin Panofsky to Cinzia Bianchi and beyond, is in itself a subjective, partial analysis, in any case filtered by the view and experience of the scholar, however much he or she tries to draw inspiration from studies, techniques and methodologies. Regarding the language of the commercial in particular, the analysis must be multidisciplinary, because each element has its own value: voice, music, shots, cinematographic language, iconography, iconology, semiotics, words, language, silences, pauses and colours. This is the first limitation found not only in this research, but in any research that approaches the study of images and creative representations: the extreme relativity of what one has tried to read, decipher, describe and narrate.

Another central point is the *medium*: television is now defined as a residual medium. Who watches it? When? Is the world of communication now only that of the web? So what does advertising on the web look like instead? Does the Barilla brand use a different image on the web from the one it uses on television? Do the commercials, the images, the narratives change? The analysis of each major Made in Italy food and beverage brand between television, the web and social networks could be another possible focus of the research: that is, their way of declining advertising according to the medium that shares it.

It would also be interesting to study viewers' *engagement* in social media advertisements. How do consumers comment on advertising images? How do they comment on YouTube videos? What is the *feedback* on viewers? How do 'prosumers' behave?

Furthermore, it would be stimulating to study how the communication of a Made in Italy brand is abroad: how do narratives change? Do they adapt to the cultures they address? Or do they take up the same stylistic features intended for the Italian public?

Remaining in the medium of television, it would also be engaging to carry out an analysis based on the times of day and the seasons of the year. Here, the focus was on the evening time in the month of April. Would anything change in the morning time? Or in the afternoon? What are the programmes at that given time? And what is the main target of a chosen time? And what if we changed the month of the year instead? What would the findings be?

This area of research is indeed wide and complex, and both the limits and the new possibilities for investigation are many and unpredictable. And this is perhaps the fascination of communication: its extreme vivacity and difficulty; its layering and dimension; its consistency and inconsistency; and its concreteness and fickleness.

Bibliography

Barthes, R. (1957). *Mythologies*. French & European Pubns.

Barthes, R. (1961). Pour une psycho – sociologie de l'alimentation contemporaine, In: Annales. Economies, sociétés, civilisations, 16th ed. année, N. 5, pp. 977–986.

Baudrillard, J. (2003). Il sistema degli oggetti, Bompiani [1° ed. 1968].

Bianchi, C. (2005). Spot, Analisi semiotica dell'audiovisivo pubblicitario, Carocci Editore.

Codeluppi, V. (1996). La società pubblicitaria, Consumo, mass media, ipermodernità, Costa e Nolan

Codeluppi, V. (2013b). Storia della pubblicità italiana, Carocci Editore

Ferraresi, M. (2003). La marca, Carocci.

Floch, J. M. (1990). *Sémiotique, marketing et communication. Sous les signes, les stratégies*. Presses Universitaires de France.

Gombrich, E. H. (2003). Aby Warburg, Una biografia intellettuale, Le Comete Feltrinelli, Aprile [1° ed. 1970].

Marrella, F. M. (2019). Iconografie del Consumo, La costruzione dell'immagine femminile tra moda, arte e pubblicità, Aracne Editrice, Collana Disembedding.

Marrone, G. (2001). Panzani e Camay, Due testi esemplari nell'analisi della pubblicità, Atti del Congresso di Castiglioncelli, ottobre.

Martin, B., & Felizitas, R. (2006). *Key terms in semiotics*. Continuum Books.

Pozzato, M. P. (2013). Capire la Semiotica, Carocci Editore, Studi Superiori, Settembre.

Semprini, A. (1993). *Marche e mondi possibili, un approccio semiotico al marketing della marca*. Franco Angeli.

Semprini, A. (2003). *Lo sguardo socio semiotico, Comunicazione, marche, media, pubblicità*. Franco Angeli.

Volli, U. (2003). Semiotica della pubblicità, Laterza, Roma – Bari.

“Bold Is Beautiful” by Anouk, Myntra: An Introduction of a Social Media Campaign Against Gender Inequality

Neel Vipinchandra Naik and Ana Sofia André Bentes Marcelo

Introduction

The development of digital media has created unique opportunities for civic movements to fight for their causes. Defense of values such as women’s rights, gender equality, and women empowerment has not escaped this trend and has been on the agenda of fashion brands’ strategies, using social media to voice their activism, as Kumari (2020:626) argues, “Social media is a new tool, weapon or platform for women to discuss, communicate, voice their opinion, organize a campaign and online activism.”

On social networks such as Instagram, Facebook, or Twitter, fashion brands publish content, seeking to stimulate social debates on topics such as female homosexuality, single-parent families, among others, necessary to question the prejudices of the current society and promote the much-desired social and mentality change, networking with a whole community to support women empowerment (Lima e Casais, 2021).

Gender equality and women’s empowerment, according to UN Women, may be reached through creating a society free of bias, stereotypes, and prejudice, a society that is varied, egalitarian, and inclusive—leaving no one behind. The United Nations Development Program (UNDP) Gender Inequality Index 2020 puts India 131 out of 189 countries in terms of gender imbalance. It assesses gender disparities in three

N. V. Naik (✉)
Escola Superior de Artes Aplicadas do Instituto Politécnico de Castelo Branco (CIPEC – Centro de Investigação em Património, Educação e Cultura) & LABCOM, Universidade de Beira Interior, Covilhã, Portugal

A. S. A. B. Marcelo
Escola Superior de Artes Aplicadas do Instituto Politécnico de Castelo Branco (CIPEC – Centro de Investigação em Património, Educação e Cultura), Castelo Branco, Portugal
e-mail: ana.marcelo@ipcb.pt

A. Cruchinho, J. C. Neves (eds.), *Sustainability in Fashion: Creativity, Education, and Innovation*, https://doi.org/10.1007/978-3-031-80991-0_9

critical areas of human development: reproductive health, empowerment, and economic well-being (Mathur, 2022).

Some fashion brands such as the Indian brand Myntra, Anouk, the object of this study, have glimpsed an opportunity to develop online advertising campaigns to touch on topics of interest to women in today's society.

According to Casais (2021), brands develop advertising campaigns to promote identification with their audiences, creating messages that represent social and cultural archetypes. The phenomenon of inclusive communication includes subjects such as cultural and generational diversity, individuals with impairments or minorities, and the cause of female empowerment. Santos (2016) argues that campaigns designated as femvertising explore women's self-esteem and their self-confidence to make ambitious choices and be free from any social pressures. Advertising thus assumes a decisive role: to be a catalyst for social change, by including in the contents of the campaign's values and attitudes that are the mirror of society, following its evolution.

India is embracing a more digital society in line with the global trend. Utilizing a variety of platforms, including Facebook, Instagram, Twitter, and YouTube, among others, serves as a catalyst for India's overall economic growth.

The number of online shoppers is increasing, and three out of every four are anticipated to originate from medium-sized cities. With Internet penetration predicted to reach 60% by 2022 and a rapidly expanding "netizen" population, these customers are looking for a closer connection with brands and a definite preference for traditional/ethnic content.

Brands have more freedom in this new digital reality to choose from a wider range of diversified and current subjects that satisfy the expectations of the target audience. Advertising stands out as a key component of digital marketing strategies for strengthening this bond between brands and their customers.

Every so often, a groundbreaking advertising film makes headlines for its sheer audacity or for defying tradition. "Bold is Beautiful," the ad film campaign for Myntra's ethnic clothing line Anouk, honors the uncommon Indian woman who breaks all unwritten conventions and, as a result, garners attention as well as plaudits.

The new generation of Indian women is featured in three different scenarios in this three-film ad campaign. The first film, named "The Visit," focuses on the problem of homosexuality, which is at the heart of this chapter's examination. The Visit is not India's first lesbian advertisement, but it is the first to come out openly. From a methodological standpoint, we chose an exploratory approach, first conducting a quick bibliographical search on the topic and then conducting a case study of Anouk's advertising campaign.

The Brand: *Myntra*

Myntra is India's biggest e-commerce company. They are dedicated to making fashion and leisure products affordable to everybody. From their humble origins as a customization company in 2007, they've progressed to become technological and fashion trailblazers.

Mukesh Bansal, Ashutosh Lawania, and Vineet Saxena founded Myntra in February 2007. Myntra is based in Bangalore and was founded with the help of venture capital investors (Bapna, 2011).

Personalization of products, such as T-shirts, mugs, and caps, was the company's first foray into the B2B (business to business) market. In 2010, the company changed its approach to become a B2C (business to customer) company, increasing its product line to include fashion and leisure items.

Myntra.com is a multi-brand aggregator. In 2011, Myntra moved away from personalization and started selling fashion and leisure products. By 2012, Myntra had 350 Indian and international brands under its umbrella to choose from (Bazala, n.d.).

Flipkart purchased Myntra in 2014 for a reported Rs. 2000 crore (US$270 million). Under Flipkart's ownership, Myntra continues to operate as a separate brand, focusing mostly on "fashion-conscious" customers (Sen, 2018). In 2014, Myntra had over 150,000 products from over 1000 brands in its portfolio. Adidas, Nike, Puma, Catwalk, Inc. 5, United Colors of Benetton, FCUK, Timberland, Avirate, FabIndia, and Biba are just a few of the well-known international and national brands. Myntra's private label brand portfolio, which includes Roadster, All About You, Mast & Harbour, DressBerry, Kook N Keech, and others, currently accounts for 25% of the platform's income (Chengappa, 2018). Anouk is an ethnic wear label that is a part of Myntra's private label brand portfolio.

The Label: *Anouk*

Anouk: Materializing a Girl's Heart, Mind, and Soul!

Anouk is a contemporary ethnic wear brand from Myntra's fashion portfolio that promotes the existence, beauty, and, most importantly, independence of women. It was launched in 2013. It is presently the most popular ethnic clothing brand on Myntra (Sharma, 2015).

Manish Aggarwal, VP Marketing at Myntra Fashion, says, "Anouk presents a range of apparel for women that are stylish, contemporary and modern. The brand is all about celebrating women, their beauty, their existence, their grace and their freedom. According to prevalent thinking, ethnic wear is seen as the safe option as it makes them look less sensuous and less appealing to the male gaze (Neogy, 2015)."

Anouk's designs are made for women who are always evolving, discovering, charming, and adventurous. Anouk is the epitome of a cosmopolitan Indian woman's vivacious liveliness. The excellent printing, dyeing, beading, and gorgeous thread embroidery, as well as the stunning artwork featured in this clothing, have a major influence on customers.

Bold Is Beautiful

India is regarded as one of the world's fastest-growing economies, and this is reflected in its modernized society and advertising community (Gupta, 2015a, b). As Indian society becomes more comfortable discussing topics like live-in relationships and homosexuality, the branding world is adding fuel to the fire by including these concepts into its advertising campaigns.

In 2015, Myntra launched a new digital campaign for its ethnic apparel brand Anouk called "Bold Is Beautiful." There have been several discussions about the campaign, and it has received a lot of attention on social media. The three videos, created by Ogilvy & Mather and produced by Hectic Content, depict three different aspects of modern Indian women's choices. The topic stands out because of the issues it addresses, such as "homosexuality," "single-parenting," and the "pleasure of remaining single."

Manish Aggarwal, VP Marketing, Myntra Fashion Brands, describes the campaign as follows, "A plethora of ethnic brands occupy this safe space of ethnic wear, and are mutually indistinguishable. We wanted to give Anouk a sharply defined identity in this crowded category—describing it as ethnic wear that is bold, vibrant and stylish and even edgy at times. Our intention was to highlight that wearing Anouk is not a compromise, but a choice (Neogy, 2015)." The videos are approximately 3 min long and address subjects that society does not discuss openly. The Anouk woman is depicted in the advertisement as a free spirit who is yet connected to her heritage.

There has been a shift in the advertising industry's attitude to conceptualization, particularly when it comes to addressing such delicate social topics. Many recent campaigns, such as Dove's "Real Beauty" and Jabong's celebration of the theme of "Being You," are good examples of difficult themes being addressed.

"The purpose of the ad was to promote Indian-style garments. Both the client and us, we believe that just because certain women wear Indian clothes, does not mean that they are not empowered or have the right to think about society as they would like to," Ogilvy & Mather's Executive Chairman and National Creative Director, Piyush Pandey, stated (Gupta, 2015a, b).

Nicole Braganza, Associate Creative Director, BC Web Wise, commented on the new digital campaign, "The Anouk series is a take on bold problems, the ones that women are often taught not to make a huge issue out of. It works well to position the brand as one for the bold and unconventional Indian woman. What is refreshing is that the brand message "Bold Is Beautiful" is nonintrusive but instead is delicately interwoven through a very real, slice-of-life story. We're increasingly seeing brands go the "social message" route to connect with audiences through online films, so the idea may not be entirely novel. But it's tastefully done and deserves all the praise for that (Neogy, 2015)."

The ad campaign's motto, "Bold is Beautiful," has not only sparked debate but also gone viral on social media. On YouTube, the ad has received over 1.73 million views. Anouk has attempted to appeal to the new-age modern consumer who believes in breaking societal taboos in India's oversaturated ethnic wear market. YouTube, Facebook, Amnet, Vuclip, Vdopia, and Affle were the social media platforms that increased the campaign's reach. Storypick, ScoopWhoop, and Twitter were the channels that ensured the continuance of the theme's conversation after the campaign's quick impact and viral propagation. *#AnoukBolIsBeautiful* became a trending hashtag and a nationwide conversation subject. The campaign received extensive media coverage and instantly became a topic of discussion in the creative and corporate worlds. At a total production cost of $50,000, the campaign generated 16 million paid reach, 4 million video views, and 11,000 tweets that reached 32 million people, resulting in a threefold increase in Anouk's sales (Mandrawadkar, 2015).

The Visit

In India, homosexuality has always been discussed in hushed tones in society. The "bold and beautiful," contemporary women living life on their own terms, are projected in the trio ad film series developed and implemented by Ogilvy & Mather (Fig. 1). "The Visit," the first of these three ad films, has piqued everyone's interest by displaying the homosexual relationship in an advertisement. The first film, "The Visit," has been praised for making headway by openly presenting what is thought to be India's first lesbian relationship in advertising—without going into humorous or sleazy terrain. Deepa Mehta's film "Fire" was one of the few instances of such a relationship that one may recall in the Indian media (Kohli, 2015).

"The Visit" was made to portray a lesbian relationship as something relatable, normal, and everyday rather than something unique (Gupta, 2015a, b).

The movie follows the two ladies as they transition from a cramped, stuffy bedroom to an open, breezy living room as they get ready for one of their parents to arrive. The two ladies "come out" to the audience before coming out to their parents as their connection and relationship are gradually made clear. By the end of the movie, the audience is asked if they have altered their opinion of them after seeing their metamorphosis. Or do they think of this pair as a reflection of themselves who have fallen in love with their better halves? Are they engaged in a union that is just as legitimate as yours? The courage with which these two ladies acknowledge their relationship in a society that still rejects lesbian love is what makes it extraordinary.

According to Shamik Sen Gupta, a member of the ad's creative team as well as the film's director, an early iteration of the script featured only one woman speaking off-camera to her partner. In the end, it was revealed that her lover is a woman. His choice as a director was to show both women and put the audience right in the heart of their daily lives. He planned to introduce them to the audience without revealing

Fig. 1 Stills from the ad film "The Visit." (Source: YouTube video Myntra, 2015)

their sexual orientation or relationship. It's a decision that eliminates the automatic divide between "them" and "us" that occurs when watching a lesbian film (Gupta, 2015a, b). The film's realism and candidness are enhanced by acclaimed cinematographer and documentary filmmaker Ranjan Palit's handheld camerawork and minimal artificial lighting.

Analysis

The goal of this digital campaign is to challenge society's established expectations for women and emphasize their real-life experiences. The long-format advertisements tell the tale of three women who made conscious decisions in their lives. As the story progresses, they are depicted defending their decisions, confronting society's questions, and emerging victorious.

Gunjan Soni, Chief Marketing Officer and Head, International Brands Business, Myntra, commented on the campaign's conception "Anouk's campaigns have always exemplified how the Indian woman can voice her opinion and craft her own destiny. Through these campaigns, the intent has been to deliver the brand message in an impactful but non-intrusive manner" (exchange4media Staff, 2016).

Ogilvy and Mather were tasked with launching Anouk as a line of bold and contemporary ethnic clothes for the modern Indian woman, and they took advantage of the undifferentiated ethnic wear market.

Without any explanations, double entendre, or teary monologues, Myntra's ads beautifully reflect the new India. The ads, which were launched on the digital platform, went viral and received a lot of positive feedback on social media. None of the characters appears to be as obnoxious or over-the-top as they are in a Hindi soap opera or film. Furthermore, the actors' performances provide a serious tone to the storytelling. Overall, the three advertisements give the impression that we live in a new, progressive India (Bhattacharyya, 2015).

According to Manish Aggarwal, VP Marketing, Myntra Fashion Brands, digital is now the only medium being used to build the brand. He adds that it is the best medium for storytelling, "This medium allows you to create interesting storylines, engage with consumers and get a feel of consumer responses in real-time. Given that we wanted to encapsulate real-life situations in a short film format and showcase three short stories simultaneously, digital became the most apt choice" (Sharma, 2015).

"The Visit" is an excellent illustration of the fear that a person feels upon meeting their partner's family for the first time. The campaign has eloquently conveyed how a woman may put a stop to all social stereotypes and be fearlessly "Bold and Beautiful"! The statement "let's not hide it anymore" aims to weave ethnic clothing and one's personal decision together. Ethnic apparel has often been associated with showing the traditional way of life; however, this advertisement attempts to deviate from this idea by incorporating a delightful aspect of color and choice.

This campaign is revolutionary in the sense that it has approached the subject sensitively, incorporating the required emotional component. Creating beautiful layers of a relationship, full of love and, more importantly, honesty. Genesis Film Production's founder, Prahlad Kakkar, claims "The ad has been dealt in a very sensitive way and has portrayed today's Indians who are ready to come out of the closet and declare their preference. The point is that our conservative society, institution, and the Right Wings are having certain reservations in accepting the reality. But since it has come into the limelight and people are ready to accept it, media like television, the internet is bound to take on issues resulting in communication in that direction (Fig. 2). As long as it's not offensive I think it is fine" (Adgully Bureau, 2015).

Jabong has recently been in the spotlight for launching ads that address a variety of concerns. Their most recent campaign, "Be you," has made an impression on the

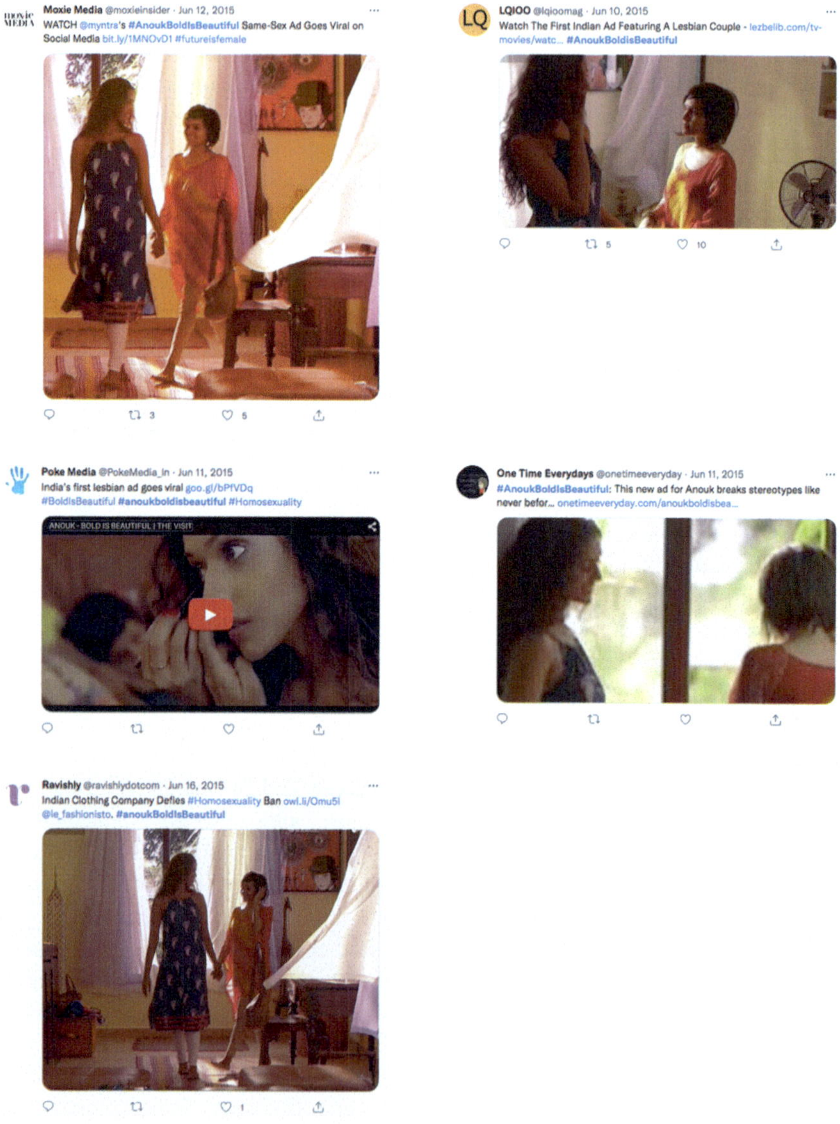

Fig. 2 Examples of a Twitter post thread for the campaign #AnoukBoldisBeautiful. (Source: Twitter @AnoukEthnicWear Myntra, 2015)

audience. Praveen Sinha, MD and Co-founder of Jabong says: "As a brand, we are not hesitant in talking about or touching sensitive issues. It helps us connect better with the current generation who believes in embracing their uniqueness and is unescorted by the thoughts of others. Likewise, Jabong too believes in & stands for 'Be you.' Nevertheless, it is imperative that such sensitive issues should be taken up with a lot of care & responsibility" (Adgully Bureau, 2015).

Kiran Khalap, Co-founder and MD of Chlorophyll Brand and Communications Consultancy, says that it is critical for an ad to achieve the brand's goal, which Anouk's digital campaign delivers. "These ads effectively communicate the new attributes of boldness. There is a faint connection with the boldness of the product, and I only hope this gets strengthened over time for the brand to get differentiated," he says (Sharma, 2015).

Conclusion

Although many brands have experimented and altered their strategies in response to consumer demands, it is still unclear if they have as a whole adjusted to the evolving market landscape. While a variety of political and economic concerns predominate in the news, there are other social issues that call for discussion and debate. However, most of urban India remains wary of such discussions. Some of these issues include homosexuality, single parenting, and women's rights. Such social "upheavals" would make us desire to be deaf and blind. However, India is undergoing such societal changes discreetly but steadily. And, in recent years, advertising has acted as a mirror for these changes.

The industry's reaction to the "Bold is Beautiful"campaign is mixed. The campaign has also offered many people a platform to express their opinions on delicate matters, while also allowing other brands to step forward and address even more sensitive issues. These films demonstrate how vital it is to be courageous in one's life decisions and how to stand up for one's beliefs by challenging deeply held values. There's a reason why companies are interested in such topics. Not only do such advertising help them stand out in the crowd and get seen, but they also appeal to today's audience—particularly the youth that every brand aspires to reach. Such issues are a goldmine for brands looking to push the boundaries.

Aditya Birla Group—Financial Services CMO Ajay Kakar remarks, "India is changing and so are consumers. In this changing world, brands have to be contemporary, but the definition of contemporary should depend on the needs, demands and preferences of a brand's target audience. Very simply, can a financial service brand do something like this, the answer is "no," but a fashion brand might use such a concept for marketing. So, having said that, it all depends upon the brand's core essence and value and whether or not it will fit with its target audience. The advertisement is good and progressive" (Kohli, 2015).

Bibliography

Adgully Bureau. (2015). *Anouk Myntra Campaign edgy yet sensitive, Adgully.com*. Available at: https://www.adgully.com/anouk-myntra-campaign-edgy-yet-sensitive-61944.html

Al Akayleh, F. (2021). The influence of social media advertising on consumer behaviour. *Middle East Journal of Management, 8*(4), 344. https://doi.org/10.1504/mejm.2021.10037485

Anouk Bold is Beautiful. (2015). Twitter. https://mobile.twitter.com/hashtag/anoukboldisbeautiful?src=hashtag_click

Bapna, A. (2011). *Overall Lifestyle category will cross $100 billion in 2015 in India: Mukesh Bansal, Founder-CEO, Myntra.com, Economic Times*. Available at: https://economictimes.indiatimes.com/opinion/interviews/overall-lifestyle-category-will-cross-100-billion-in-2015-in-india-mukesh-bansal-founder-ceo-myntra-com/articleshow/9187245.cms?from=mdr

Bazala, H. B. (n.d.). *History of Myntra, Scribd*. Available at: https://pt.scribd.com/doc/111773096/History-of-Myntra

Bhattacharyya, A. (2015, June 23). Straight talk. *The Financial Express*. https://www.financialexpress.com/industry/straight-talk/88683/

Bhavsar, R. (2018). A study on social advertisements and their effect on preference of consumer. *(Bhavsar, 2018). Meida Mimansa.*

Casais, B. (2021) Pode o femvertising ser um boomerang para as marcas? Meios & Publicidade – Meios & Publicidade (meiosepublicidade.pt). Accessed 24 Jan 2022.

Chauhan, G. S., & Shukla, T. (2016). Social media advertising and public awareness: Touching the LGBT chord! *Journal of International Women's Studies, 18*(1), 145–155.

Chengappa, S. (2018, October 17). Myntra to invest in 15 emerging fashion brands. *The Hindu Business Line*. https://www.thehindubusinessline.com/companies/myntra-to-invest-in-15-emerging-fashion-brands/article25250732.ece

Corporate Reporter. (2011, September 7). *Myntra.com to expand operations*. The Hindu. https://www.thehindu.com/business/companies/myntracom-to-expand-operations/article2433066.ece

Das, S. (2015, June 14). Video: Myntra's Anouk Lesbian ad is a good step but fails to look real. *The Indian Express*. https://indianexpress.com/article/trending/video-myntras-anouk-ad-on-being-a-lesbian-in-india-is-a-good-step-but-fails-to-look-real/

Exchange4media Staff. (2016). Myntra's in-house brand Anouk gets bolder with new ad film. Exchange4media.Com. https://www.exchange4media.com/advertising-news/myntra's-in-house-brand-anouk-gets-bolder-with-new-ad-film-65708.html

Govindarajan, A. (2018). Indian advertisements breaking stereotypes in the perspective of erving goffman.

Gupta, H. (2015a, June 19). Bold and beautiful: Anouk's ad featuring a lesbian couple. *Digital Media Marketing News*. http://www.digitalmarket.asia/bold-and-beautiful-anouks-ad-featuring-a-lesbian-couple/

Gupta, S. S. (2015b, June 17). Why we made the "lesbian ad" that went viral. Dailyo.In. https://www.dailyo.in/arts/anouk-lesbian-ad-bold-is-beautiful-behind-the-scenes-myntra-lgbtq-section-377/story/1/4402.html

Johri, V. (2015, June 10). India's first lesbian ad is actually wonderful. Dailyo.In. https://www.dailyo.in/lifestyle/myntra-anouk-ad-bold-is-beautiful-lesbians/story/1/4243.html

Kohli, A. (2015, December 6). Myntra's ad film breaks norms, creates ripples with a lesbian storyline. Bestmediaifo.Com. https://bestmediainfo.com/2015/06/myntras-ad-film-breaks-norms-creates-ripples-with-a-lesbian-storyline/

Kumari, M. (2020). Social media and women empowerment. *International Journal of Scientific & Technology Research, 9*(03), 626–629.

Lima, A. M., & Casais, B. (2021). Consumer reactions towards femvertising: A netnographic study. *Corporate Communications An International Journal, 26*(3), 605–621. https://doi.org/10.1108/ccij-02-2021-0018

Mandrawadkar, K. (2015). *Myntra case study Anouk Campaign.*

Mathur, B. (2022, March 15). Breaking the gender bias: A look at the determinants of well being of women in India. NDTV-Dettol Banega Swasth Swachh India. https://swachhindia.ndtv.com/breaking-the-gender-bias-a-look-at-the-determinants-of-well-being-of-women-in-india-67168/

Melo, E. (2018). *Gestão de Marketing e Branding: a arte de desenvolve e gerenciar marcas*. Editora Alta Books.

Myntra. (2015). *ANOUK, BOLD IS BEAUTIFUL – A wonderful video titled "THE VISIT."* Youtube. https://www.youtube.com/watch?v=Ef27m5ocK6Q

Neogy, S. (2015). *Myntra unveils its "Bold Is Beautiful" campaign for ethnic wear brand Anouk*. Exchange4media.Com. https://www.exchange4media.com/advertising-news/myntra-unveils-its-bold-is-beautiful-campaign-for-ethnic-wear-brand-anouk-60340.html

PricewaterhouseCoopers. (n.d.). *Propelling India towards global leadership in e-commerce*. PwC. Retrieved January 31, 2022, from https://www.pwc.in/research-insights/2018/propelling-india-towards-global-leadership-in-e-commerce.html

Santos, L. (2016). Femvertising: o empoderamento da mulher na propaganda sob a ótica das receptoras. Santos_Lisangela_Deni_dos_2016_TCC.pdf. Accessed 24 Jan 2022.

Saxena, A., et al. (2011, March 14). *Updated: Myntra gets $14M in second round funding*. MediaNama. https://www.medianama.com/2011/03/223-myntra-gets-14m-in-second-round-funding/

Sen, A. (2018, January 9). *Flipkart fashion business catches up with Myntra*. Mint. https://www.livemint.com/Companies/crgAFNgipSWuf9reSeNDLJ/Flipkart-fashion-business-catches-up-with-Myntra.html

Sharma, A. (2015, June 10). Myntra: Redefining Tradition. Afaqs! https://www.afaqs.com/news/digital/44490_myntra-redefining-tradition

Singh, P. (2017). At home with their queerness: Same-sex relationality and the Indian family in advertising media. *Feminist Media Studies, 17*(5), 721–736. https://doi.org/10.1080/14680777.2017.1298144

Scenes of Affection in the Advertising of the Brazilian Brand Reserva: Contemporary Reflections on Fashion, Body, and Aging

Henrique de Souza Goulart, Daniela Novelli, Icléia Silveira, Lucas da Rosa, and José Alfredo Beirão Filho

Introduction

This chapter unfolds from the Valentine's Day 2022 campaign presented by the Reserva brand on its social networks, reflecting on the presence and/or perception of the old body in the fashion advertising in Brazil. For this, it proposes to understand the market of the "silver economy"as well as the possible intersections between fashion, body, and aging, directing the gaze toward research methods that offer a holistic understanding of the reality of this public and its real needs.

Thus, this research is classified as basic, qualitative, and descriptive, in which a narrative literature review is articulated, carried out in different databases, such as Scopus and Google Scholar, using the following search terms: "old bodies," "fashion advertising," "aging," and "silver economy." Based on the results obtained, a selection of texts that supported this research was prepared, including journal articles and event proceedings, books, dissertations, and doctoral theses. Furthermore, data made available by institutions such as the World Data Lab, among others, via their virtual addresses, were used.

Reflect on the visual production of the Valentine's Day 2022 campaign of the Reserva brand through the relationship between fashion, body, and aging. To do so, the specific objectives were: (a) to understand the silver economy market; (b) to present the intersections between fashion, body, and aging; (c) to investigate the presence of the old body in fashion advertising; (d) to develop a semiotic analysis of the Valentine's Day 2022 campaign of Reserva.

The relevance of this research is justified based on the professional relationship of the researcher, established since the first semester of 2020, with men and women over 60 years in the Extension Program "Eterno Aprendiz" of the Faculdade Metodista Centenário —FAMES of Santa Maria, Rio Grande do Sul, an opportunity

H. de Souza Goulart (✉) · D. Novelli · I. Silveira · L. da Rosa · J. A. B. Filho
Santa Catarina State University, Florianópolis, Santa Catarina, Brazil

A. Cruchinho, J. C. Neves (eds.), *Sustainability in Fashion: Creativity, Education, and Innovation*, https://doi.org/10.1007/978-3-031-80991-0_10

that generated the personal motivation for the development of the research, considering the repeated complaints about the lack of representation of old people in fashion advertisements, supported by studies presented by authors such as Lima (2020), Loiola (2021), Lopes et al. (2012), Minichiello et al. (2000), Motta (2012), Oliveira (2004), Palacios (2004).

With the result of this research, we hope to broaden the discussion regarding the perception of old people in the fashion market, especially in advertising, moving away from the image of dependent elderly people, limited in the sense of challenging the normativity of beauty standards, based on youth, illuminating the potential of the older consumers and its different faces.

Silver Economy: Fashion, Body, and Aging

According to Américo (2021) from the portal VC S/A, population aging happens mainly in wealthier countries, where the number of births decreases and life expectancy increases. In the early 1990s, individuals over 60 years old represented 6% of the world population, and in 2021, this figure had already amount to 13%. It can be observed that since 2018, the world population has changed, and for the first time in history, the global number of elderly people was greater than children. That is, there were 705 million people over 65 years old and 680 million under 5 years old. Brazil, according to the author, jumped from 7.1 million people over 60 in 1990 to 30 million in 2021, or 15% of the total population (Américo, 2021). According to the Serviço Brasileiro de Apoio às Micro e Pequenas Empresas (Sebrae), the purchasing power of people over 60 years should exceed R$ 30 trillion worldwide (SEBRAE, 2021) and, according to the *World Data Lab* (2020), the group of consumers that will see the greatest growth by 2030 is the over-65s.

Contrary to what one might imagine, if fashion repeatedly acts as a mechanism to combat aging, to what extent do the elderly have autonomy of choice in the construction of their appearance? We consider the silver market to be in full growth and seek to identify gaps in the fashion market regarding the representativeness of the public of people over 65. In this sense, addressing the relationship between fashion, body, and aging contributes to the understanding of the aesthetic and sociocultural aspects involved in the historical process of growth of the silver economy.

As of the twentieth century, the field of gerontology began to gain contours in Brazil, and the images surrounding old age began to be driven by the involvement of various social agents in the country, shifting from the idea of decay, or frailty and dependence, to aspects linked to the "experiences of aging," as presented by Lopes et al. (2012) *apud* Mesquita and Castilhos (2012, p. 45). The experience of aging is understood as something that arises from a confluence of manifestations, which are both external and internal to the physical body; i.e., it involves the self-perception of old age, which goes through both issues of the organic body (appearance, energy, and health) and the interactive age, which can be defined by the age that others see (Minichiello et al., 2000).

New ways of aging have been proposed, especially seeking to combat the mentality inherited from the past, anchored in the idea of decadence and dependence. However, according to Lopes et al. (2012), the excessive production of overly positive images of old age weakens the possibility of legitimizing the various possibilities of being old and being in old age, typical of the heterogeneity of aging.

Supported by Motta (2012), this study proposes the use of the terms "mature" and/or "old" to designate people over 60 years, since, according to the author, one tries to rehabilitate the word old(o) to banish it due to the desire of consumer society to pasteurize old age and dissimulate its antipathy to this stage of life. And, in this sense, fashion has acted as one more among many agents fighting aging.

Grounded also in the writings of Barnard (2003), this research recognizes that fashion and clothing are cultural, that is, part of how a group constructs and communicates its identity. Crane (2006), in turn, writes about clothing being indispensable to the social construction of identity, and clothing can be considered a way of expression of subjects and as cultural identification and traits of each individual's identity. According to Oliveira (2004, p. 10), "in fashion, and through it, subjects show themselves, showing their ways of being and being in the world, which positions them in it."

When it comes to fashion, it can be said that the body is seen as a medium through which fashion communicates, because this is a means of expression much exploited by the Media in consumer society. To understand fashion's preference to portray only what is young/new as being the valid ideal of beauty, Tonarque (2012) explains that the limitations perceived by mature people at the time of composing the "look" are due, especially, to the fact that fashion industries disregard the changes that the body goes through when aging. In this sense, we find again in Svendsen (2010) the support for what was exposed, in which the author presents the body of postmodern society as a "trapped" body, taken as an object to show fashion.

Villaça (2007) presents the bodies molded in gyms, submitted to diets and all forms of alternative therapies that reiterate the idea of a healthy mind and body. The young, fit and molded body is sanctioned according to the canons of the consumer society, which massifies the ideals of beauty through various means of communication. The body here is understood as a domesticated body, or docile (Foucault, 1999) to the determinisms of fashion, which adorns and modifies its contours, that is, remodeling bodies is something that does not apply only to artistic representation but corresponds to a real experience, experienced by all who are at the mercy of the prevailing ideals of beauty.

Therefore, the old bodies must be "camouflaged", the exhibition of them is not allowed because they do not correspond to the values and sociocultural dispositions of this post-modern body. The image considered ideal is that of mannequins with thin arms and legs, in an adolescent style. In this sense, it can be stated that the body, victimized by the appeals of consumption, is culturally affected by the various spheres of power.

If, on the one hand, consumer goods industries increasingly create and reproduce novelties with the promise of perfect bodies, which are obtained quickly and without much personal effort (Jordão, 2021), on the other hand, we notice new statutes being created in an attempt to establish the "new old age" or "active aging," terms used to refer to the new understanding of aging. In this sense, this new archetype differs from the one presented at the beginning of this topic, precisely regarding issues linked to behavior, which end up expanding, according to Teixeira et al. (2018, p. 3):

One of the implications of the Longevity Revolution is the construction of the image of the elderly by the media, which does not happen without purpose. In the view of Bezerra (2006), an industry entirely focused on this target audience has been assembled and expanded with the help of media appeal and the adoption of a positive view of aging that is favorable to the modern capitalist state, because to achieve this standard, an arsenal of products and processes is created and directed to the consumption of this genre.

Starting from a succinct but necessary historical cut, to understand how the old body was seen in advertising pieces over time, evidence shows values linked to the periodization of life, and until 1950, the image of the old person was related to aspects such as health and disease (Teixeira et al., 2018); however, from this decade on, the advertising representations of old age move away from this axis and approach a more realistic idea, even while preserving the role of grandparents played by mature people who presented ads for personal hygiene, cosmetics, clothing, food, and even financial institutions.

According to Vasconcelos (2001) *apud* Teixeira et al. (2018), there was another, even more radical, change in the way of representing old age from the 1980s and 1990s: old people started being portrayed more positively, even when occupying roles linked to subversive practices, social activities, and modernity.

Palacios (2004) states that, in an attempt to pasteurize old age, in 1990 there was also a change in the advertising discourse, in which the word "old age" is replaced by the expression "third age"—which then assumes a preponderant role in society as a positive ideal of old age. Thus, according to Noronha and Correia (2021), it is precisely in this period that the older people market gains recognition, that is, campaigns are created by marketing professionals to conquer this public. Cirillo (2012), Drigo and Perez (2015) and Moura and Souza (2021) *apud* Teixeira et al. (2018, p. 4) state that in the twenty-first century, although old age is associated with family dependence, old men and women have often abandoned situations of negativity and are increasingly assuming positive attitudes, confidence, power, attitude, security, motivation, relaxation, and transgression of behavioral patterns considered pertinent to this period of life.

Those who managed to avoid aging are the ones who are associated with these positive values in advertising campaigns, because they complied with the consumption imperative, especially aimed at the elderly. According to Loiola (2021), advertising has the habit of portraying the elderly always as asexual beings, often sick and dependent, lonely, emphasizing concepts related to health care and, above all, how to delay aging. "The simulacra of the images built by the cosmetics industry, or the media objectify beauty, simplifying it, taking away the real sense of beauty" (Loiola, 2021, p. 93).

Valentine's Day 2022 Editorial from the Reserve Brand

The conception of a visual message brought by Martine Joly (1996) translates into a complex system of meaning and communication, drawing attention to plastic, iconic, and linguistic elements. In this sense, according to Novelli (2009, p. 115), "advertising discourses are validated by the image society," with the conception of the advertising image being essentially communicative, aiming at reading by the public and thus reinforcing all the potential attributed to the privileged field observation of the mechanisms of meaning production through the image.

For Joly (1996), "defining the objective of an analysis is essential to install your tools, remembering that they determine a large part of the object of the analysis and its conclusions" (Joly, 1996, p. 49). The first step proposed by Joly's (1996) image analysis methodology, the latter being in dialogue with Roland Barthes, is precisely focused on the fact that the perception of shapes and objects is cultural, that is, an image appears similar to us because the units we highlight in it have cultural affinities, determined by our habit of detecting these same units in the world (Joly, 1996).

Next, the careful description of the image stands out, as only through it can it be possible to differentiate the different types of messages that compose it. Although simple, this step proves to be fundamental, considering that it constitutes the translation of visual perceptions into verbal language (Joly, 1996). The verbalization of the visual message manifests processes of perceptual choices and recognition that govern its interpretation. This passage from "perceived" to "named," this crossing of the border that separates the visual from the verbal, is determined in both directions (Joly, 1996, p.72).

The third aspect points to the fact that the image promotes the interweaving of different materials together, to compose a visual message. The heterogeneity of the image and the multiplicity of materials that articulate meanings to produce the global message are then evident (Joly, 1996). According to the author [...] the objects represented refer to the custom of "shopping" in a certain type of society; [...] the presentation of the ad, its place in the magazine, advertising. In other words, in addition to the literal or denoted message, evidenced by the description, there is a "symbolic" or connoted message, linked to the pre-existing and shared knowledge of the advertiser and the reader (Joly, 1996, p.75).

It is also essential to observe the diversity of linguistic supports since the visual message is made up, as already stated, of three other messages: plastic, iconic, and linguistic (Novelli, 2009). Textual elements, such as captions or slogans that accompany an image, become important linguistic signs as they can function as an anchor or relay (Joly, 1996) and must be interpreted in the exercise of semiotic analysis.

In this area, Joly (1996) highlights the main elements that make up a plastic message: (i) the support: paper, format, page; (ii) the dimensions of the advertisement: as well as the layout and type of characters used; (iii) frames: which represent the physical limits materialized by a frame, following different periods and styles; (iv) framing: represents the size of the image, the distance between the photographed object and the lens; (v) the angle and choice of objective; (vi) composition:

geographical composition of the interior of the visual message, plays an essential role in the hierarchization of vision and, therefore, in guiding the reading of the image; (vii) forms: anthropological and cultural interpretation of plastic forms and tools; (viii) colors and lighting; and (ix) texture.

Regarding the iconic message, it can be said that it resides in the interpretations generated by the connotation process, related to the viewer's knowledge. In other words, figurative signs are present in the image through the connotations they evoke, also considering "the metonymic character of such signs, as it is generally only possible to visualize parts of elements that designate the 'whole' by contiguity, through a kind of displacement of meaning organized in a certain way" (Novelli, 2009, p. 129). In this sense, Joly (1996, p. 106) puts as an example the postures observed in the models' poses: "they often put characters on stage, and part of the interpretation of the message is, then, determined by the scenography that takes up postures that are also culturally coded."

Based on previous examples of women over 65 who are not subject to the marginalization that is often imposed on them, and supported by Lima (2020), it can be said that advertising is in a process of transformation to expand the representativeness of the female population with the inclusion of more real women, including black, obese, and elderly women. According to Teixeira et al. (2018), these are women who are looking for comfortable clothes and with fashion information that escapes from the stereotypes of "granny clothes," which shows that the new reality experienced demands new solutions in various sectors of society, including the clothing and fashion industry.

On May 26, 2022, the Reserva brand published on its Instagram profile (@ reserva) an editorial alluding to Valentine's Day and a 31-second video in black and white on its YouTube channel, where two old people, one cis white man and a cis white woman star in scenes of affection that, to the sound of violins, unfolds until both are left wearing nothing but the brand's underwear (Fig. 1).

In this publication published on Reserva's brand social networks, it is noted at the outset that the brand used the following caption: "Finally, love is in the air. This Valentine's Day, Reserva celebrates passion, love, and sensuality with the launch of the Reserva Underwear line, starring those who know that there are things that don't change with time." In this sense, it is understood through plastic, iconic, and linguistic signs how the campaign explores, in a sensitive and at the same time innovative way, the affection and love between an old couple.

The constant use of black-and-white photography connotes, in the context of this proposal, the desire to immortalize the moments lived, emphasized by the same chromatic choice of the lingerie pieces and the contrast of light and shadow, which adds a dose of drama and intensity, but also tranquility and security at certain times. The variation in framing planes gives movement to the scenes, contributing to the iconic meaning of a dynamic sexual and affective interaction, also reinforced by the mutual play of desires, gestures, and looks. In other words, the poses, especially when exchanging affection and more intimate interactions, refer to a poetic and delicate language, highlighting the idea that the flame has not gone out with time, that love has no age, or that even with more than 65 years, it is possible to celebrate

Fig. 1 Scenes from Reserva's Valentine's Day 2022 campaign. (Source: https://www.ocarafashion.com/2022/05/27/reserva-dia-dos-namorados-2022/. Accessed on July 05, 2022)

passion and sensuality. And yet, the interaction of hands, arms, and legs, as well as their bodies on display naturally and spontaneously, with all the marks of time, are expressive elements of an authentic old age, whose intimacy is shown to the public without shame or insecurity.

It is contextualized that this advertising action is linked to the launch of the brand's underwear collection, expanding the discussion and putting on the agenda issues such as the visibility of older people, the deconstruction of ageism, and the naturalization of sex in old age. According to Lima (2022), for Rony Meisler, Co-founder of Reserva and CEO of Ar&Co, the inclusion of older people in advertising campaigns is an important movement against ageism, which is one of the pillars of the Reserva brand. However, it is possible to notice that people over 50 often appear in the brand's campaigns, which prioritize real people in their content, as seen in the campaign under analysis, which featured a real elderly couple wearing only underwear, photographed under the eyes of Mauricio Nahas.

According to the model, Solange Nakad, the campaign aimed to show intimacy in old age. The model reveals on her social networks that, "we are aging healthily (sic) and with consumption in all (sic) segments [...] We want to have pleasure. In maturity, we seek to touch, feeling, the pleasure of giving and receiving" (Nakad, 2022). She was photographed with her partner Lelio Nakad—with whom she lives for 55 years—she presents herself as a senior model on her social networks, hoping to have instigated other couples with her participation in the Reserva brand's Valentine's Day Campaign.

The semiotic analysis revealed that this representation of the old body meets the conceptions of gerontology, which, according to Debert (2003), in Brazil, invests its efforts in reinforcing the idea that old age can no longer be seen as a "moment of loss" and seeks to highlight the positive side of aging. It is clear in all the exchange of gestures and looks between the couple that there is a search for the pleasure of touch in maturity, supported by the complicity built over time. It is worth noting that, since 2018, the Reserva brand has been developing the "Cara e Coroa" Project, which offers opportunities for professionals aged 50 or over to enter its stores. In this sense, Meisler clarifies that it is not uncommon to find job offers for older professionals for "minor" positions; however, the company proposes that the "coroas," as defined by Reserva's CEO, feel part of the family and understand that their importance within the organizational context is equal to that of younger employees.

It can be noted, then, that advertising has been adopting new strategies to reach this audience, given the criticality of this niche of consumers, even if these paradigm shifts in the way of "looking and advertising" follow interests focused on selling their products, he points out. It is an initial path towards a new perception of the old person in the media and/or advertising world. However, this inclusion of older people in fashion content and editorials is accomplished, and not just a "performative inclusion, that is, it is included through standardization and erasure, disregarding the heterogeneity of old age" (Teixeira et al., 2018, p. 74). It is necessary to free oneself from thinking enslaved by the incessant search for an illusory youth.

Final Considerations

When one seeks to understand the historical, sociocultural, and aesthetic aspects related to the presence of "new old age" bodies in visual fashion production, one consequently reconsiders the value of beauty historically rooted in youth, especially in Brazil—considered one of the countries that most perform plastic surgeries around the world, often motivated by the desire to alleviate aging and correct physical defects in the search for the perfect body.

Based on the authors and concepts presented throughout this research, as well as the semiotic analysis carried out, it is understood that when the Reserva brand takes the position of including older people in its campaign images that deal with affective and sexual issues, this greatly contributes to promoting important ruptures, destabilizing social constructions about the old body, especially in a country where the cult of beauty, youth, and sensuality is recurrent in advertising. In this sense, the increase in the presence of fashion images featured on Instagram by old people such as Iris Apfel and Vovó Izaura testifies to this movement, even in different contexts.

Initiatives like these may reflect how the population over 65 represents an important market niche, the silver economy as mentioned, which demands from fashion brands a more attentive vision regarding body aesthetic requirements, seeking to understand more in-depth the new interests and needs linked to the comfort and practicality of this public. The visual message of the campaign revealed an innovative approach to the perception of the old body, as the brand proposes to naturalize affection and sex in old age, as well as to expose real old bodies, outside the idealized beauty standard, to convey your products.

Finally, from a broader point of view, the conceptual and aesthetic valorization of the old body and the use of the old person's labor by the Reserva brand instigates the promotion of the re-education of the entire society about overcoming prejudiced social constructs around old age, overcoming the idea that beauty resides in smooth skin and a thin body, that it is necessary to desperately paralyze the natural changes of life. Thus, bodies with wrinkles, expression lines and marks from the lives of old lovers invite the celebration of passion, love and sensuality, wrapped in genuine, sensitive and authentic feelings that are revealed in black-and-white advertising.

References

Américo, J. (2021). *Economia prateada: A ascensão dos consumidores 60+*. Disponível em: https://bit.ly/3ce1plp (acesso em 05 de jul. 2022).

Barnard, M. (2003). *Moda e Comunicação* (2nd ed.). Ed. Racco.

Bezerra, A. K. G. (2006). *A construção e reconstrução da imagem do idoso pela mídia televisiva*. Dissertação, Universidade Federal de Campina Grande, disponível em: http://www.bocc.ubi.pt/pag/guedes-ada-imagemidoso-midia-televisiva (acessado em 1 de agosto de 2022).

Cirillo, M. A. (2012), *A vez da terceira idade*: o discurso da publicidade na construção da imagem do idoso nas revistas. Dissertação (mestrado em Comunicação Social) - Faculdade de Comunicação, Universidade Metodista de São Paulo, São Bernardo do Campo, 2012. 252 f. Disponível em http://tede.metodista.br/jspui/handle/tede/621. Acesso em 5 Jun. 2022.

Crane, D. (2006). *A moda e seu papel social: classe, gênero e identidade das roupas*. Senac.

Debert, G. G. (2003). O velho na propaganda. *Cadernos Pagu*, Campinas, *21*(1), 133–155, disponível em: https://bit.ly/3K0O5gR (Acessado em 05 jul. 2022).

Drigo, M. O., & Perez, C. (2015). Marcas, publicidade e valores relativos à periodização da vida: Possibilidades de ressignificação da maturidade adulta. Signos Do Consumo, 7(1), 22-38. https://doi.org/10.11606/issn.1984-5057.v7i1p22-38. Acesso em 05 Jun. 2022.

Foucault, M. (1999). *Vigiar e Punir: o nascimento da prisão*, 20ª ed, Vozes, São Paulo.

Joly, M. (1996). *Introdução à análise da imagem*. Papirus.

Jordão, M. E. (2021). *Vamos falar de envelhecimento? a invisibilidade do idoso na publicidade*, trabalho de conclusão de curso de graduação, Pontifícia Universidade Católica de Goiás, Goiânia, disponível em: https://bit.ly/3A3PRt1 (acessado em: 05 jul. 2022).

Lima, F. R. A. (2020). *Construção do sentido do feminino na publicidade de marcas de beleza*, dissertação, Pontifícia Universidade Católica de São Paulo, São Paulo, disponível em: https://sapientia.pucsp.br/handle/handle/23688 (acessado em 05 de jul. de 2022).

Lima, C. (2022). Reserva aposta em equipe com colaboradores 60+ em suas lojas. *Vogue*, Disponível em: http://glo.bo/3STI6i8 (acessado em: 05 jul. 2022).

Loiola, R. F. (2021). Análise sociossemiótica dos modos de vestir na velhice. *Dobra[S] – Revista da Associação Brasileira de Estudos de Pesquisas em Moda*, [S.L.], n. 31, p. 88–102, disponível em: https://dobras.emnuvens.com.br/dobras/article/view/1290 (Acessado em 05 jul. 2022).

Lopes, A. et al. (2012). Envelhecimento e Velhice: Pistas e reflexões para o campo da moda. In: C. Mesquita, K. Castilho (Org.), *Corpo, Moda e Ética* (pp. 45–53). Estação das Letras e Cores, São Paulo.

Mesquita, C., & Castilho, K. (Org.) (2012). *Corpo, Moda e Ética*. Estação das Letras e Cores, São Paulo.

Minichiello, V., Browne, J., & Kendig, H. (2000). Perceptions and consequences of ageism: Views of older people. *Ageing And Society,* [S.L.], *20*(3), 253–278, Cambridge University Press (CUP), Cambridge, disponível em: https://bit.ly/3caljOF (acessado em 05 jul. 2022).

Motta, A. B. (2012). Mulheres Velhas: elas começam a aparecer. In C. Pinsky (Ed.), *Nova História das Mulheres no Brasil*. Contexto.

Moura, T., & Souza, S. (2021). A velhice pelo olhar do adulto contemporâneo: Uma análise das imagens visuais de idosos na mídia. *Latin American Journal of Development*, 3(1), 117–133. https://doi.org/10.46814/lajdv3n1-014. Acesso em 05 Jun. 2022.

Nakad, S. (2022). Namorados Reserva – 2022, Instagram: @solangemaria.nakad1, disponível em: https://www.instagram.com/solangemaria.nakad1/ (acessado em 15 ago. 2022).

Noronha, I. A. (2021). A relação da moda com a terceira idade: As vovós fashionistas (TCC de graduação). Instituto de Cultura e Arte, Universidade Federal do Ceará. Disponível em http://www.repositorio.ufc.br/handle/riufc/61419. Acesso em 5 abril 2022.

Novelli, D. (2009). Juventudes e imagens na revista Vogue Brasil (2000–2001). Dissertação (Mestrado em História do Tempo Presente) – PPGIH-UDESC: Florianópolis, 275f.

Oliveira, A. C. (2004). Entre as Plásticas da Moda e o Corpo, o Sujeito. In K. Castilho (Ed.), *Moda e Linguagem* (2nd ed., pp. X–X). Anhembi Morumbi.

Palacios, A. R. J. (2004). *As marcas na pele, as marcas no texto: sentidos de tempo, juventude e saúde na publicidade de cosméticos em revistas femininas durante a década de 90*. Tese de doutorado, Universidade Federal da Bahia, Salvador, disponível em: https://repositorio.ufba.br/handle/ri/11292 (acessado em 05 jul. 2022).

SEBRAE. (2021). *Conheça o potencial do mercado da terceira idade*. disponível em: https://bit.ly/3RAMbab (acesso em: 05 de jul. 2022).

Svendsen, L. (2010). *Moda: uma filosofia*. Zahar.

Teixeira, D. P. et al. (2018). Construção da imagem do velho na mídia: Representatividade ou estímulo ao consumo? *Revista Vértices*, [S.L.], *20*(3), 385–397, disponível em: https://www.redalyc.org/journal/6257/625767802011/625767802011.pdf (acesso em: 05 jun. 2022).

Tonarque, S. A. (2012). *Velhice e moda: incursões históricas e realidade atual*, dissertação de mestrado, Pontifícia Universidade Católica de São Paulo, São Paulo, disponível em: https://tede2.pucsp.br/handle/handle/12404 (acesso em 05 de jul. 2022).

Vasconcelos, S. M. (2001). *O "velho" na Publicidade Brasileira*, dissertação de mestrado, Universidade Metodista de São Paulo, São Bernardo do Campo.

Villaça, N. (2007). *A edição do corpo: tecnociência, artes e moda*. Estação das Letras.

WORLD DATA LAB. (2020). *The world is not only getting richer but also much older*, disponível em https://worlddata.io/blog/silver-economy-richer-older (Acesso em: 05 de jul. 2022).

Online Influencer Marketing: Defining the New Cool

Pedro Dourado and Vasco Ribeiro

Introduction

With the evolution of the Web in recent years, digital technologies have proven to be an important means of production and consumption to meet in a more intimate and informal space (Brewer, 2019; Gazzola et al., 2020), with users becoming increasingly connected (Tafheem et al., 2022). New ways of interaction have also enabled new modes of consumption and marketing strategy development (Leung et al., 2022).

The power of social media is now widely known (Gazzola et al., 2020) and is critical for brands and businesses to promote their offer (Shrivastava et al., 2021). We are witnessing the growth of online influencer marketing (Belanche et al., 2021; Bu et al., 2022; Kim & Kim, 2021; Leung et al., 2022; Lou & Yuan, 2019) as a viable alternative to the ineffectiveness of traditional advertising (de Veirman et al., 2017) and to combat the skepticism towards marketing that comes directly from brands (Leung et al., 2022).

Marketing communication via digital influencers is now widely accepted as an important tool of persuasion (Jung & Im, 2021), with the size of the global influencer market tripling since 2019 (Statista, 2025). The positive outcomes are mirrored in the industry's professionalization: companies know which types of influencers will provide a return (von Mettenheim & Wiedmann, 2021), and influencers create the right content to be selected (Jorge et al., 2018; Leung et al., 2022).

The role of digital influencers in the fashion industry is noteworthy: perceived as close to the general public (Gazzola et al., 2020), influencers have gained an important voice in dictating consumer demand (Orminski et al., 2021).

P. Dourado (✉) · V. Ribeiro
Universidade do Porto, Porto, Portugal
e-mail: vribeiro@letras.up.pt

A. Cruchinho, J. C. Neves (eds.), *Sustainability in Fashion: Creativity, Education, and Innovation*, https://doi.org/10.1007/978-3-031-80991-0_11

This popularization of digital influencers, as well as the trust and credibility that followers place in them (Johnstone & Lindh, 2022), is reminiscent of how marketing appropriated the relationships observed between teenage consumers and their peers at the turn of the millennium. Indeed, this study examines the documentary Merchants of Cool, which aired on PBS in 2001, and exposes the main findings of an investigation into how marketers and media industry executives create and sell pop culture to American teenagers.

In this chapter, we look for similarities and differences between how marketing in the late 1990s/early 2000s and today's online influencer marketing work to find the next cool thing (Goodman & Dretzin, 2001). Are digital influencer marketers the new Merchants of Cool?

Although the effects of content produced and disseminated by online influencers have received special attention in recent years from academia and industry (Bu et al., 2022), the truth is that the relationship between marketers and influencers is little addressed (Kim & Kim, 2021), and guidelines that take cases like the early millennium's under the radar marketing as an example are nonexistent.

Following a framework that considers the technological paradigm, the evolution of social networks, and the evolution of influencer marketing, this chapter exposes the path taken by journalist Douglas Rushkoff in his search for marketing tactics and techniques in "Merchants of Cool" in parallelism with today's reality. Finally, the conclusion synthesizes the work done, presenting reflections as well as the study's main theoretical and practical contributions.

Technological Advancement and Social Networking

Since the mid-twentieth century, Information and Communication Technologies (ICTs) have proven indispensable to the functioning of our societies, transforming all aspects of human activity. Castells' Network Society has enabled the "rise of new patterns of social interaction" (Castells, 2004, p. 145). With the introduction of Web 3.0 and mobile devices and applications, the popularity of social media platforms has grown, allowing any user to build solid networks of followers (Leung et al., 2022).

In reality, digital technologies, in addition to facilitating peer communication, foster new forms of interaction between production and consumption (Bulovic & Covic, 2020), resulting in the reconfiguration of several industries and sectors. Consumers actively exchange information about brands, products, and shopping experiences via digital platforms and virtual communities (Colliander & Hauge Wien, 2013).

As a result, consumers have become valuable assets in the creation process (Tafheem et al., 2022). The co-creation of value generates capital in the form of knowledge and opinions (Kohler et al., 2011), rather than providing value to the passive consumer (Zhang et al., 2015). Indeed, we are currently witnessing a disintermediation process in which intermediaries are being replaced and producers and consumers are aligned in terms of knowledge, power, and authority (Crewe, 2013).

This is also true in the fashion industry, where users actively exchange information as a common practice in social interactions in the digital environment, which can enhance their consumer experiences and co-create value with brands (Quach & Thaichon, 2017).

The Internet, as a technology of freedom (Castells, 2004), is a critical component of this paradigm shift. The channels that have existed for decades, such as the press, are being neglected, to the detriment of the Internet's ubiquity, which takes shape through digital social networks, such as Facebook or Youtube, and which eliminates gatekeepers (Piazza & Abrahamson, 2020).

The Effectiveness of Online Influencer Marketing

Digital technologies enable the acquisition of new customers as well as improved interactions with existing ones (Bulovic & Covic, 2020; Kim & Kim, 2021). In this context, social media marketing emerges as an interdisciplinary and cross-functional process that employs social media to achieve organizational objectives by providing value to stakeholders (Felix et al., 2017). It should also be noted that, in addition to utilizing cutting-edge technological advances to facilitate extremely frequent interactions, this type of digital platform facilitates content creation, both upstream (marketer-generated content) and downstream (user-generated content) (Zhang et al., 2015).

Indeed, over the last decade, online influencer marketing has grown in popularity (Phua et al., 2017), with user-generated content proliferating across social networks and some users emerging as opinion leaders, actively producing personal content and product and service reviews (Masuda et al., 2022). Today, consumers interact daily with thousands of contents via social media (Lou & Yuan, 2019), in direct contact with these opinion leaders and in an environment accessible to all users (Marwick & Boyd, 2011).

With brands increasingly entrusting digital influencers to present and promote their products and services, online influencer marketing has emerged as a global phenomenon critical to the long-term viability of business marketing (Leung et al., 2022). It is presented by Lou and Yuan (2019, p. 59) as "a marketing strategy that uses the influence of key individuals or opinion leaders to drive consumers' brand awareness and/or their purchasing decisions."

In practice, companies choose or encourage specific users to interact with their social media followers (Leung et al., 2022). Then, these users share content (in various formats, such as videos, text, or images) that their followers find useful, enjoyable, and engaging. Finally, followers show their interest and satisfaction by liking, commenting on, or subscribing to pages or events (O'Donell, 2018). Thus, using the influencers' reputation and power, online influencer marketing uses digital platforms to increase the exposure of the product or service being promoted, expand its reach, and establish a stable online relationship with its audience (Thaichon et al., 2020).

This strategy enables brands to add value to their products or services by placing them in a specific context and through a personal narrative (Khamis et al., 2017). Influencer positioning and content are now used to promote a company's offerings (Leung et al., 2022). Furthermore, influencers are proving to be simple assets with which to collaborate with brands (Masuda et al., 2022), and they enable marketing objectives to be achieved at low costs (von Mettenheim & Wiedmann, 2021), often with nonmonetary compensation, such as free products or experiences (Leung et al., 2022).

This is clearly noticeable in the evolution of fashion communications: where brands and retailers used to set trends and persuade consumers in their fashion consumption, people now follow influencers who inspire them via social media platforms (Vehmas et al., 2018).

One of the most significant outcomes of these new forms of interaction is the consolidation of value co-creation processes that emerge as a result of the increasingly common cooperation between production and consumption (Quach & Thaichon, 2017). Online influencer marketing has proven to be an important tool for marketers in obtaining trustworthy recommendations from their target audience (Jorge et al., 2018), demonstrating that social media is indeed a space to understand consumer needs, collaborate, and co-create (Bu et al., 2022).

Nonetheless, brands' lack of message control (Bar & Pedreira, 2021; Leung et al., 2022) ends up being both a great strength and a potential risk. Indeed, influencers' considerable creative freedom is variable, and it should entail management and coordination aligned with company objectives (Leung et al., 2022). Finding the right influencer for each brand—"a well-fitting influencer" (von Mettenheim & Wiedmann, 2021, p. 1277)—is therefore critical for risk reduction. In this regard, the selection process should be thorough, with issues such as the authenticity of the influencer and the veracity of their results (follower analysis and statistics) taken into account (Belanche et al., 2021).

The truth is that marketing communication via influencers is now widely accepted and practiced, owing primarily to their effectiveness in persuading their followers (Jung & Im, 2021). Of note, online influencer marketing has proven to be more effective than traditional marketing efforts at generating sales and returns (Leung et al., 2022): according to the Influencer Marketing Hub (Santora, 2022), every dollar invested in influencer marketing can result in a \$5.20 benefit, quintupling the amount invested in return.

Digital Influencers: From Intimacy to Dependence

Although the term "influencer" was already mentioned in the context of online influencer marketing, it is worth noting that, according to Abidin (2015, p. 1), influencers are:

> Everyday, ordinary Internet users who accumulate a relatively large following on blogs and social media through the textual and visual narration of their personal lives and lifestyles, engage with their following in digital and physical spaces, and monetise their following by integrating 'advertorials' into their blog or social media posts.

To this, Leung et al. (2022) add that influencers should be considered opinion leaders in their network, while Bu et al. (2022) emphasize the accessibility of these actors due to the ease of relationships with their audience. In this regard, it is necessary to distinguish influencers from celebrities: Belanche et al. (2021, p. 187) place influencers "between celebrities and friends" and argue that they represent a new version of the celebrity endorsement marketing tactic.

Although characteristics such as influencers' activity history and network size can forecast the reach of their actions (Shrivastava et al., 2021), influencers' success is determined by other factors (von Mettenheim & Wiedmann, 2021). Online influencers and their attributes have been the focus of scientific investigation (Jung & Im, 2021), primarily because the information they distribute is based only on their personal tastes and interests (Masuda et al., 2022) and because they have proven to be an important voice in determining product and service demand and supply (Orminski et al., 2021). The literature identifies five aspects that can be studied: (1) authenticity, (2) accessibility, (3) intimacy, (4) trust, and (5) congruence.

For starters, by presenting products and services that are integrated into their personalities and lifestyles (Bu et al., 2022), influencers expose more organic content, which can reach the brands' audiences more effectively than brand-generated content (Lou & Yuan, 2019). In fact, these actors have an impact on audiences because they are perceived as equals and genuine (Jorge et al., 2018; Kim & Kim, 2021). Add to that their ease of access and direct communication (Bu et al., 2022), which motivates interactions as if they were friends (Belanche et al., 2021).

These first two aspects pave the way for intimacy to be exposed, which in this case occurs in the various ways people do common tasks, such as drinking coffee, putting on makeup, getting ready for a party, or studying (Marôpo et al., 2020). This intimacy develops through the promotion of forms of sociability and self-revelation close to the followers, and it is entirely up to the influencers to set the limits on the topics they wish to address or not.

Influencers' dissemination of commercial messages to their followers is only effective if there is a pre-existing and cultivated trust relationship between them (Kim & Kim, 2021). According to Johnstone & Lindh (2022), there is a growing direct trust in digital influencers, who are more important than actors such as friends, family, and colleagues in influencing consumption.

It is also worth noting that congruence between brands and their influencers has been shown to affect affinity and intimacy (Xiao et al., 2021). Belanche et al. (2021) investigated the effects of product, consumer, and influencer congruence and discovered that when a follower comes across an influencer who promotes a product that appears congruent with their personality—and reflects their own values or image—they tend to align their perceptions of the product with the influencer's implicit perceptions.

This congruence leads to consumers imitating influencers they regard as "role models" (Ruvio et al., 2013, p. 61). In other words, the better the fit between the consumer and the influencer, the more likely the imitation (Casaló et al., 2010). According to Ruvio et al. (2013), the consumer's doppelganger effect indicates that some people intentionally imitate the consumption behavior of others in order to look or behave like their influencers.

Indeed, the role of social networks in connecting their users is undeniable (Xiao et al., 2021), and by sharing personal information and attractive characteristics, influencers can even come to gain dependence and social power among followers (Kim & Kim, 2021).

It is critical to introduce the concept of homophily in this context. Homophily, which has been identified as an important dimension of source characteristics (Ladhari et al., 2020), can be defined as "the degree to which people who interact are similar in beliefs, education, social status, and the like" (Eyal & Rubin, 2003, p. 80). Indeed, by displaying their tastes and interests, digital influencers attract like-minded people, resulting in self-segmentation, with followers looking for influencers who resemble them (Mcquarrie et al., 2013).

The New Merchants of Cool

The documentary Merchants of Cool, produced by Barak Goodman and Rachel Dretzin, was broadcast on PBS in 2001. Douglas Rushkoff conducted a series of interviews with marketers, media industry executives, and cultural critics to learn the ins and outs of American marketing, which was popular at the turn of the millennium for creating and selling pop culture to teenage consumers. The documentary presented us with the relentless search for the next cool thing (Goodman & Dretzin, 2001) by traveling through the secrets of marketing and exploring its tactics and tools.

Twenty years later, it is critical to observe, analyze, and comprehend the phenomenon confronting the marketing strategies presented in Merchants of Cool and the online influencer marketing, which is the focus of this chapter's research. The ubiquity of the Internet (Piazza & Abrahamson, 2020) and the establishment of social media (Gu & Ye, 2014) represent continuous and overwhelming changes in the way production and consumption interact today (Bulovic & Covic, 2020).

The content analysis performed—a method of analysis that extracts meaning from textual data (Bardin, 2011) without necessarily referring to a pre-established theoretical framework (Ghiglione & Matalon, 1997)—allows us to take a fresh look at Douglas Rushkoff's journey through the "complex world of buying and selling cool" (Goodman & Dretzin, 2001, 1:10).

It is critical to understand one of the main paradigmatic differences between the two perspectives right away: the technological evolution that separates them. At the turn of the millennium, data such as 75% of teenagers having a television in their bedroom and one-third having a personal computer where they spend an average of

2 h a day online (Goodman & Dretzin, 2001, 4:20) reflected the massive consumption of entertainment. Today's most important statistics are clear: in 2019, the number of cell phones in use surpassed the number of people on our planet (Statista, 2021b), and in 2020, the average daily consumption of social networks by Internet users worldwide reached 145 min (Statista, 2021a).

Not so different is the need for companies and brands to know their target audience. Marketing executive Rob Stone emphasizes the importance of understanding consumers in the documentary: "If you don't understand and recognize what they are thinking and what they are feeling, and then be able to take that in and come up with a really precise message (…) in their terms, you gonna lose" (Goodman & Dretzin, 2001, 5:00). Indeed, we have seen that finding the right influencer for a brand has become a critical task, prompting marketing teams to engage in extensive selection processes (Belanche et al., 2021; von Mettenheim & Wiedmann, 2021). According to Leung et al. (2022, p. 15), "only when firms select influencers whose audience, positioning, and abilities provide a good fit with their brands could firms reap the benefits of OIM [*online influencer marketing*] to enhance marketing communication effectiveness."

The truth is that, just as there were many brands competing for the same audience in the past (Goodman & Dretzin, 2001, 4:50), the number of companies recognizing and adopting the digital influencer strategy is increasing (Bu et al., 2022; Masuda et al., 2022), with 94% of marketers believing in the effectiveness of online influencer marketing (Lou & Yuan, 2019) and looking to collaborate with these actors (Masuda et al., 2022).

Nonetheless, gaining knowledge of trends and who feeds them has changed significantly—the search for "what's going to happen next" (Goodman & Dretzin, 2001, 6:22). Brands like MTV had researchers on their teams in the 1990s and early 2000s who conducted rigorous ethnographic analyses, studying the audience—young people in this case—in their natural habitat. The concept of "cool hunting" (Goodman & Dretzin, 2001, 6:07) has been simplified: nowadays, influencers, as well as the final consumer, share their daily lives and frequently lead their content to mundane topics (Marôpo et al., 2020), exposing their interests and tastes (Masuda et al., 2022).

Marketing, both past and present, realized that once it had gathered information about its targets, it needed to give them control and some of the power. Just as MTV allowed viewers to choose what they wanted to watch and listen to (Goodman & Dretzin, 2001, 19:49), brands now engage in true cooperative and co-creation activities with influencers and consumers (Quach & Thaichon, 2017; Zhang et al., 2015). Furthermore, as previously stated, online influencer marketing requires brands to relinquish a significant amount of control: "an inherent tension in OIM [*online influencer marketing*] is that it provides firms with the creativity benefits of leveraging influencers' content, but it also creates content control problems, due to sourcing from a group of influencers" (Leung et al., 2022, p. 13).

The great advantage of influencer marketing is that it allows companies and brands to focus on a specific group of the audience—their influencers—and attribute that power to them alone, not least because they are the ones the audience looks to (Mcquarrie et al., 2013), in a kind of search for advisors with whom they share interests (Belanche et al., 2021).

The significance of both strategies has resulted in the professionalization of these industries. In 2001, agencies such as Cornerstone specialized in "under the radar marketing" (Goodman & Dretzin, 2001, 12:30), offering tactics that their target audiences were unaware of. Sprite was a key example in this regard: brand activation through hip-hop events communicated to young consumers that the goal was not to sell the product but to integrate it into their culture (Goodman & Dretzin, 2001, 11:44). Online influencer marketing has become increasingly profitable, professionalized, and even industrialized in recent years (Jorge et al., 2018). Huynh et al. (2021) demonstrate the power of technology by presenting a system for collecting data from social networks and detecting potential influencers for a given marketing campaign. Today, these tactics and tools are used to sell a product to a consumer through specific intermediaries that bear no resemblance to traditional advertising: "the influencers (...) need to create campaigns that seem natural to their audience, rather than eliciting suspicions of covert advertising, such that followers believe the promoted message and adopt positive responses" (Belanche et al., 2021, p. 187).

With the realization of the power that comes with being a teenager on one hand and an influencer on the other, young people and users of various social networks have become professionalized in order to integrate the "20%, the trendsetters, that are going to influence the other 80%" (Goodman & Dretzin, 2001, 6:50).

Rushkoff introduces us to the concept of the "midriff" (Goodman & Dretzin, 2001, 28:38), a media-created character based on a stereotype that captivated the teenage female audience: sophisticated, highly sexual, and confident. Teenage girls competed in contests and conventions to prove that they, too, had what it took to be the next Britney Spears and become famous and successful in the media industry (Goodman & Dretzin, 2001, 30:30).

Today, the world of digital influencing is not very distinct: influencers like the Portuguese SofiaBBeauty understand how "circuits of global culture play out on digital platforms" (Jorge et al., 2018, p. 93) and how to become profitable in collaboration with brands (Jorge et al., 2018; Leung et al., 2022).

We are dealing with a large "feedback loop" in both scenarios (Goodman & Dretzin, 2001, 42:10). Media and brands observe today's influencers and attempt to sell their products through their image and lifestyles. The public looks up to them and aspires to be like them. The two cultures are so inextricably linked that it is impossible to tell which came first: "Who is mirroring who?" (Goodman & Dretzin, 2001, 41:33).

Conclusion

The Internet gave rise to new and varied ways of communicating and consuming (Bulovic & Covic, 2020; Leung et al., 2022), which had a significant impact on brand and company marketing strategies, forcing them to adapt in order to remain active and profitable (Gazzola et al., 2020; Kim & Kim, 2021; Masuda et al., 2022).

When discussing technology in the twenty-first century, we must consider the evolution and stability of social networks (Gu & Ye, 2014), particularly the proximity of contact that they have enabled between producers and consumers (Bulovic & Covic, 2020). Today, we are witnessing co-creation processes (Zhang et al., 2015), and for some time now, we have begun to disregard long-established intermediaries such as the press (Piazza & Abrahamson, 2020) or traditional advertising (de Veirman et al., 2017).

Online influencer marketing has won over brands all over the world (Leung et al., 2022) and continues to grow in popularity (Bar & Pedreira, 2021; Kim & Kim, 2021; Lou & Yuan, 2019). Furthermore, businesses are becoming more aware of what they need to do to be successful with this marketing strategy (von Mettenheim & Wiedmann, 2021), and influencers have comfortably taken over the role that was previously held by actors in promotional campaigns (Lou & Yuan, 2019) or even celebrities (Belanche et al., 2021).

In the fashion industry, influencers are now regarded as a voice for fueling users' fashion addiction (Orminski et al., 2021). The truth is that social media platforms allow the fashion industry to connect with their customers and learn about their fashion preferences (Shrivastava et al., 2021).

Indeed, in this digital age, the fashion industry is witnessing an increase in the role of influencers, as consumers perceive their lifestyle to be more authentic and appealing than traditional advertisements (Jin et al., 2019), making the online presence of designers and brands even more necessary (Ahmad et al., 2015).

The documentary Merchants of Cool (Goodman & Dretzin, 2001) provided us with a fresh perspective on how marketing communication evolves today, always in contrast to the one designed by brands and marketers between the late 1990s and early 2000s. The recent literature and journalistic work of Douglas Rushkoff presented in this chapter show that structural differences have occurred in the last two decades. Although the efforts differ, the intensity and purpose of brands in getting to know their customers, reaching them, and remaining profitable are similar.

The purpose of this study was to determine whether digital influencer marketers are the new Merchants of Cool, and the parallelism between the two strategies confirms that they are—despite the two very different contexts (primarily social and technological) and significant differences between the players involved. Even though the actors and targets of the strategies have changed over the last two decades, the relentless pursuit of cool and trendsetting remains unwavering.

In fact, both observed scenarios raise the same concerns: if marketing seeks to mirror reality, and if true reality is altered by its actors to please marketing, who is replicating what? Authenticity and true cool are caught in a feedback loop that affects the final consumer, who is overly receptive to content manufactured by the media industry or influencers, prompting us to consider issues such as dependency, which is compounded by the consumer's constant need to receive signals that tell them what to think and how to behave.

This study makes significant contributions to academia and marketing. First and foremost, because it emphasizes the relationship between marketers and influencers—contrasting with the usual approach by consumption—then by the confrontation between two strategies apparently so unequal but that meet in the use they make of a kind of "new intermediaries" that ended up subverting the rules and taking control.

In Merchants of Cool, Jimmy Lovine, co-chairman of Interscope Records, explains that "there is no way to stop a movement in popular culture" (Goodman & Dretzin, 2001, 47:00). However, there will always be room for a fresh perspective in the complex world of buying and selling cool (Goodman & Dretzin, 2001, 1:10), whether it is analyzing teenagers, influencers, or the next cool thing.

Bibliography

Abidin, C. (2015). Communicative intimacies: Influencers and perceived interconnectedness. *A Journal of Gender, New Media & Technology, 8.*

Ahmad, N., Salman, A., & Ashiq, R. (2015). The impact of social media on fashion industry: Empirical investigation from Karachiites. *Journal of resources development and management, 7.*

Bar, C. M., & Pedreira, J. F. (2021, June 23). Is influencer marketing still influential? *Iberian conference on information systems and technologies, CISTI.* https://doi.org/10.23919/CISTI52073.2021.9476344.

Bardin, L. (2011). *Análise de conteúdo* (4th ed.). Edições 70.

Belanche, D., Casaló, L. V., Flavián, M., & Ibáñez-Sánchez, S. (2021). Understanding influencer marketing: The role of congruence between influencers, products and consumers. *Journal of Business Research, 132*, 186–195. https://doi.org/10.1016/j.jbusres.2021.03.067

Brewer, M. K. (2019). Slow fashion in a fast fashion world: Promoting sustainability and responsibility. *Laws, 8*(4), 24. https://doi.org/10.3390/laws8040024

Bu, Y., Parkinson, J., & Thaichon, P. (2022). Influencer marketing: Homophily, customer value co-creation behaviour and purchase intention. *Journal of Retailing and Consumer Services, 66*, 102904. https://doi.org/10.1016/j.jretconser.2021.102904

Bulovic, V., & Covic, Z. (2020). The impact of digital transformation on sustainability in fashion retail. *SISY 2020 – IEEE 18th international symposium on intelligent systems and informatics, proceedings* (pp. 149–154). https://doi.org/10.1109/SISY50555.2020.9217087.

Casaló, L. V., Flavián, C., & Guinalíu, M. (2010). Antecedents and consequences of consumer participation in on-line communities: The case of the travel sector. *International Journal of Electronic Commerce, 15*(2), 137–167. https://doi.org/10.2753/JEC1086-4415150205

Castells, M. (2004). *A Galáxia Internet – Reflexões sobre Internet, Negócios e Sociedade.* Fundação Calouste Gulbenkian.

Colliander, J. & Hauge Wien, A. (2013), Trash talk rebuffed: consumers' defense of companies criticized in online communities, *European Journal of Marketing*, *47*(10), 1733–1757. https://doi.org/10.1108/EJM-04-2011-0191

Crewe, L. (2013). When virtual and material worlds collide: Democratic fashion in the digital age. *Environment and Planning A, 45*(4), 760–780.

de Veirman, M., Cauberghe, V., & Hudders, L. (2017). Marketing through instagram influencers: The impact of number of followers and product divergence on brand attitude. *International Journal of Advertising, 36*(5), 798–828. https://doi.org/10.1080/02650487.2017.1348035

Eyal, K., & Rubin, A. M. (2003). Viewer aggression and homophily, identification, and parasocial relationships with television characters. *Journal of Broadcasting & Electronic Media, 47*(1), 77–98. https://doi.org/10.1207/s15506878jobem4701_5

Felix, R., Rauschnabel, P. A., & Hinsch, C. (2017). Elements of strategic social media marketing: A holistic framework. *Journal of Business Research, 70*, 118–126. https://doi.org/10.1016/J.JBUSRES.2016.05.001

Gazzola, P., Pavione, E., Pezzetti, R., & Grechi, D. (2020). Trends in the fashion industry. The perception of sustainability and circular economy: A gender/generation quantitative approach. *Sustainability (Switzerland), 12*(7), 1–19. https://doi.org/10.3390/su12072809

Ghiglione, R., & Matalon, B. (1997). *O inquérito: teoria e prática* (3rd ed.). Celta Editora.

Goodman, B., & Dretzin, R. (2001). *The merchants of cool*. Public Broadcasting Service. https://weta.org/watch/shows/frontline/merchants-cool

Gu, B., & Ye, Q. (2014). First step in social media: Measuring the influence of online management responses on customer satisfaction. *Production and Operations Management, 23*(4), 570–582. https://doi.org/10.1111/poms.12043

Huynh, T., Nguyen, H. D., Zelinka, I., Nguyen, K. V., Pham, V. T., & Hoang, S. N. (2021). Advo: A system to manage influencer marketing campaigns on social networks. *Applied Sciences (Switzerland), 11*(14). https://doi.org/10.3390/app11146497

Jin, S. V., Muqaddam, A., & Ryu, E. (2019). Instafamous and social media influencer marketing. *Marketing Intelligence and Planning, 37*(5), 567–579. https://doi.org/10.1108/MIP-09-2018-0375

Johnstone, L., & Lindh, C. (2022). Sustainably sustaining (online) fashion consumption: Using influencers to promote sustainable (un)planned behaviour in Europe's millennials. *Journal of Retailing and Consumer Services, 64*, 102775. https://doi.org/10.1016/j.jretconser.2021.102775

Jorge, A., Marôpo, L., & Nunes, T. (2018). I am not being sponsored to say this: A teen youtuber and her audience negotiate branded content. In *Observatorio (OBS*)*. http://obs.obercom.pt

Jung, N., & Im, S. (2021). The mechanism of social media marketing: Influencer characteristics, consumer empathy, immersion, and sponsorship disclosure. *International Journal of Advertising, 40*(8), 1265–1293. https://doi.org/10.1080/02650487.2021.1991107

Khamis, S., Ang, L., & Welling, R. (2017). Self-branding, 'micro-celebrity' and the rise of social media influencers. *Celebrity Studies, 8*(2), 191–208. https://doi.org/10.1080/19392397.2016.1218292

Kim, D. Y., & Kim, H. Y. (2021). Trust me, trust me not: A nuanced view of influencer marketing on social media. *Journal of Business Research, 134*, 223–232. https://doi.org/10.1016/j.jbusres.2021.05.024

Kohler, T., Fueller, J., Matzler, K., & Stieger, D. (2011). Co-creation in virtual worlds: The design of the user experience. *Management Information Systems, 35*(3), 773–788.

Ladhari, R., Massa, E., & Skandrani, H. (2020). YouTube vloggers' popularity and influence: The roles of homophily, emotional attachment, and expertise. *Journal of Retailing and Consumer Services, 54*, 102027. https://doi.org/10.1016/J.JRETCONSER.2019.102027

Leung, F. F., Gu, F. F., & Palmatier, R. W. (2022). Online influencer marketing. *Journal of the Academy of Marketing Science, 50*, 226. https://doi.org/10.1007/s11747-021-00829-4

Lou, C., & Yuan, S. (2019). Influencer marketing: How message value and credibility affect consumer trust of branded content on social media. *Journal of Interactive Advertising, 19*(1), 58–73. https://doi.org/10.1080/15252019.2018.1533501

Marôpo, L., Jorge, A., & Tomaz, R. (2020). "I felt like I was really talking to you!": Intimacy and trust among teen vloggers and followers in Portugal and Brazil. *Journal of Children and Media, 14*(1), 22–37. https://doi.org/10.1080/17482798.2019.1699589

Marwick, A. E., & Boyd, D. (2011). I tweet honestly, I tweet passionately: Twitter users, context collapse, and the imagined audience. *New Media and Society, 13*(1), 114–133. https://doi.org/10.1177/1461444810365313

Masuda, H., Han, S. H., & Lee, J. (2022). Impacts of influencer attributes on purchase intentions in social media influencer marketing: Mediating roles of characterizations. *Technological Forecasting and Social Change, 174*, 121246. https://doi.org/10.1016/j.techfore.2021.121246

Mcquarrie, E. F., Miller, J., & Phillips, B. J. (2013). The megaphone effect: Taste and audience in fashion blogging. *Journal of Consumer Research, 40*(1), 136–158. https://doi.org/10.1086/669042

O'Donell, E. (2018). *Instagram influencers: When a special relationship with fans turns dark.* The Conversation. https://theconversation.com/instagram-influencers-when-a-special-relationship-with-fans-turns-dark-100543

Orminski, J., Tandoc, E. C., & Detenber, B. H. (2021). #sustainablefashion–a conceptual framework for sustainable fashion discourse on twitter. *Environmental Communication, 15*(1), 115–132. https://doi.org/10.1080/17524032.2020.1802321

Phua, J., Jin, S. V., & Kim, J. (2017). Gratifications of using Facebook, Twitter, Instagram, or snapchat to follow brands: The moderating effect of social comparison, trust, tie strength, and network homophily on brand identification, brand engagement, brand commitment, and membership intention. *Telematics and Informatics, 34*(1), 412–424. https://doi.org/10.1016/J.TELE.2016.06.004

Piazza, A., & Abrahamson, E. (2020). Fads and fashions in management practices: Taking stock and looking forward. *International Journal of Management Reviews, 22*, 264–286.

Quach, S., & Thaichon, P. (2017). From connoisseur luxury to mass luxury: Value co-creation and co-destruction in the online environment. *Journal of Business Research, 81*, 163–172.

Ruvio, A., Gavish, Y., & Shoham, A. (2013). Consumer's doppelganger: A role model perspective on intentional consumer mimicry. *Journal of Consumer Behaviour, 12*(1), 60–69. https://doi.org/10.1002/cb.1415

Santora, J. (2022). *Key influencer marketing statistics you need to know for 2022.* Influencer Marketing Hub. https://influencermarketinghub.com/influencer-marketing-statistics/

Shrivastava, A., Jain, G., Kamble, S. S., & Belhadi, A. (2021). Sustainability through online renting clothing: Circular fashion fueled by instagram micro-celebrities. *Journal of Cleaner Production, 278*, 123772. https://doi.org/10.1016/j.jclepro.2020.123772

Statista. (2021a). *Daily time spent on social networking by internet users worldwide from 2012 to 2020.* https://www.statista.com/statistics/433871/daily-social-media-usage-worldwide/

Statista. (2021b). *Number of mobile (cellular) subscriptions worldwide from 1993 to 2021.* https://www.statista.com/statistics/262950/global-mobile-subscriptions-since-1993/

Statista. (2025). *Influencer marketing market size worldwide from 2015 to 2025.* https://www.statista.com/statistics/1092819/global-influencer-market-size/

Tafheem, N., El-Gohary, H., & Sobh, R. (2022). Social media user-influencer congruity: An analysis of social media platforms parasocial relationships. *International Journal of Customer Relationship Marketing and Management, 13*(1). https://doi.org/10.4018/IJCRMM.289213

Thaichon, P., Liyanaarachchi, G., Quach, S., Weaven, S., & Bu, Y. (2020). Online relationship marketing: Evolution and theoretical insights into online relationship marketing. *Marketing Intelligence and Planning, 38*(6), 676–698. https://doi.org/10.1108/MIP-04-2019-0232

Vehmas, K., Raudaskoski, A., Heikkilä, P., Harlin, A., & Mensonen, A. (2018). Consumer attitudes and communication in circular fashion. *Journal of Fashion Marketing and Management, 22*(3), 286. https://doi.org/10.1108/JFMM-08-2017-0079

von Mettenheim, W., & Wiedmann, K. P. (2021). The complex triad of congruence issues in influencer marketing. *Journal of Consumer Behaviour, 20*(5), 1277–1296. https://doi.org/10.1002/cb.1935

Xiao, L., Saleem, A., Tariq, S. M., Haq, J. U., & Guo, M. (2021). I wish i could be like her/him! How self-congruence stimulates a desire to mimic. *Journal of Theoretical and Applied Electronic Commerce Research, 16*(7), 3025–3042. https://doi.org/10.3390/jtaer16070165

Zhang, H., Lu, Y., Wang, B., & Wu, S. (2015). The impacts of technological environments and co-creation experiences on customer participation. *Information & Management, 52*, 468–482.

Glasses as a Fashion Accessory in Enhancing Styling Creativity and Their Communication in Sustainable Fashion Brands

Mónica Lopes, Manuel J. Damásio, and Alexandra Cruchinho

Introduction

Fashion is directly associated with an individual's attitude and behaviors, both in relation to oneself and to others (Loschek, 2009).

To create closeness within their own group and to differentiate themselves from others, individuals with similar lifestyles and routines invented fashions that were uniquely theirs, thus forming strong connections among individuals in each group (Campos & Rech, 2010).

It is indisputable that glasses are a significant ally for doctors in the pursuit of better vision quality, even though initially they were used as an adornment due to the lack of lens manufacturing technology. From the twentieth century onward, there was a need to better adapt them to consumers' faces, as aesthetics became a more prominent requirement for those who wore them. Currently, glasses are also seen as a fashion accessory, essential in shaping the look of the wearer, as well as a fashion accessory in reinforcing the creative concepts of brands (Robertson & Jacobson, 2011).

According to Robertson and Jacobson (2011), this object has moved beyond its functionality to be perceived as an element whose symbolism reflects a specific lifestyle in a market that is constantly changing and transforming the end consumer.

The use of products with their own identity that simultaneously allow individuals to better express their identity will always be a strategy to consider (Bucci, 1998).

Glasses provide a better quality of life in the sense that individuals gain greater visual comfort by using this accessory, which also serves to convey a certain image of themselves (Robertson & Jacobson, 2011).

M. Lopes (✉) · M. J. Damásio · A. Cruchinho
ULHT – CICANT, Lisbon, Portugal
e-mail: monica.lopes@ulusofona.pt; mjdamasio@ulusofona.pt; alexandra.cruchinho@ulusofona.pt

A. Cruchinho, J. C. Neves (eds.), *Sustainability in Fashion: Creativity, Education, and Innovation*, https://doi.org/10.1007/978-3-031-80991-0_12

New frames tailored to each face type and personality, recent technology that continuously creates varied materials, and new designers crafting distinct and appealing models for each collection have reversed this trend. Additionally, digital communication has increased drastically in recent times, encompassing different channels than those used previously, causing shifts in consumption patterns. The fact that today's consumers are continuously online and use social media daily demonstrates how the way information and fashion content are consumed has changed, marking a new cycle that allows fashion brands to be much closer to their audience.

The way consumption occurs, the lifestyle associated with it, both products and services, directly impacts the environment and is a result of the depletion of our planet's nonrenewable resources. Fashion is one of the major contributors to this depletion (Schulte et al., 2013).

The increasing awareness of consumers puts growing pressure on fashion brands and industry distributors to reduce surplus waste more responsibly, not only through recycling and reusing these surpluses but also by planning a much more precise production (Bird, 2018).

Some luxury fashion brands, such as Stella McCartney, Gucci, Balenciaga, Bottega Veneta, and Hermès, have been working on the development of circular economy practices and have been producing all their products within sustainable production practices (Schilling, 2017).

As a Fashion Accessory

From Kawamura's perspective, fashion can be differentiated into two main categories: the first concerning clothing and the second as a resource or a general method that encompasses the entire clothing area. Fashion applies primarily to areas where taste is a central concept. Fashion refers to clothing but also to music, poetry, cinema, and architecture. Fashion also has an influence on morality, although in that field, this influence is less. Fashion focuses on general changes in human lifestyles, and all fashions are constantly changing (Kawamura, 2005).

In Lipovetsky's point of view, fashion arises from individual tastes and attitudes, resulting from the ideas and thoughts of the individual in their quest to assert themselves in a particular group. All the rapid changes and transformations in fashion come from ornaments and accessories, such as glasses, as changes in clothing are much slower. It is these trivial details that distinguish and mark the person who wears them (Lipovetsky, 2010).

According to Rosa (2019), Simmel early on observed that fashion could be used as an indicator of the civilization process because an awareness of fashion indicates self-awareness in the sense of egocentrism, and a faster development in fashion indicates an increasing complexity of self-image, self-identity. At the extreme of this development lies the completely decentralized subject. The search for identity thus results in its opposite: the total dissolution of identity. That is the path approaching.

In the current market, in every season and in all segments, glasses are present in both the portfolios of luxury brands and the fashion of urban subcultures. This accessory, which has maintained its presence over time, is synonymous with a cool lifestyle (Brown, 2015).

Glasses, as a fashion accessory, allow individuals to express themselves individually and thus showcase their personality, their personal identity. Numerous public figures from the artistic world, including music and cinema, use this accessory to convey a specific message about themselves: Audrey Hepburn, John Lennon, Karl Lagerfeld, and Lady Gaga make glasses a part of their artistic identity, inspiring the public to wear them as well (Brown, 2015).

Some authors, such as Handley (2011) or Brown (2015), point out that glasses have the power to instill confidence and make people feel cool, regardless of the situation in which they are worn.

It is undeniable that the fact that glasses are a fashion accessory, always present in the runway shows of the most prestigious fashion weeks worldwide, in advertisements and editorials in the field, or even in globally successful films, significantly contributes to the transformation and acceptance of glasses. They were once used only for the correction of refractive errors or visual difficulties and are now considered a fashion icon.

Glasses as a Styling Element in Reinforcing Creative Concepts in Fashion Brands

Watching fashion shows, productions, magazine editorials, or even fashion films, we can observe that glasses are an increasingly used styling element by designers and creatives to convey a particular message in many collections and throughout all seasons. This fashion accessory, used by both creatives and designers in presenting their collections, as well as by the public, has remained in fashion over time, and its demand is growing.

In the latest edition of the fashion event Portugal Fashion, fashion designer Katty Xiomara also incorporated the use of glasses as a styling choice to creatively reinforce the concept of her collection titled "Propaganda" for the SS/23 proposals. In this collection, the designer aims to convey the message of contradicting the sense of belonging that individuals have to their group by wanting to feel equal to others, thus nullifying individual thought. Focusing on the feeling and desire for individual freedom, this Venezuelan designer presented proposals with austere yet delicate lines, using translucent fabrics like tulle, with very feminine details.

Katty Xiomara's Fashion Show at Portugal Fashion SS23: Glasses by Bolon Eyewear

The designer believes that glasses are considered a fashion accessory that is incorporated, meaning they are not just a decorative piece of the look but also serve a functional purpose and harmonize with the face (Fig. 1). It is a form of individual identification for those who use this object (Xiomara, 2022).

The brand Duarte Brand aims to innovate and transform streetwear style, creating and originating a cool lifestyle. Known for exclusive prints created by the brand's designer and creative, Ana Duarte, the brand emphasizes sustainability from its conception. In 2021, Ana Duarte won the C.L.A.S.S. Icon award. This competition recognizes creatives who can combine design, innovative techniques, and materials along with communication to raise awareness and sensitize consumers to the new values of sustainable fashion, contributing to an active lifestyle while working towards a better future. "Street Swell" is the name of the collection that the creator presented at the 59th edition of Lisbon Fashion Week 22, which took place from October 6 to 9, 2022. This collection is inspired by the culture of skaters, influencing urban culture and the carefree style that characterizes this community, which does not worry about demonstrating who they are.

Fig. 1 Ugo Camera Portugal Katty Xiomara for Fashion SS23. (Source: www.viewfashionbook.com)

Duarte Brand Fashion Show at Lisbon Fashion Week SS23: Glasses by Bolon Eyewear

Just like Brown (2015), Ana Duarte also mentions that in her collections, she always incorporates glasses (Fig. 2) as they are a complement to the overall look—a component that reflects a lifestyle and a cool aspect she aims to communicate with her collections. The increasing use of glasses by the public aligns with the common and identifying element, as her collections are designed to be worn on the streets. Each collection has its concept, and glasses are another styling element to reinforce the message conveyed by the collection itself (Duarte, 2022).

Sharing a similar perspective with designer Ana Duarte is Carlos Gil. This designer, with a solidly built professional career, has been increasingly conquering international markets, always emphasizing elegance and his impactful way of creating pieces of unquestionable quality.

In the last edition of Lisbon Fashion Week for the SS/23 season, Carlos Gil presented the collection "Time." For this creative, time is a theme that everyone should reflect on—from how we manage our time to how that management affects our future.

Carlos Gil Fashion Show at Lisbon Fashion Week SS23: Glasses by Bolon Eyewear

An elegance and eccentricity that have always characterized the brand were once again present in this collection full of color and movement (Fig. 3).

Fig. 2 Ugo Camera Duarte Brand for Lisbon Fashion Week Oasis 22. (Source: www.viewfashionbook.com)

Fig. 3 Ugo Camera Carlos Gil for Lisbon Fashion Week Oasis 22. (Source: www.viewfashion-book.com)

Carlos Gil is another designer who also relies on glasses to creatively reinforce the concept of his collections. For this designer, accessories are an integral part of an outfit, and glasses are an indispensable element to communicate a specific message, where everything is meticulously considered, from the materials of the glasses to the color of the lenses (Gil, 2022).

Sustainability in Fashion

In a definition of fashion as creation or language that fosters identity, encourages artistic development, and results in consumer products for various market segments, both national and international, we have, because of this process, products that are consumed in different markets (Moura & Almeida, 2013; Takamitso et al., 2019).

All stages of a fashion product, from production to communication, are somehow linked to working with textile materials (Santana & Pessoa, 2022).

Sustainable consumption and production are relevant factors in preserving limited natural resources and addressing climate change. In the textile sector, the increase in consumption is attributed to fast fashion, resulting in an increase in textile waste at a much faster rate (Elander et al., 2015; Fletcher & Williams, 2013).

The fashion industry has significant consequences for the environment, and considering a circular economy, designed from the outset, can be part of the solution to the current situation. With this approach, a nonexistent waste cycle could be created if designed from the beginning. As a result, there would be greater exclusivity and innovation (Moorhouse & Moorhouse, 2017; Nishimura & Triska, 2021).

The primary function of a fashion designer is to make creative thoughts tangible and add value to those thoughts so that the end consumer can relate to and identify with the products they create. Acting from such an early stage, the designer plays a highly active role throughout the process of creating a specific product, and if the primary focus is on sustainability, it can significantly contribute to improving the fashion production system, which still has its main objective focused on profit (Schulte et al., 2013).

Berlin (2012) states that consumer reflections on the environmental and fair-trade impact caused by the fashion business began in the 1960s. It was only in the late 1980s that consumers started paying attention to the origins of raw materials, their production, how they are treated or transformed, and their impact on both the environment and society. Once these issues become a concern for the consumer, they will be attentive to a production that is part of an ethical production chain, completely opposed to the inherent production of fast fashion, where these values are completely set aside, aiming only for profit, and significantly increasing environmental impact (Fletcher & Grose, 2011; Vehmas et al., 2018).

From the industry's perspective, fabrics are being viewed and analyzed in a completely unique way than before. There is concern, especially among designers and factories, that these fabrics have a more sustainable origin, such as organic cotton and hemp, which give them a quality seal. It is no longer sufficient for them to be practical to clean or to have a good drape (Lee, 2009; Fletcher & Grose, 2011).

Another increasingly prevalent trend, regardless of the area, is the return to traditions, handmade products, what is unique and exclusive, and the work of artisans where mass production does not apply. Such products, resulting from controlled production, are highly valued by the end consumer for several reasons: undeniable quality, exclusive design, and a much more appealing (small collections). In these types of collections, the production process becomes clear to the consumer, adding a lot of value to both the producer and the consumer. Designers are already embracing this handmade trend and vintage inspirations, where the customer can actively participate in the creative, design, and production process. This is a strategy highly valued by the end consumer, who ends up with a product designed and personalized only for them, exactly as they want it, making it completely unique and fostering loyalty as the consumer feels more connected to the brand (Broach & Kirkby, 2011; Nurcahyanti & Septiana, 2018; Mazzarola et al., 2019).

Clothing accounts for a considerable portion of the products we consume, and the end consumer has realized that how they consume will dictate the future of the planet (Lee, 2009).

Sustainability has become a general concern, and even luxury brands make a point of showing that they are fully aware and working with this focus. An example of this is the fact that luxury brands have stopped having so much stock, communicating that modern luxury involves being socially and environmentally responsible (Bird, 2018; Rodriguez et al., 2021).

Brands need to reinvent themselves from design, creating sustainable solutions when creating and launching products, to their own organizational structure. To attract consumer attention and add value to the brand identity, they increasingly need to be exclusive and capture attention, given the speed and how competitors' content is consumed today through social media (Williams et al., 2018; Moorhouse & Moorhouse, 2017).

Currently, driven by the shift towards sustainability, designers are already working together with agricultural producers, raw material suppliers, artisans, and other stakeholders in the fashion product production chain to work towards a more sustainable future (Williams, 2010; Motta, 2022).

Communication of Fashion Brands in the Digital Era

According to Lipovetsky (2010), what characterizes the essence of fashion is the desire for the new. The author states, however, that we are in hypermodernity, and this period we live in is a complete deconstruction of various sectors of our society: economy, politics, education, and culture. This deconstruction, resulting from modern lifestyles, is characterized by an immense proliferation of products for consumers, altering human relationships and even how they perceive the environment (Costa et al., 2020).

To find viable solutions for this deconstruction, it is essential to continue investing in research focusing on conscious design and consumption, social responsibility, and sustainable social development driven by innovation and technology. Innovation appears as a great ally in creating value for a product and a brand. Companies seek to gain a competitive advantage in a reality marked and awakened by sensitivity to new things. To innovate, one must somehow be connected to networks to better understand how to connect effectively (Goldsmith, 2015).

The incessant growth and change in the market where brands operate and compete require them to adapt and align new strategies, management methods, and business models that fit into these new scenarios. It is the Internet that encompasses these opportunities that allow companies to reach a much broader consumer base, reduce costs, and determine how they connect with their brands, and consequently, create value for consumers (Lu, Wang & Kennon, 2017).

The current fashion market is the result of the constant change in the final consumer's way of thinking and its influence on the industry, which has been exponentially increased with the use of digital social networks (Vehmas et al., 2018).

The strategy with which a company positions itself and manages its presence on a particular social network is of utmost importance since it represents its influence and power on that social network (Tidd & Bessant, 2015).

For there to be relevance in the results of a particular market, it must integrate into its business the omnichannel strategy: the use of online stores, digital payments, online product launches, and fashion shows, as well as other initiatives focused on social networks are some of the approaches used by brands (Barreto et al., 2022).

Fashion communication aims to persuade the consumer through the message of a brand or designer, which is constantly changing, considering social changes and the constant search for new communication methods, along with these changes in consumer culture (Lee et al., 2011; Wu et al., 2013). The twenty-first century is the era when the digital technology revolution allows bidirectional communication, and emotional values are emphasized, thus interfering in the communication and fashion industry. Along with fashion shows, advertising, and other initiatives, many fashion brands have increased the use of fashion films to launch and communicate their products (Wu et al., 2013; Huh et al., 2016; Díaz Soloaga & García Guerrero, 2016; Gibson, 2017; Rodriguez & García, 2018).

Analysis of Communication by Brands with Glasses in Sustainable Fashion

Stella McCartney

It is a luxury brand named after the designer and creator of the brand. Stella McCartney's main value is the trust of consumers who are using a brand focused on sustainability and the environment. It aims to awaken the sexy and feminine side of the woman who wears the brand. The company is committed to ethical values, and it believes it is responsible for the resources it uses and their impact on the environment. It is created from bio-based materials, at the forefront of sustainable innovation (Fig. 4). From entirely ecological packaging to displays made from recycled industrial waste and sustainably sourced wood, the brand maintains its luxury perspective while protecting forests, reducing carbon emissions, and reducing dependence on petroleum-based plastics. Frames are created from bioacetate using over 68% organic carbon derived from wood seeds and cotton, along with plant-based plasticizers.

The use of biolenses also helps reduce the use of oil-based plastics, as they are made from natural and renewable materials such as castor oil.

The brand has a UL Certification (the leading global authority on testing, inspection, and certification of sustainable products, providing customers with additional clarity and transparency), which validates the percentage of bio-based content (Fig. 5).

Fig. 4 Displays Stella McCartney 2019. (Source: https://www.stellamccartney.com)

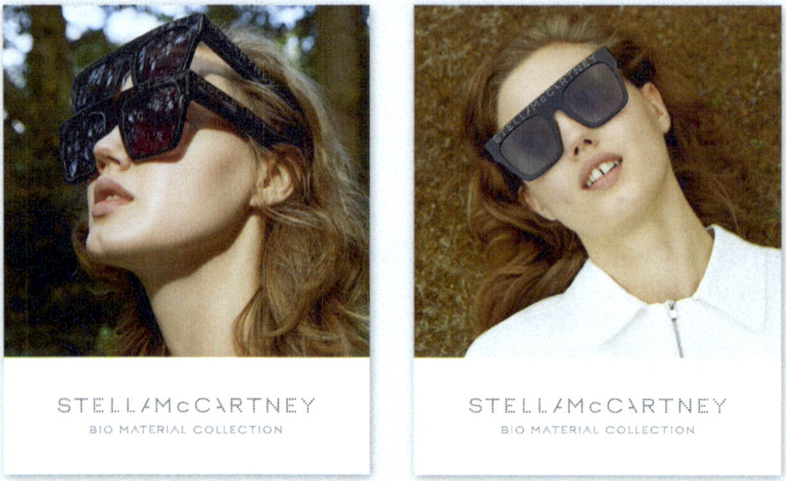

Fig. 5 Displays Stella McCartney 2019. (Source: https://www.stellamccartney.com)

For packaging, the brand uses a vegan-inspired, crocodile-patterned glasses case, inspired by its prêt-à-porter handbag collection. Each case contains a cleaning cloth, leaflet, and label created with recycled materials, as well as a mini-grip and temple caps made from recycled paper.

In the communication of its campaigns in retail stores, its displays with images show glasses from a sustainable perspective; these displays are made of acrylic from industrial waste and wood wool panels, a durable material used in sustainable architecture that combines wood from sustainably certified forests and mineral binders made from sawdust.

Karün

This sustainable eyewear brand is based in Patagonia, and its concept revolves around creating products with a focus on nature conservation and rural development. "Karün" means "to be nature" in the Mapuche language, an ancestral indigenous culture of Chile. The Patagonian Desert, a remote and pristine region where nature flourishes and inspires the brand's designs, values, and way of working, has shaped how it envisions the world—striving to see beyond the obvious and reflecting on the essence of human beings. The brand defines its mission and core value as inspiring the world to contemplate the question: "What would the world be like if we understood that we are all nature?" They value indigenous cultures and their legacy, which, according to the brand, has the wisdom to live in harmony while always respecting biodiversity.

Karün Eyewear is part of the Karün World Group. The brand's creation process begins with Karün collectors from the coast of Chile. These collectors gather several types of discarded waste that affect the natural ecosystem (such as ropes, fishing nets, and metals) and sell them to the brand. With this additional income, these collectors can boost their own micro-businesses. Thus, waste becomes capital for this population to develop their businesses and raw material for the brand's high-quality sunglasses. With a Circular Production Process (Fig. 6), all materials delivered by Karün collectors are then recycled by a recycling partner of the brand. They are transformed into regenerated ECONYL® nylon, a material with virgin technical characteristics that is 100% recycled. Once transformed into raw material, these waste materials are sent to Italy or Turkey to produce the final pieces. The company is offsetting the residual carbon emissions produced by its operations and coordination, making it a CarbonNeutral® Company. Also focusing on sustainability, one of the strategies used by the company at the end of the product's life is to encourage customers to return them to the brand, thereby avoiding new plastic waste. In return, the brand offers the end consumer a 25% discount on their next purchase.

To close the cycle in terms of regeneration and conservation, the brand, with a substantial part of its revenue from sales, is financing a 4-year program involving more than 600 rural entrepreneurs and their families in the Cochamó and Río Puelo Valley, located in Northern Patagonia and known as the "Yosemite of South America." This program is being developed with a social partner, Balloon Latam, and aims to protect 400,000 hectares of wilderness through work with the local community.

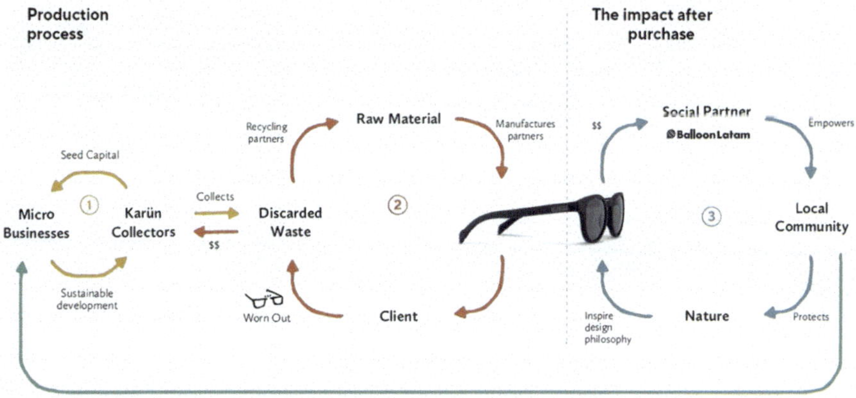

Fig. 6 Imagem adaptada da marca. (Fonte: www.karunworld.com)

Process of Eyewear Production

Brand Certification Seals

One of their sunglass collections was designed in partnership with activist and actress Shailene Woodley, who reinterprets the classic styles of the 1960s and 1970s, combined with a sustainable aspect. The campaign, named "Wear What You Stand For," invites consumers to use style as a personal expression (Fig. 7).

"Wearing what we believe in means using fashion, clothing, and accessories that we wear every day as a tool for self-expression and a means to achieve our common goals," argues Shailene Woodley in May 2022, in an interview with LuxWoman magazine.

The group believes it can change the way we interact with our planet (Fig. 8).

Methodological Procedures

The adopted research methodology is a mixed methodology: non-interventionist and interventionist. In the first phase, the non-interventionist methodology involved a bibliographic search. This type of non-interventionist methodology involves checking references already published in books, scientific articles, dissertations, or theses. The first step in any scientific research begins with a bibliographic search (Salge et al., 2021).

To advance the research, exploratory research with a qualitative approach was developed, using bibliographic and documentary research as a data collection method (Gil, 2002).

Fig. 7 Imagem adaptada da marca. (Fonte: www.karunworld.com)

Fig. 8 Adapted image from karunworld. (Source: www.karunworld.com)

In the second phase, using an interventionist approach, telephone interviews were conducted with three Portuguese fashion designers who used glasses in the presentation of their collections at fashion events, as a creative reinforcement in their styling, at the Oasis edition of Lisbon Fashion Week, held from October 6 to 9, 2022, and at the Portugal Fashion SS23 event, which took place from October 11 to 15, 2022, in Porto.

The sample selection was random, a convenience sample, and the designers who showed willingness to participate in the study and who use glasses to complement their collections were chosen.

In the final phase, case studies of two eyewear brands intentionally chosen for using sustainable raw materials in the creation of these fashion accessories were analyzed. Two brands were chosen from a diverse international universe: Stella McCartney and Karün. The choice was made because these are the brands with which the researcher has greater proximity and that meet the requirements for selection.

The main objectives of this study are to understand the relevance of glasses as a fashion accessory in the creative reinforcement used by creators, to understand the importance for brands of having this styling element in their product portfolios. Another objective of this research is to understand how glasses are communicated in sustainable fashion brands.

Final Considerations

A fast-paced lifestyle has become routine, and the emergence of social media and new B2B and B2C platforms has completely transformed the fashion industry, both in terms of consumption and how companies develop and launch products. Companies face the challenge to be fast to anticipate competition and execute it for their own survival. The promotion of consumption promotes the rapid production and consumption of fashion products, and the tools of neuromarketing and trend forecasting, which are so perfected in this industry, have a greater impact on this process. This situation has significant consequences, caused by excessive consumption, not only for the environment but also for society and the economy, jeopardizing human longevity. To find new practices and technological and innovative processes for a more sustainable future, movements in this direction are already noticeable worldwide. Consumers are responsible for the impact generated by the products they consume and for how they are used and discarded. Consumers' interest in the recycling process and more sustainable solutions has increased considerably. It has become important for all product information to be explicit and visible to consumers, even through multiple communication channels. These digital services that accompany the product add value to it for the consumer. It is essential to determine a fair price that aligns with environmental standards, as well as with all the people included in the product value chain. All information regarding the origin, production, manufacture, and maintenance of the piece must accompany that piece for transparency from companies to consumers. This situation would bring many benefits, such as making prices more justifiable, holding consumers more accountable for their choices, and even educating the population for reduced consumption. In order to be part of the solution that aims at sustainability, many designers are already actively participating in this process so that they can contribute positively to the transformation of the fashion system in that direction. On the other hand, companies that want to progress and remain in the market must focus on sustainability. Businesses depend on nature for absolutely everything, and if the situation is not reversed, we are compromising future generations with the strategies we use in the present. Education has the role of stimulating in students a vision that contributes to a better future, with initiatives of the present that have the potential to cross the boundaries between theory and practice, thus linking conception to production for success. Adding value to fashion involves creating something unique, based on new knowledge about sustainability, supporting biodiversity among design choices, and sharing ideas, sharing curiosities. Glasses, as a fashion accessory, can add this value, accompanying the constant metamorphosis of the individual, allowing them to express all sides of their personality. According to Robertson and Jacobson (2011), this product is no longer focused solely on functionality but is also perceived as a component, a symbol of a lifestyle in a constantly changing market and the evolution of the end consumer.

Based on the conducted research, it can be concluded that glasses are indeed a widely used fashion accessory by creators to enhance the creative concept of the messages they want to convey through their collections. The future is increasingly digital and focused on omnichannel approaches; this situation thus opens opportunities for sustainable initiatives and circular economy practices in the fashion sector. Practices such as fiber recycling, working on innovative product planning and development result in reduced waste and costs. Millennials are the largest consumers, born with social media; therefore, it is relevant to pay attention to this consumer group and reach them through the same networks and technologies. Slow fashion and sustainable brands are favored in this process because, through these networks, brands can highlight their values, processes, and ecological ways of working, thus reaching their target audience more effectively.

References

Barreto, L., Freitas, V., & Paula, V. (2022). Omnichannel em Tempos de Pandemia: Tendência da Estação ou Moda Atemporal? *Revista Ciências Administrativas., 28*. https://doi.org/10.5020/2318-0722.2022.28.Esp.e11837

Berlim, L. (2012). *Moda e Sustentabilidade: uma reflexão necessária*. Estação das Letras e Cores Editora.

Bird, J. (2018). Fashion's dirty little secret and how it's coming clean. *Recuperado em, 20*.

Broach, Z., & Kirkby, B. (2011). *Fashioning the future 2011. The international student awards for sustainability in fashion. Centre for sustainable fashion*. London College of fashion.

Brown, V. (2015): Cool shades: The history and meaning of sunglasses. .

Bucci, A. (1998). *L'impresa guidata dalle idee. Management della moda e del design* (2nd ed.). Domus Academy.

Campos, A. Q., & Rech, S. R. (2010). Considerações sobre moda, tendências e consumo. *IARA-Revista de Moda, Cultura e Arte, 3*(3).

Costa, A., Lima, C., & Lima, L. (2020). La tomada de consciencia en las relaciones de consumo: ciudadanos conscientes y sociedades sostenibles. *Rev. Geogr. Académica, 14*(1).

Díaz Soloaga, P., & García Guerrero, L. (2016). Fashion films as a new communication format to build fashion brands. *Communications Society, 29*(2), 45–61. https://doi.org/10.15581/003.29.2.45-61

Elander, M., Palm, D., Whatson, D., Kiorboe, N., Salmenperä, H., Dahlbo, H., Rubach, S., Hanssen, O., Gíslason, S., Ingulfsvann, A., & Nystad, O. (2015). A Nordic strategy for collection, sorting, reuse and recycling of textiles.

Fletcher, K., & Grose, L. (2011). *Moda & Sustentabilidade: design para mudança*. Editora Senac São Paulo.

Fletcher, K., & Williams, D. (2013). Fashion education in sustainability in practice. *Research Journal of Textile and Apparel, 17*(2), 81–88.

Gibson, C. P. (2017). The fashion narratives of tom ford: Nocturnal animals and contemporary cinema. *Fashion Theory, 21*(6), 629.

Gil, A. (2002). *Como elaborar projetos de pesquisa* (4th ed.). Atlas.

Goldsmith, E. (2015). *Social influence and sustainable consumption*. Springer.

Huh, Y., Chun, J., & Ha, J. (2016). A research on the communication characteristics of fashion film-focused on Chanel, Prada, Kenzo, Alexander Wang. *Journal of the Korean Society of Clothing and Textiles, 40*(2), 315–329.

Kawamura, Y. (2005). *Fashion-ology an introduction to fashion studies*. Berg.

Lawson Handley, L. J., Estoup, A., & Evans, D. M. (2011) Ecological genetics of invasive alien species. Bio Control 56, 409–428. https://doi.org/10.1007/s10526-011-9386-2

Lee, M. (2009). *ECO CHIC: O guia de moda ética para a consumidora consciente* (1st ed.). Larousse.

Lee, G., Lee, J., & Kwon, S. (2011). Use of social-networking sites and subjective well-being: A study in South Korea. *Cyberpsychology, Behavior, and Social Networking, 14*(3), 151–155.

Lipovetsky, G. (2010). *O Império do Efémero, A moda e o seu destino nas sociedades modernas*. Publicações Dom Quixote.

Loschek, I (2009), When clothes become fashion: design and innovation systems. Berg.

Lu, R., Wang, Y., & Kennon, R. (2017). Cross-cultural research for luxury fashion brands in the chinese market: A review of long-versus short-term orientation in national culture dimensions. *Business models and ICT technologies for the fashion supply chain: Proceedings of IT4Fashion 2016*. Springer Nature. https://doi.org/10.1007/978-3-319-48511-9_3.

Mazzarellaa, F., Storeya, H., & William, D. (2019). Counter-narratives towards sustainability in fashion. *Scoping an academic discourse on fashion activism through a case study on the centre for sustainable fashion*. EAD 2019.

Moorhouse, D., & Moorhouse, D. (2017). Sustainable design: Circular economy in fashion and textiles. *The Design Journal, 20*(sup1), S1948–S1959. https://doi.org/10.1080/14606925.2017.1352713

Motta, M. (2022). Palestras sobre um sistema de moda e têxtil sustentável: uma análise qualitativa dos stakeholders. *Moda Palavra, Florianópolis, 15*(35), 97–128.

Moura, M., & Almeida, M. (2013). A relação entre a sustentabilidade e o design de moda contemporâneo: uma análise sobre o segmento jeanswear. *Comunicação e Sociedade, 24*, 232.

Nishimura, M., & Triska, R. (2021). Design de Serviço e Moda: relações para o desenvolvimento de plataformas colaborativas. *ModaPalavra, Florianópolis, 14*(34), 111–136.

Nurcahyanti, D., & Septiana, U. (2018). Handmade eco print as a strategy to preserve the originality of ria Miranda's designs in the digital age. *MUDRA Journal of Art and Culture, 33*(3), 395–400.

Robertson, P., & Jacobson, D. (2011). *Knowledge transfer and technology diffusion. E.U.A.* Edward Elgar Publising, Inc.

Rodríguez, A. F., & García, H. S. (2018). Advertising communication in the fashion industry: Branded content, the case of fashion films. *Revista de Comunicación, 17*. https://doi.org/10.26441/RC17.1-2018-A1

Rodríguez, C., Mcquillan, K., & Segarra-Saaavedra, J. (2021). La comunicación de la sostenibilidad como herramienta de ventas y de cambio social: fast fashion vs slow fashion. *Revista de Comunicación de la SEECI, 54*, 17–37. https://doi.org/10.15198/seeci.2021.54.e667

Rosa, A. (2019). *Imitação e Desejo: Moda, Marcas e Copyright*. Mediaxxi Publishing, Research and Consulting.

Salge, E., Oliveira, G., & Silva, L. (2021). Knowledge for construction of documentary research. *Revista Prisma. Rio de Janeiro, 2*(1), 123–139.

Santana, C., & Pessoa, A. (2022). Sustentabilidade ilustrada. *ModaPalavra, Florianópolis, 15*(35), 157–184.

Schilling, M. K. (2017). *Stella McCartney is weaving a new way forward*. https://www.fastcompany.com/40474789/stella-mccartney-is-weaving-a-new-way-forward (acedido em 05 dezembro 2022).

Schulte, N., Lopes, L., Alessio, M., & Freitas, B. (2013). Fashion in the sustainabilit context. *ModaPalavra e-Periódico Ano, 6*(11), 194–211.

Takamitso, H. T., & Junior, J. A. (2019) See now buy now: Desafios e oportunidades no desenvolvimento de novos produtos de moda sustentáveis do mercado de luxo na era das mídias sociais. Future Studies Research Journal Trends - FIA Business School. https://doi.org/10.24023.

Tidd, J., & Bessant, J. (2015). *Gestão de empresas – Inovação*. I. Bessant, John. II. Bookman. 5ª edição.

Vehmas, K., Raudaskoski, A., Heikkilä, P., Harlin, A., & Mensonen, A. (2018). Consumer attitudes and communication in circular fashion. *Journal of Fashion Marketing and Management: An International Journal, 22*(3), 286–300.

Williams, D. (2010). *Fashion education in sustainability: Change through experimental crossing*. Centre for Sustainable Fashion. London College of Fashion/University of the Arts London.

Wu, J., Won Ju, H., Kim, J., Damminga, C., Kim, H. Y., & Johnson, K. P. (2013). Fashion product display: An experiment with Mockshop investigating colour, visual texture, and style coordination. *International Journal of Retail & Distribution Management, 41*(10), 765–789.

Entrevistas Realizadas

Duarte, A. (2022). Entrevista em.
Gil, C. (2022). Entrevista em.
Xiomara, K. (2022). Entrevista em.

Product Design: Contribution to User's Comfort

Paula Mercedes Neves and Fernando Moreira da Silva

Introduction

This research project is directly related to the act of traveling. Everything that involves travel has a considerable impact on the world economy and the environment.

The context of travel has evolved throughout history, as well as the needs of the type of luggage the traveler carries. This action of traveling has always involved carrying luggage, so we focus not only on what is necessary for the day-to-day of today's traveler but also on minimizing the user's physical efforts when transporting all their hand luggage.

The main objective of this project is to contribute to the comfort of the traveler in air travel, as well as in other types of transportation. Through previous doctoral research in Design, an innovative product was achieved that aims to contribute to optimizing the management, organization, and transport of hand luggage. The product makes it possible to speed up accessibility to the objects needed during the trip, as well as allowing the entire range of luggage to be transported, grouped into the cabin wheels suitcase, freeing the traveler from carrying weights on their shoulders and hands.

In this chapter, we present the evolution of the final product, in which some significant changes were introduced in the design of the initial functional prototype.

These changes have led to a new type of travel bag, which is intended to be disseminated to academia and industry. Since it is a new type of bag, it can take on

P. M. Neves (✉)
ESART, Escola Superior de Artes Aplicadas, Instituto Politécnico de Castelo Branco, Castelo Branco, Portugal

F. M. da Silva
CIAUD, Research Centre for Architecture, Urbanism and Design, Lisbon School of Architecture, Universidade de Lisboa, Lisbon, Portugal

A. Cruchinho, J. C. Neves (eds.), *Sustainability in Fashion: Creativity, Education, and Innovation*, https://doi.org/10.1007/978-3-031-80991-0_13

different versions in terms of its appearance. In the specific case of this chapter, we present just one of the possible looks that this product may have in the future.

Methodology

The methodology followed in this study is a theoretical-practical mixed one, essentially empirical (through active research). The theoretical component began with a literature review, focused on the research of authors related to different themes within the scope of the investigation.

We then proceeded to a brief contextualization of the reasons that led human being to travel throughout history and on the act of traveling itself, with emphasis on the most used transport. An approach was made to its evolution up to the present time and, also, to the importance of nowadays baggage restrictions.

The results of the doctoral research, "Hand Luggage Organizer Design for Airplane Travel Contribution to Traveler Comfort" (2018), support this study: a questionnaire was implemented to target groups of travelers where the problems and questions they encounter were identified, both when they pack their suitcase as during its transport.

The exploratory work is based on studying aspects of inclusive design, ergonomics, and sustainability.

The exploratory and laboratory phases are fundamental for the development of this project. It started with the development of the MOOV.22 design, where the studies of form, function, and materials are preponderant. The changes made to the 2018 functional prototype promote improvements in its usability through a new design, and later, it will be formally applied in a final prototype designed for series production. We aim this product to bring real added value to the comfort of today's travelers.

The Context of Travel and Baggage Transport

The reasons for traveling vary from person to person. Each of us will be able to travel for work, leisure, study, or for other reasons, in the same year, month, or week.

John Urry (2012) talks about the universe of human mobility, mentioning that, in 2012, an average of 300,000 passengers were flying over the United States of America, the equivalent of a city in the air. The transport of luggage on air travel is increasingly directed so that the traveler only uses hand luggage, that is, that he only travels with cabin luggage.

After the impact that Covid-19 has had on air travel, other restrictions have been added for travelers, such as they have to present Covid-19 tests and vaccination certificates.

The Covid-19 pandemic has had a negative impact on the global travel industry. According to Dube et al. (2021), possible recovery paths for the global aviation sector are examined. Currently and from now on, airports and airlines must find solutions that prevent the spread of diseases and reduce the environmental footprint caused by pollution in this sector. "The study recommends that as the sector opens up, it does so in a responsible manner, which puts measures that protect travelers, reduce costs, increase efficiency, and ensure a quality customer experience anchored on employees' health and customer safety" (Dube et al., 2021, p.1).

The Suitcase and Its Evolution Over Time

The invention of the suitcase at the beginning of the twentieth century is related to the transport revolution. We highlight its first creators:

In 1920, Louis Vuitton created a suitcase for air travel with a weight limit of 26 kg. In the 1970s, Bernard D. Sadow invented the suitcase with wheels. This idea was patented as "rolling luggage," due to the fact that the bag was pulled horizontally; however, balance and ergonomic problems still needed to be solved. A few years later, Northwest Airlines pilot Bob Plath created a new wheeled luggage system that allowed him to transport his suitcase more quickly through the terminals of busy major airports. In 1987, Plath designed this suitcase upright, with two wheels and an extendable handle. He patented the Rollaboard® suitcase (Fig. 1) and marketed it through sales to fellow pilots and flight attendants (Yukimura, 2014).

Today, in the market, we can find bags with two and four wheels, which are more ergonomic and functional. The four-wheel suitcases with telescopic handles are more stable and slide alongside the user without great effort, providing comfort. "An extremely useful element and present at different times in life, suitcases are also a product of a long tradition in the fashion world. A simple idea [...] to take the goods from one place to another. [...]" (Higa, 2015, s.p.).

Since the beginning of this investigation in 2012, there has been a growing market for bags and travel accessories, innovations, and contributions to the comfort of today's traveler.

Cabin Baggage in the Context of Travel

When traveling by plane, passengers usually face several problems and inconveniences, such as the following: (1) Sometimes, it is impossible to carry more than one bag per passenger; (2) difficulty in organizing and accessing personal objects during the journey at the airport; (3) difficulty in organizing and accessing personal items during the flight; (4) transport of personal effects at the destination of the trip; (5) too much luggage in the hands and shoulders to carry; and (6) setbacks during the trip, from leaving home to the destination.

Fig. 1 Pilot Bob Plath carrying his invento Rollaboard®, 1987. (Source: http://criandoeinovando.blogspot.pt/2014_07_01_archive.html [2 Nov. 2014])

All of the above can cause physical and psychological discomfort in the traveler. If the traveler's needs are met, the level of travel comfort can increase. Thus, keeping the necessary objects on hand can be a contribution to increasing that level.

The list of objects that make up the luggage that is taken into the cabin is different for each individual, depending on variables such as gender, age, health and food, types of travel, social level, professional activity, cultural level, religion, traveling with a partner or not, change of climate from departure to return, type of accommodation, and duration of the trip.

Some restrictions on hand luggage are: the number of bags, weight and size, which vary according to the requirements of airlines and customs.

Concept Moov.22 and Its Validation

In this project, the exploratory phase essentially focuses on the study of inclusive design, ergonomics, and sustainability. It's inclusive, because it's customizable; it allows you to satisfy the particular needs of users. As Julia Cassim (2012) explains, according to the inclusive design approach, a project can be conceived from extreme situations to average situations, the reverse being not possible.

It is ergonomic, as it integrates aspects that facilitate the well-being of the traveler, before, during, and after the trip, minimizing the physical and psychological fatigue of the user. According to Neves (2018), it is important to develop interventions that reduce the lack of tranquility and situations that cause stress. In ergonomics, users' discomfort levels are usually measured. It is also sustainable, as it is designed with noble materials, such as genuine leather, quality of manufacture with superior quality details and accessories, allowing this product to have a long-life period.

This concept was achieved through the contribution (online questionnaire survey in 2015) of travelers in the following target groups: tourists, executives, and crew (sample of 402 respondents). The concept is related to the organization of hand luggage, from the inside to the outside of the cabin bag, making the list of objects necessary during the trip accessible, thus facilitating its transport. We can consider hand luggage divided into two parts: what you carry inside your cabin bag—hard-to-reach luggage—and what you need in transit and during your trip (outside your cabin bag)—easily accessible luggage.

The Concept

The organizer is a cover with enough elasticity to adapt to most cabin bags whose width, height, and depth measures must not exceed 115 cm in total, and in accordance with the rules in force in most airlines for luggage cabins.

When placed on the cabin bag, MOOV.22 allows you to carry your luggage as a whole. For example, attaching a bulky coat to MOOV.22 that takes up almost the entire cabin bag, making it advantageous to take it out of the bag and bundled with it (Figs. 2 and 3), and also being able to add a free shopping bag shop, a bottle of water, and an umbrella, through the compartments, webbing straps, and carabiners.

Inside the cabin, the MOOV.22 can be easily removed from the cabin bag: the user can easily remove it from the bag, just by pulling the main handle, before it is placed in the place intended for these.

When removed, the MOOV.22 appears to be just a small bag that contains the items the passenger needs during the journey (Fig. 2).

At the airport, the document bag is accessible from check-in to boarding (Fig. 4).

When carrying all the hand luggage grouped in the MOOV.22, the user can easily extract it from the bag, just by pulling the handle.

Fig. 2 Information on how to use MOOV, 2022

Fig. 3 Mobility and portability, 2022

Prototipe (Functional Model) and Its Validation

The prototype (functional model) was presented in the north of Portugal at CENTI (Center of Nanotechnology for Technical, Functional, and Intelligent Materials) to a panel of experts and specialists to be evaluated and possibly validated.

The expert panel was composed of specialists from different areas, the main areas present in our study: Design | Professor Ana Cristina Broega & Professor Joana Cunha; Ergonomics | Professor Paula Carneiro & Professor Doctor Nelson Costa; Technologies | Engineer Domingos Moreira; Industry | Mr. Domingos Lima;

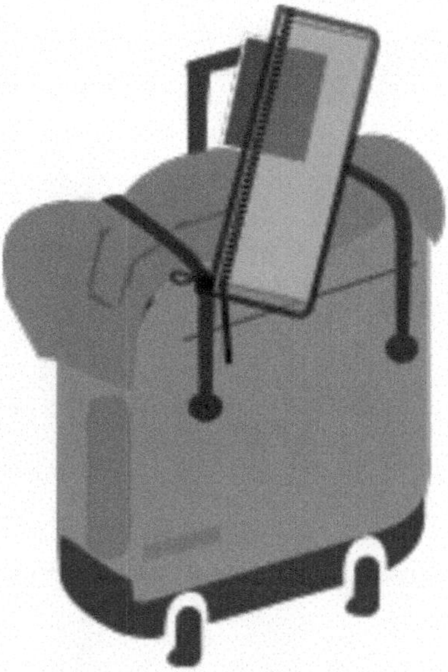

Fig. 4 Documentation is accessible, 2022

Fashion and Industry Design | Mr. António Simões. The researcher was accompanied by the research supervisor team.

For that specific purpose, we had to travel to the North of the country by plane, from Lisbon airport to Porto airport (Fig. 5). During the flight, the prototype had also its first usability test on a plane trip.

After the detailed presentation of the product, the panel members had the opportunity to interact with the product and contribute with their knowledge, giving their opinions and suggestions to improve the product, which were later introduced in the final prototype (Fig. 6).

Moov.22 Illustrated Drawing (Render)

MOOV.22 (Fig. 7) is adjustable to most cabin bags, with dividers to organize the luggage by sectors, identifiable by the asymmetrical placement of the compartments on two of its main sides (A and B). Mobility is maximized (focused on ergonomic aspects), helping the user travel lighter, as it allows transporting all the luggage grouped in the cabin suitcase with wheels, avoiding weight on the hands and shoulders. It is composed of materials with elasticity on the sides. One of the sides has an opening to access the side handle of the cabin bag.

In MOOV.22, all cabin bag handles are accessible through the openings at the top and on one of the sides. It is also possible to place a third bag grouped in the telescope, counting this as the second bag, as the MOOV.22 is aggregated with the bag like a cover.

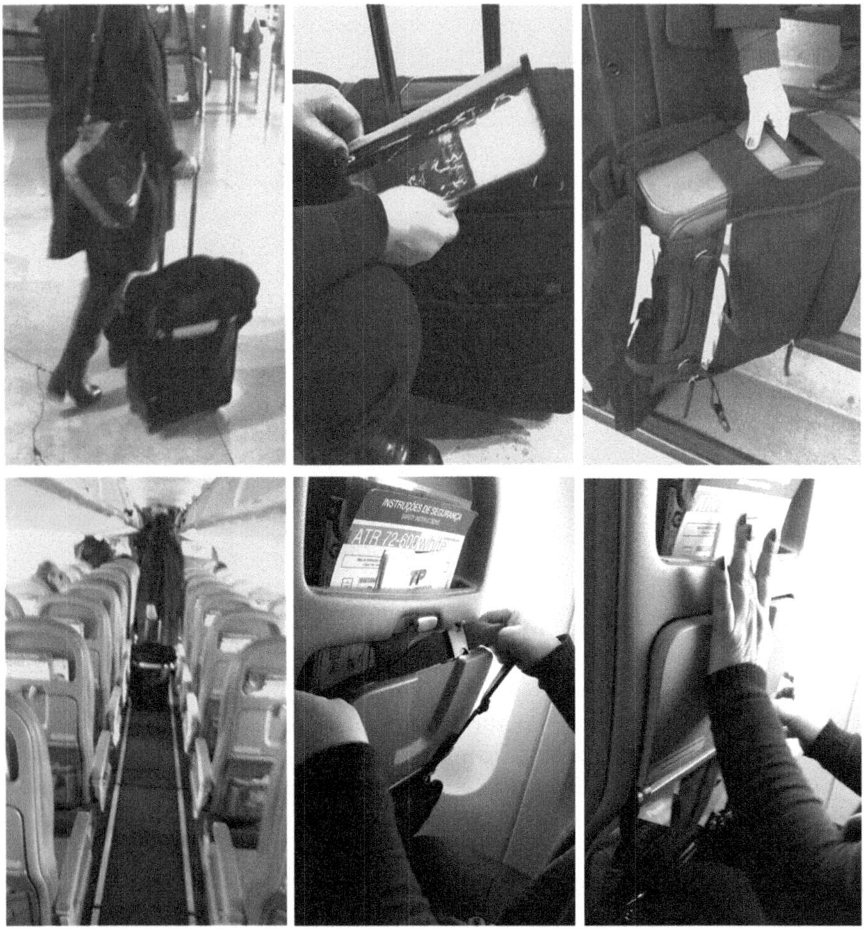

Fig. 5 Arrival at the airport; check-in; on the stairs for embarkation; arrival at the plane cabin; prototype adjusted to the seat tray in front of the passenger, 2017

At the top of the MOOV.22, there are two strap loops with carabiners that attach to half-moons to group luggage, which we do not want to carry in our hands.

Face A is designed to carry a small laptop, in the upper and lower compartments for the necessary gadgets, such as chargers, etc.

The visible improvements focus on formal and functional aspects. Thus, still on face A, we chose to introduce ergonomic backpack straps attached to the top, next to the handle: these straps hold one on each side, with half-moons placed at the bottom of face A.

When MOOV.22 is not being used as a backpack, the straps can be collected in a pocket next to the handle.

Fig. 6 Specialists panel | Prototype evaluation (functional model) CENTI, Vila Nova de Famalicão, 2017

Fig. 7 MOOV.22 applied on the suitcase, face B and face A, 2022

On face B, the compartment at the top is designed to store the mobile phone and contains a document pouch, with this pouch attached and grouped to the MOOV.22 by a snap or carabiner.

Two hand grips were also added, one on each side (A and B), to use it as a suitcase. These changes contribute to making its usability more intuitive and comfortable. The option of using genuine leather can make the product more sophisticated in appearance and increase its longevity.

The main handle has been designed to have triple functionality: it converts the MOOV.22 into a crossbody bag, adjustable to the size of each individual; bundles extra luggage in the suitcase when on the move, freeing up hands and shoulders; and adjusts the MOOV.22 to the seat tray in front of the passenger.

So, with these changes, we simplified the product, making it more minimalist and more flexible in its use, that is, more inclusive, comfortable, and empathetic.

The product has an on-going international patent application—Patent Cooperation Treaty (PCT), denominated SUITCASE COVERING ACCESSORY, WIPO WO2021075989, (2021).

Conclusions

With the prototype (functional model) completed and the MOOV.22 model under study and development, some issues still need to be addressed before this product can be launched on the market.

Here are topics for future tasks: Completion of International Patent Registration; Small Series of parts for usability testing; Feasibility study of the commercial prototype executed by APICAPES; Negotiations with Partners and Investors; Dissemination in Industry and Academia.

As the MOOV.22 Project is a new type of case/cover, in the future it may have numerous versions based on other Design proposals and different Brands.

However, we can already underline the important improvements that have been added to the initial product, making it even more inclusive, ergonomic, sustainable, and comfortable.

Acknowledgments The authors would like to acknowledge CIAUD, Research Centre for Architecture, Urbanism and Design of the Lisbon School of Architecture, Universidade de Lisboa; IPCB, Polytechnic Institute of Castelo Branco; LMA; BELCINTO; APICAPES, Portuguese Footwear, Components, Leather Goods Manufacturers' Association; DGL; SEPOL; RCF, PROTECTING INNOVATION; INP; and to everyone who has contributed from the beginning until now and from now on to this project.

References

Cassim, J. (2012, January 10). *The 3rd 24 hour inclusive design challenge: Oslo on Vimeo [video file]*. Retrieved from: http://vimeo.com/44306386

Dube, K., Nhamo, G., & Chikodzi, D. (2021). COVID-19 pandemic and prospects for recovery of the global aviation industry. *Journal of Air Transport Management, 92*, 1–12. https://doi.org/10.1016/j.jairtraman.2021.102022

Higa, F. (2015). *Uma viagem pela moda: a história das malas mais icônicas até os dias atuais*. http://www.laparola.com.br/uma-viagem-pela-moda-a-historia-das-malas-mais-iconicas-ate-os-dias-atuais. http://criandoeinovando.blogspot.pt/2014_07_01_archive.html.

Neves, P. M. (2018). *Hand luggage organizer design for Airplane travel: Contribution to traveller comfort*. (Doctoral dissertation). Repositório Institucional da Universidade de Lisboa. http://hdl.handle.net/10400.5/15425

Urry, J. (2012). *What is the mobility turn?* http://en.forumviesmobiles.org/video/2012/12/10/what-mobility-turn-467

WIPO WO2021075989. (2021). *Suitcase covering accessory, PI Portal*. https://patentscope.wipo.int/search/en/detail.jsf?docId=WO2021075989&tab=PCTBIBLIO&_cid=P10-KOZS5A-81091-1.

Yukimura, C. (2014). *Momento Eureka: Inovação – Invenção – Lead Users – JTBD. Disponível em:* http://criandoeinovando.blogspot.pt/2014_07_01_archive.html

Textile Waste Bank: Business Model Implementation in Selective Cooperative and Survey for Replicability

Suzana Barreto Martins, Fernanda de Oliveira Massi, Claudio Pereira de Sampaio, and Tiffany Maria Pimenta Silva

Introduction

According to data from the World Bank (Hoornweg & Bhada-Tata, 2012), the annual average generation of textile waste per individual in Brazil is 9 kg per person (average in Latin America), which allows inferring an annual generation of about 1,900,000 tons of this type of waste in the country, about 2.6% of urban solid waste (MSW). This represents a major environmental challenge, since textile waste from clothes, rugs, shoes, furniture, and other sources is not considered recyclable material in the municipal selective collection systems; in the case of clothing, although a good part of these products are donated to new users, there is still a significant fraction of products in poor condition—around 20%—that end up being sent to municipal landfills.

Using the city of Londrina as an example and considering an average of 9 kg per inhabitant/year, it can be estimated that the population of around 560 thousand people who live there generates approximately 5040 tons/year of waste textiles.

In another research, Lourenço (2018) obtained an estimate of about 837 tons/year of textile waste/year that pass through the conveyor belts and separation tables of selective collection and recycling cooperatives in the city of Londrina destined for the municipal landfill as they are considered tailings. This value does not include, therefore, textile waste that is donated or disposed of incorrectly by the population, and it can be inferred that textile waste sent to the municipal landfill represents about 16% of the total generated annually in the city.

Another survey carried out by the university's research group Desin pointed out that only one cooperative received around 18 tons/year of clothes alone (which are only part of the textile waste), even without being able to receive this material, since

S. B. Martins (✉) · F. de Oliveira Massi · C. P. de Sampaio · T. M. P. Silva
State University of Londrina, Londrina, Brazil
e-mail: suzanabarreto@uel.br; claudiopereira@uel.br; tiffany.maria.pimenta@uel.br

A. Cruchinho, J. C. Neves (eds.), *Sustainability in Fashion: Creativity, Education, and Innovation*, https://doi.org/10.1007/978-3-031-80991-0_14

its selective collection contract with the public administration does not include textile waste as recyclable, but as waste.

This fact represents an environmental problem that manifests itself in several ways: inadequate disposal of textile waste in public places, mainly valley bottoms, and in vacant lots in the city; burning of these residues in these places; and sending them to the city landfill, causing the reduction of its expected useful life and the eventual clogging of leachate drainage ducts.

On the other hand, waste also represents an opportunity once its economic value is recovered. This can occur through strategies such as reuse, recycling, and energy recovery through incineration. This project is mainly in line with the first two strategies, based on the proposition of implementing a system for collecting, separating, cleaning, disassembling, processing, and manufacturing new products from textile waste from both post-consumer (of city residents) and industrial waste.

In turn, it represents the continuation of a set of research activities previously developed, in which the research group Desin sought to understand the phenomenon of textile waste generation at a global, national level, and in the city of Londrina (2018–2019 period). This understanding resulted in the proposition and detailing of a system and business model aimed at the revaluation of textile waste (period 2019–2020), which has already been evaluated and is in the evaluation and implementation phase with a selective collection and recycling cooperative in the city of Londrina (period 2021–2022).

This model is being tested through a pilot implementation of a complete prototype of the system on a small scale, including all the steps of the process already developed in the previous edition of the project. In this way, we seek to systematize this prototyping stage to implement a socio-environmental business model, including training cooperative members to operate the system so that it can be replicated in similar contexts in other selective collection and recycling cooperatives in the state of Paraná and other states in the country. Thus, the present work also aims at a qualitative-quantitative survey of selective collection cooperatives in the state in order to identify the replication potential of a socio-environmental enterprise for the recovery of waste.

The expected impacts of this project include the social impact, resulting from the creation of new job and income opportunities; environmental impact, due to the reduction in the amount of textile waste that is sent to the municipal landfill or discarded in inappropriate places; economic impact, resulting from the creation of a new value chain for a material until then considered rejected; and scientific impact, with the production of new knowledge on the creation of socio-environmental businesses that can be replicated in other cooperatives in the region, and even in the country.

Methodology

To achieve the aforementioned objective, which includes the effective implementation of the Textile Waste Bank in recycling cooperatives, the methodology of this project is interventionist, exploratory, and qualitative, using a generative approach in the form of Action Design Research (ADR). According to Petersson and Lundberg, ADR was proposed by Sein et al. (2011) to combine Design research with Action Research, with the objective of generating design knowledge in a prescriptive way from the learning obtained in the construction and evaluation of an artifact in a given organizational environment in order to solve a problem. Therefore, ADR combines the strengths of the two approaches that gave rise to it: Action research, which allows for a deeper immersion in the social reality studied (in this case, a selective collection cooperative), and Design Science Research, which uses the creation of an artifact (in this case, a system) as a means of producing knowledge. In this way, the activities carried out were:

- Complementary survey of information about the BRT, as well as its history and theoretical-methodological bases;
- Discussion of contents, proposition, and analysis of ideas with the interdisciplinary research group and with participants of the cooperative;
- Planning and monitoring of initial implementation stages;
- Survey of cooperatives in the state of Paraná to verify the project's replicability potential.

Development

For 10 years, the research group Desin has been studying textile waste and solutions for these materials, which until then had not been properly disposed of. The initial proposal of the project was to propose innovative and sustainable solutions for products and processes in the clothing sector, in order to promote the reuse of textiles in the city of Londrina. Although textile waste is not included in the National Solid Waste Policy, research has been carried out, and the group has found an opportunity for a business model for textile waste through the application of concepts such as Distributed Economies, Circular Economy, Design for Sustainability, and Systems Design.

The group started by researching the themes to be addressed and then began the development of a system model for the collection, separation, and processing of textile waste. Initially, the model would be operated in partnership with cooperatives but not within them. With the development of research, it was defined that the model would be operated by the cooperative itself, since they are the actors who receive the textile materials along with the other waste. In addition, the cooperative members already have expertise in receiving waste, sorting, processing, and sale, thus facilitating the implementation of the system.

The first business model was prototyped at the beginning of 2020 with funds from the competition of companies Leroy Merlin and Tomie Ohtake within the partner cooperative and with the participation of two cooperative members, a chemist, fellows, and project coordinators. After prototyping the system for a month, it was possible to refine the model, solve problems, and address bottlenecks that were identified during the process, thus completing the development of the so-called Textile Waste Bank, BRT.

After receiving the funds from the SIVEPAR public notice, the BRT was officially launched for society in an event that took place on July 7, 2022, at the Shopping Aurora in the city of Londrina. The event was a conversation circle with stylists Ronaldo Fraga and Dudu Bertholini on the theme "Stories of Fashion, Art, Culture and Behavior," organized by a clothing and apparel retail company in the city and with the sponsorship and support of several companies and institutions in the clothing sector. Before the start of the conversation with the stylists, the launching ceremony of the Textile Waste Bank took place with the participation of the coordinating professors of the research group, some cooperative members, and the project's fellows (Fig. 1). The objective of the launch was to present to the city community the new system of collection and processing of textile waste in the city, since the model counts on the reception of waste after use by the population, and to start the collection at points of the city through the built-in collectors.

Fig. 1 BRT launching ceremony. (Photo: Desin Group)

The launch of the BRT had an excellent impact on society and resulted in several journalists interested in writing about the new textile waste collection system, in companies interested in sending their industrial waste to the cooperative, and in companies interested in implementing the model in their cities. The population was also interested in the model and in the new way of correctly disposing of post-use textile waste.

The first step in the implementation was the placement of collectors throughout the city. The collection points were defined together with a cooperative member responsible for the selective collection logistics in the city, spaces were defined in shopping malls in the city and at the State University of Londrina. The collectors were made of reinforced cardboard, 1.20 m high and 45 cm on the sides, with stickers with information about the BRT arranged in the form of a storyboard (Fig. 2).

The openings of the first collectors were cut in different formats in order to compare the best way. In addition to the cuts, protections were inserted in E.V.A to ensure that there is no removal of material from the collector (Image 3), since that is not their purpose. The collector cover was sealed with a nylon tie tape. The collectors were made in greater quantity than the number of collection points so that the cooperative could remove the full collector to take to the BRT space and leave a void in the place.

After defining the collectors, the next questions prior to implementation were delimited. The first of these was the selection of the space where the BRT would be allocated within the cooperative. The shed that would be used for the implementation, which was used as a basis for the previous layouts, presented problems with regard to the electrical wiring and the energy capacity of the place, since the cooperative is already operating at its energy limit, and it would not be possible to insert one more equipment of high energy consumption. To solve this problem, the

Fig. 2 BRT collectors. (Photo: Desin Group)

cooperative suggested the relocation of the BRT to the headquarters they have in another neighborhood of the city, as it is a place with greater energy capacity and free space of 8 meters by 7 meters. Thus, a new layout was designed for the BRT in this new space (Fig. 3).

Despite solving the energy issue, the new space required the construction of two walls and a door to protect the equipment and items from rain and wind. To visualize the space, the wall design was previously carried out in the place to be built, as shown in Fig. 4.

Another important decision point was the choice of cooperative members to assume the operationalization of the system. Initially, there was a need for three jobs

Fig. 3 BRT layout in the cooperative. (Photo: Desin Group)

Fig. 4 New walls in the BRT. (Photo: Desin Group)

to operate the BRT, defined through conversation between the cooperative members and the availability and interest of each of them. The form of training for these jobs was also defined, as the delay in implementation caused a lack of time for the cooperative members to be trained, and a specific technical scholarship was defined for the training of the cooperative members, together with the BRT coordinator, from July 2022. Such training considered aspects related to the sale of the material, its composition, colors, weights, qualities, among other classifications, which have an outlet market. In this way, the cooperative members are being instructed to correctly sort the material for its later commercialization, as well as the separation of pieces and fabrics for thrift stores and for the partner social enterprise. They are also being trained on how to dismantle the product without damaging the accessories so that they can be marketed.

In addition to the correct sorting of the materials, the cooperative members were instructed on the care with the cleaning of the materials, which is a requirement mentioned by all possible buyers. Thus, materials need to be well stored in places away from moisture and other materials that can attract vectors.

One of the cooperative's concerns was in relation to possible buyers for the materials processed by the BRT. To solve this issue, a partnership was signed with Centro Brasil Design, together with a fellow specifically to prospect customers and buyers for each type of material offered by the BRT, in addition to simulating values for these materials. Furthermore, with the disclosure of the BRT system, companies interested in buying the material emerged.

The tissue shredder was delivered to the cooperative in September 2022, starting the operation of the BRT, which is in the initial phase of implementation, with a processing capacity of 30 tons per month. The functioning of the model in its entirety will be monitored and evaluated in the coming months and in subsequent research.

Survey of Cooperatives to Verify Replicability

The survey of cooperatives, on the other hand, had as its initial focus the geographical area of the interior of Paraná, excluding the metropolitan region of Curitiba and the coast, with a population between 50 thousand inhabitants and 500 thousand inhabitants; this cut was made because, for the viability of the BRT, it is estimated that a minimum monthly processing of 20 tons of waste is necessary, and it is estimated that cities from 50 thousand inhabitants would be able to produce this amount. In Paraná, excluding Londrina as it is the city in which the project was first developed, there are 25 cities that meet these criteria. The table below lists the selected cities and their population data (Fig. 5).

In these 25 cities, 47 collectors' cooperatives were identified. There is a great variety in terms of their size, both in terms of cooperative members and in terms of monthly production volume. These cooperatives were significantly impacted during the Covid-19 pandemic because, due to the crisis, many people started to informally collect recyclable materials, reducing the volume of material sorted in the sheds. Given this reality, one of the biggest difficulties was finding updated information,

Fig. 5 Cities selected for study and number of inhabitants. (Source: Authors (2022))

City	Population
Maringá	436.472
Ponta Grossa	358.838
Cascavel	336.073
Foz do Iguaçu	257.971
Guarapuava	183.755
Toledo	144.601
Apucarana	137.438
Arapongas	126.545
Umuarama	113.416
Cambé	108.126
Sarandi	98.888
Campo Mourão	96.102
Franscisco Beltrão	93.308
Paranavai	89.454
Pato Branco	84.980
Cianorte	84.779
Telêmaco Borba	80.588
Castro	72.125
Rolândia	68.165
Irati	61.439
União da Vitória	58.298
Ibiporã	55.688
Marechal Cândido Rondon	54.031
Prudentópolis	52.776
Palmas	52.503

which makes it difficult to understand their current reality in the post-pandemic context.

The main methodology of this study was desk and documentary research using secondary sources, with the collection and analysis of information and documents available on the websites of city halls of cities in the State of Paraná, as well as reports on the subject, in order to elaborate a description of the reality of the collectors' cooperatives identified in the study. An interview was also carried out with an expert who provided some information about cooperatives in the area. No direct contact was made with these cooperatives.

The results obtained from the survey of information about the cooperatives were divided into three distinct categories for a better understanding of the data obtained, namely:

- Category 1—Number of members in cooperatives in cities between 100,000 and 500,000 inhabitants (medium cities, according to IBGE classification);
- Category 2—Number of members in cooperatives in cities between 50 thousand inhabitants to 100 thousand inhabitants (small cities, according to IBGE classification);
- Category 3—Amount of waste in cooperatives.

The objective of category 1 was to observe the size of cooperatives in relation to the number of cooperative members in **medium-sized** cities in Paraná, and it was possible to verify that there is a big difference in size between them: there are cooperatives with 150 members and others with only 16, generating very different realities and forms of work organization, which represents a first important aspect to consider in order to replicate the BRT in collectors' cooperatives.

Category 2 also seeks to analyze the size of cooperatives in relation to the number of cooperative members, but in **smaller cities**. A less significant difference was observed between them than in the first category, with the largest formed by 50 members and the smallest by 10.

Category 3 seeks to examine the **amount of waste** received, sorted, and sold, and is divided into three subcategories according to the availability of information: volume produced monthly, volume received monthly, and volume sold monthly. There is also a great diversity of volumes, and in the recent data found, there is also, in some cases, a drop in production. This category is also extremely important for the replicability of the BRT, as it is necessary to observe the monthly volume to see if the cooperative would produce enough for the Bank to be viable. The results according to the previous categorization are presented and discussed in detail below.

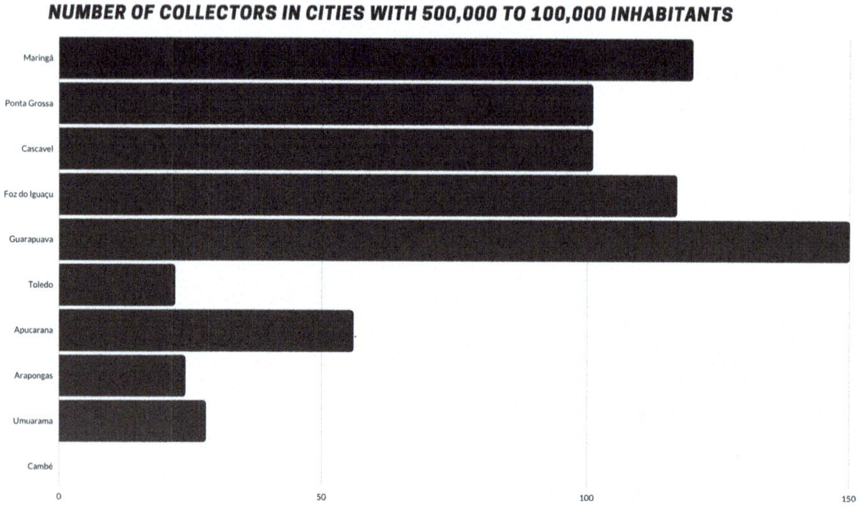

Fig. 6 Number of collectors in cities with 500,000 to 100,000 inhabitants. (Source: Authors (2022))

Category 1

Category 2

Category 3

In addition to the data related to the cooperatives presented above, additional studies were sought in the form of gravimetric analysis that would make it possible to expand the understanding of the subject (Figs. 6, 7, 8, 9, 10 and 11).

Conclusion

Finally, it was possible to follow the implementation activities of a Textile Waste Bank in a selective waste collection cooperative in the city, as well as understand the difficulties of the process and solutions for each problematic aspect. The main difficulty identified, which caused a delay in the implementation of the system, was the financial survey for the purchase of equipment and the bureaucratic process of

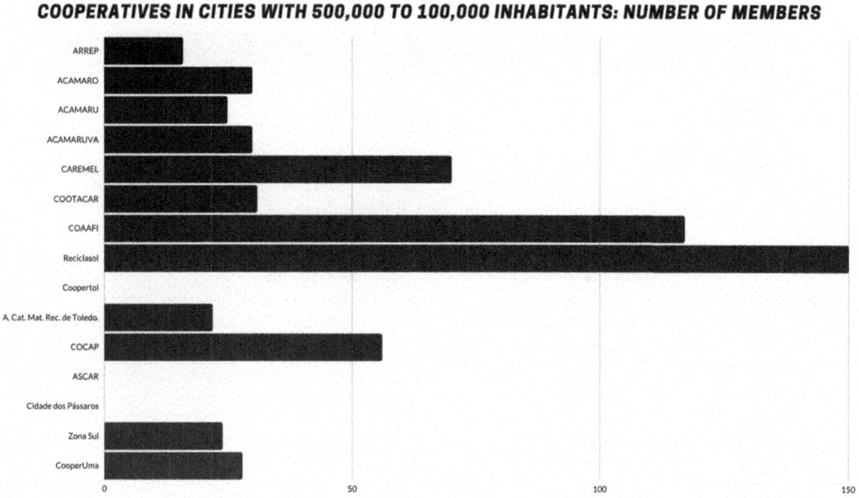

Fig. 7 Graphic of the number of members in cooperatives in cities with 500,000 to 100,000 inhabitants. (Source: Authors (2022))

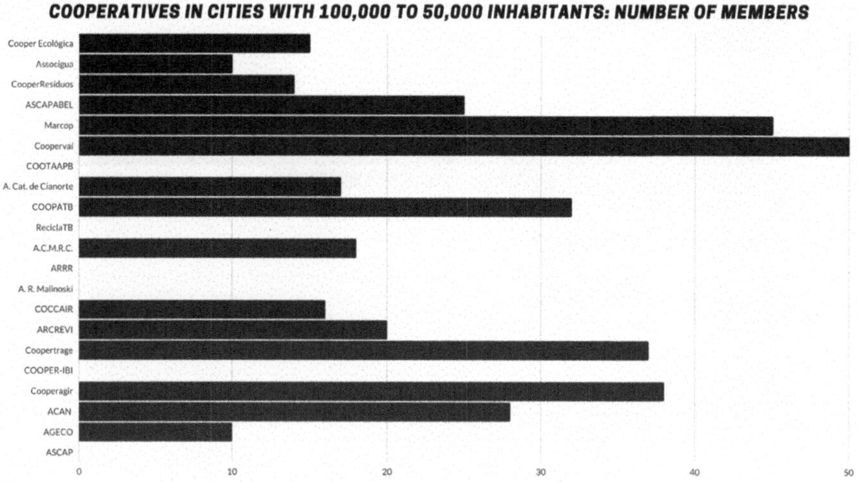

Fig. 8 Number of collectors in cities from 100,000 to 50,000. (Source: Authors (2022))

purchasing the shredding machine through the public notice of the Fundação Araucária. After approval of the project by the public notice of private institutions SIVEPAR and contemplation with the financial resource, the project advanced quickly when compared to waiting for capital from public bodies.

After the purchase of the fabric shredder, some issues were identified regarding the implementation of the system, especially regarding the cooperative's insecurity in taking over the business model; however, after meeting with the project

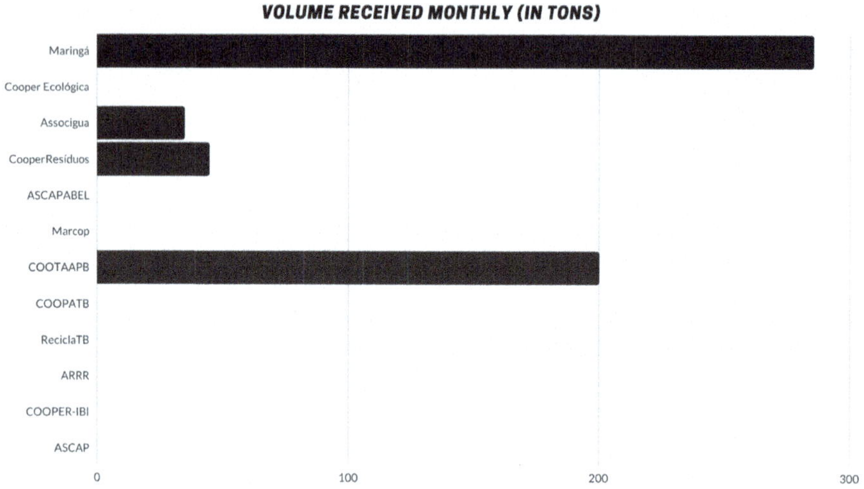

Fig. 9 Volume of waste received monthly. (Source: Authors (2022))

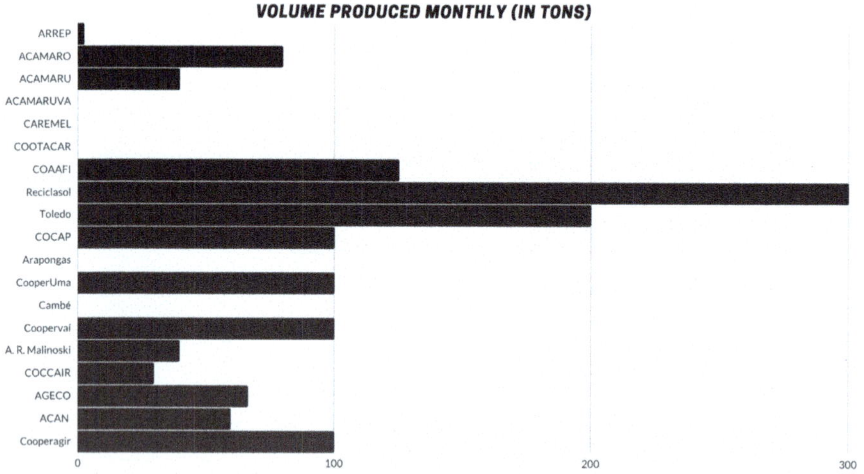

Fig. 10 Volume of waste produced monthly. (Source: Authors (2022))

coordinators, the cooperative members exposed their fears and talked to understand what the solutions would be and thus decided to undertake the BRT business model and continue with the partnership.

The effective implementation of the model showed that a lot of research and maturation of projects are necessary to achieve success in Systems Design, especially when they are related to wicked problems, as the problems that arose during the implementation were resolved due to the group's experience in research on textile waste and Systems Design. It was also possible to identify the importance of

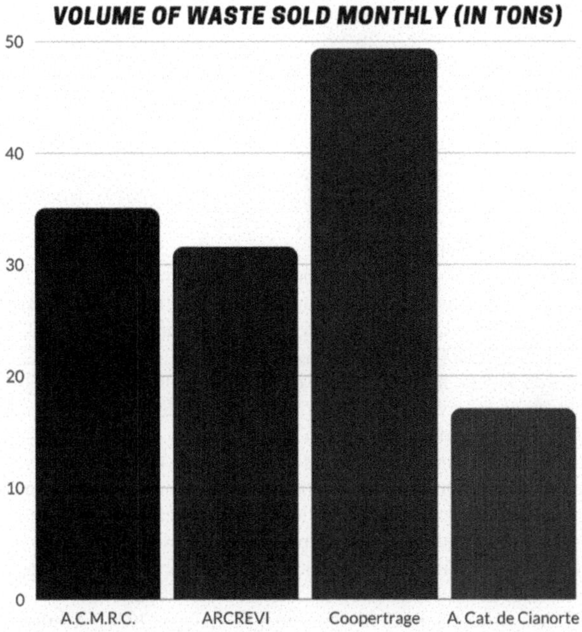

Fig. 11 Volume of waste sold monthly. (Source: Authors (2022))

multidisciplinary teamwork, since the discussions became richer with different perspectives on the same subject.

Although the implementation of the model took place assertively, it is worth noting that a period of evaluation of the functioning of the system will be necessary to ensure that the BRT operates as expected and guarantees safety to the cooperative members, who were initially apprehensive about taking on the management of the BRT system as a business model. In addition to helping with possible problems that may arise, the follow-up will contribute to market research for prospecting buyers and selling prices, as well as carrying out a quantitative survey of waste collected by the cooperative at the beginning of the dissemination of the system to the community and the beginning of collection at public points.

Furthermore, it was possible to verify the effectiveness of the Action Design Research method in projects of this scope, since it involves Systems Design, Design for Sustainability, and Circular Economy, together with the support of the Bibliographic Review.

Bibliography

ACAN e Cooperagir recolhem 390 toneladas de lixo por mês em Marechal Candido Rondon. (2019). Available at: https://www.opresente.com.br/marechal-candido-rondon/acan-e-cooperagir-recolhem-390-toneladas-de-lixo-por-mes-em-marechal-rondon/. Accessed 25 Feb 2022.

Antes refugado, isopor passa a ser reciclado pela Cooperagir. (2019). Available at: http://opresente.com.br/marechal-candido-rondon/antes-refugado-isopor-passa-a-ser-reciclado-pela-cooperagir/. Accessed 25 Feb 2022.

ARCREVI. Available at: http://uniaodavitoria.pr.gov.br/tag/arcrevi/. Accessed 23 Feb 2022.

BRAZIL. Law n° 12.305 of August 02, 2010, institutes the National Solid Waste Policy; alter the Law n° 9.605, of February 12, 1998; and takes other measures. Available at: http://www.planalto.gov.br/ccivil_03/_ato2007-2010/2010/lei/l12305.htm

COCAP registra queda de 50% na coleta seletiva com ação de atravessadores. (2021). Available at: http://www.apucarana.pr.gov.br/site/cocap-registra-queda-de-50-na-coleta-seletiva-com-acao-de-atravessadores/. Accessed 14 Feb 2022.

Cooperativa de catadores gera renda e estimula sustentabilidade com reciclagem. (2021). Available at: https://www.umuarama.pr.gov.br/noticias/agricultura-e-meio-ambiente/cooperativa-de-catadores-gera-renda-e-estimula-sustentabilidade-com-reciclagem. Accessed 16 Feb 2022.

Cooperativa Marcop pede reajuste no repasse para a coleta do lixo reciclável. (2018). Available at: https://www.jornaldebeltrao.com.br/noticia/281928/cooperativa-marcop-pede-reajuste-no-repasse-para-a-coleta-do-lixo-reciclavel. Accessed 22 Feb 2022.

Cooperativa de Reciclagem de Sarandi abrange área de serviço (2021). Available at: http://www.sarandi.pr.gov.br/web/index.php/noticias/item/cooperativa-de-reciclagem-de-sarandi-abrange-area-de-servico. Acesso em 21 fev. 2021.

Dia mundial do meio ambiente: TOORECICLANDO faz uma Toledo mais ecológica. (2020). Avaliable at: http://toledo.pr.gov.br/noticia/dia-mundial-do-meio-ambiente-tooreciclando-faz-uma-toledo-mais-ecologica. Accessed 14 Feb 2022.

Em 2019, ITAIPU repassa à cooperativa de Foz do iguaçu 87 toneladas de reciclaveis. (2020). Available at: https://www.itaipu.gov.br/sala-de-imprensa/noticia/em-2019-itaipu-repassa-cooperativa-de-foz-do-iguacu-87-toneladas-de-recicla. Accessed 9 Feb 2022.

ENVEX Engenharia e Consultoria. (2020). *Plano municipal de gestão integrada de resíduos sólidos de Guarapuava/PR*.

FOZ DO IGUAÇU, Prefeitura Municipal. Plano municipal de saneamento básico município de foz do Iguaçu, Foz do Iguaçu.

Hoornweg, D., & Bhada-Tata, P. (2012). *What a waste: A global review of solid waste management*. World Bank.

Lourenço, V. A. (2018). *Análise gravimétrica dos rejeitos da reciclagem da cidade de Londrina. Trabalho de conclusão de curso. Graduação em Administração*. Universidade Estadual de Londrina.

Prefeitura de Maringá (PR): e cooperativas de recicláveis firmam contratos de parceria. (2018). Avaliable at: https://easycoop.com.br/Noticias/39938/Informativos/. Accessed 8 Feb 2022.

Sein, M., Henfridsson, O., Purao, S., Rossi, M., & Lindgren, R. (2011). Action design research. *MIS Quarterly, 35*, 37–56.

Sarandi, Prefeitura Municipal de. (2014). *Plano municipal de gestão integrada de resíduos sólidos*.

Silva, R. (2015). *Coopervaí processa 100 toneladas de materiais recicláveis por mês*. Available at: http://semamparanavai.blogspot.com/2015/03/coopervai-processa-100-toneladas-de.html. Accessed: 22 Feb 2022.

Vezzoli, C. (2010). *Design de sistemas para sustentabilidade: teoria, métodos e ferramentas para o design sustentável de "sistemas de satisfação"*. EDUFBA.

Sustainability in Fashion and the Upcycling Movement

Paula Rosa Lopes

Introduction

The issue of sustainability in the fashion industry is not easy to address. The fashion sector implies a production cycle that encompasses the production of raw material, design, cutting, sewing, and finally its marketing. Only the first part of the process, the production of raw material, implies the consumption of a set of resources that are high. However, after millions of years of evolution, it is not acceptable for humans not to wear clothes. In this context, the European community and some authors point to solutions based on a different management of resources. In this scenario, movements such as upcycling emerge, a trend that is becoming a fashion and implies the recycling and reuse of garments, with the aim of minimizing the consumption of raw materials and the impact that their production has on the environment.

The apparel market, which encompasses every kind of clothing, from sportswear to businesswear, and from value clothing to statement luxury pieces, registered an increase in sales after a global decrease caused by the coronavirus pandemic. The revenue of the global apparel market was calculated to amount to some 1.5 trillion U.S. dollars in 2021 and was predicted to increase to approximately 2 trillion dollars by 2026 (Tighe, 2022). The United States and China generating substantially higher revenues than any other country (Tighe, 2022). In the same study, two segments should be highlighted that show notable growth. On the one hand, the secondhand clothing industry, which is expected to grow as consumer demand for more environmentally conscious clothing options increases. On the other hand, the luxury apparel market is projected to double in size from 2021 to 2026 (Xie et al., 2021).

P. R. Lopes (✉)
Lusofona University / CICANT, Lisbon, Portugal
e-mail: paula.rosa.lopes@ulusofona.pt

A. Cruchinho, J. C. Neves (eds.), *Sustainability in Fashion: Creativity, Education, and Innovation*, https://doi.org/10.1007/978-3-031-80991-0_15

At the same time, the number of pieces of clothing purchased by each person has increased in recent decades. In the European Union alone, it increased by 40% due to price reductions, which also reduced the lifespan of clothing. Once again, the figures are illustrative, and in Europe, the consumption of textile products per person is around 26 kg, of which 11 kg are thrown away on average per year.

Since the concepts about sustainable fashion can still be further clarified, the objective of this study is to understand the concept of sustainable fashion for young consumers.

Sustainable Fashion and Upcycling Fashion

Nowadays, many consumers are already aware of the sustainable fashion movement, and the upcycling movement is a reality (Park & Lin, 2020).

The concept of sustainable fashion is not yet consensual among researchers on this topic (Greco & De Cock, 2021; Orminski et al., 2021). Sustainable fashion is not a totally new theme, but at this moment, it is becoming pertinent and urgent to delve deeper into this subject due to the consumer awareness of the benefits of sustainability applied to fashion throughout its process, from the production of textiles to design and clothing production, integrating marketing and brand communication (Henninger et al., 2016).

At the same time, this movement looks to combat the large carbon footprint that the fast fashion industry has created by reducing the environmental impact such as air pollution, water pollution, and climate change.

From the point of view of the fashion industry and fashion brands, the risks and gains related to the sustainable fashion industry have been studied (Adamkiewicz, et al., 2022). This study by Adamkiewicz et al. (2022) also outlines future research opportunities in a more sustainable, holistic approach to a product's life cycle and how this can be translated into clear, transparent, or reliable certification schemes to prevent the misleading and dishonest marketing strategies, helping consumers to make a responsible choice.

Concern for the environment has led to the emergence of solutions in the fashion industry, such as recycling and upcycling. Dissanayake & Sinha (2015) highlight the benefits of the fashion recycling movement. The upcycling movement stands for reusing secondhand clothes that are then sewn again, resulting in a new piece of clothing made from the same textiles. Upcycling provides a new business opportunity for sustainable fashion designers. Some of secondhand stores have created their own line-up of upcycled clothing (Anitha, 2022; Fletcher & Grose, 2012). These movements promote the circular economy in the sector (Colasante & D'Adamo, 2021; D'Adamo & Lupi, 2021).

Sustainable Fashion Communication

Fashion brands that implement sustainability practices communicate their strategies to their stakeholders. Brydges et al. (2022) evaluate how brands communicate sustainability to consumers across three key sites: brand websites (including corporate social responsibility reports) (Jiménez-Marín et al., 2021), social media platforms (Calcagni et al., 2019), and in-store campaigns. These authors refer that not only do brands use a range of practices to define sustainability differently, but furthermore, these definitions vary depending on the context.

Castillo-Abdul et al. (2022a) study how the social networks have become crucial communication channels for brands and shows that most of the social responsibility content of the fashion brands analyzed in social networks is linked to the use of sustainable materials, the protection of natural spaces and, in the particular case of Prada, the protection of the oceans (Castillo-Abdul et al., 2022b).

Research around social networks has become increasingly important in recent years because they are a powerful means of communication between different generations, especially the younger ones. Social networks are occupying a privileged place in the marketing and communication of brands, which has been reflected in the increase in sales in various sectors of activity. Lopes et al. (2022) research aims to analyze how the relationship between brand image and clothing purchase intention is mediated by the consumers' perception of the benefits that this purchase provides them.

In Lopes et al. (2022) study, it was found that the 947 participants referred to clothing through social networks. The sample was constituted by 65.5% female and 34.5% was male. The group that expresses the greatest intention to buy clothing through social networks reveals a monthly net household income of less than €1500 and is in the 21–25 age group.

Methodology

In this exploratory study, it was applied a questionnaire in order to understand how young people see sustainability in fashion. The sample of the respondents registered 148 answers, 70.7% feminine gender and 29.3% were from masculine gender. The range ages were between 18 and 22 years old (18 years—13.7%; 19 years—22.6%; 20 years 19.9%, 21 years 15.1%, the remaining percentages were recorded in the range of 22 years to 49).

The questionnaire applied tried to understand among the group of young people what the concept of sustainability in fashion would be for them, and the results show that, despite it being proven that they have notions of sustainability in relation to the fashion brands they buy, there are still few consensual opinions regarding to the concept of sustainable fashion.

In this sense, the exhibition states that the "sustainable fashion" concept includes:

- Brands that reuse models and textiles to adapt to new collections (84.5%).
- Brands that use recycled materials in a system to make clothes or accessories (89.7%).
- Buy secondhand clothes or accessories (such as glasses or shoes) on the social media Marketplace (60.8%).
- Buying secondhand clothes or accessories (such as glasses or shoes) from friends and acquaintances (67.5%).
- Sew clothes at home, using items of clothing that we no longer use (e.g., customizing men's shirts for women's models) (81.7%).
- I prefer to buy clothes that last longer, even if they are more expensive (53.4%).

Measures and Procedures

The answers to the items that comprise them were given using a Likert-type scale that ranges from 1 (Strongly disagree) to 5 (Strongly agree), depending on the degree of agreement of the participants with each of the statements. In order to characterize the sample, a set of sociodemographic questions (e.g., gender, age) was also included.

Items that assess brand image, perceived benefits, purchase intention, and questions regarding sociodemographic characterization were introduced on the Google Forms platform. The link was shared on the researchers' social networks (e.g., Facebook, LinkedIn) and sent by email to their contact networks. All participants were aware of the research objectives and were informed that they could withdraw at any time if they wished to do so. It should be noted that all the guidelines of the General Regulation on Data Protection regarding the anonymity and confidentiality of responses were ensured.

Results

The literature review points to the need to clarify the concept of sustainable fashion, given that the main players in the industry do not always agree and end up following different approaches (Orminski et al., 2021; Adamkiewicz, et al., 2022).

The upcycling movement appears associated with sustainable fashion, giving a second life to clothes that are then sewn again, resulting in a new piece of clothing with the same textiles. Upcycling provides a new business opportunity for sustainable fashion designers (Dissanayake & Sinha, 2015; Anitha, 2022; Fletcher & Grose, 2012).

The literature review addresses online communication, on brands' websites and social networks, as a preferred channel for communicating the sustainable strategies of fashion brands, even if each brand communicates its "concept" of sustainable fashion (Brydges et al., 2022; Castillo-Abdul et al., 2022a, b). Lopes et al. (2022)

also show that social networks are a preferred channel for purchasing clothes for a sample mainly made up of young women.

The results of the questionnaire suggest that young people are aware of the importance of sustainability in fashion but need more information from brands to be able to better clarify the concept and make decisions towards it.

Conclusion

After analyzing the literature review and the quantitative study carried out through the application of a questionnaire to a sample of 148 young consumers, it was possible to conclude that sustainable fashion is a concept they understand but need to be clarified by brands (Adamkiewicz, et al., 2022).

- In a study in which a large majority of participants were women between 18 and 22 years old, the group agreed with a sustainable fashion concept that includes brands that reuse models and textiles to adapt to new collections and brands that use recycled materials in a system to make clothes or accessories, but they do not fully accept buying secondhand clothes or accessories (such as glasses or shoes) on the social media Marketplace (60.8%).

The conclusions of this study could be useful for managers of clothing brands, as well as marketing and communication professionals, when making future decisions.

References

Adamkiewicz, J., Kochańska, E., Adamkiewicz, I., & Łukasik, R. M. (2022). Greenwashing and sustainable fashion industry. *Current Opinion in Green and Sustainable Chemistry, 38*, 100710.

Anitha, B. (2022). Sustainable fashion through recycling and upcycling. *Journal of Textile Science & Engineering, 12*(9), 1.

Brydges, T., Henninger, C. E., & Hanlon, M. (2022). Selling sustainability: Investigating how Swedish fashion brands communicate sustainability to consumers. *Sustainability: Science, Practice and Policy, 18*(1), 357–370.

Calcagni, F., Amorim Maia, A. T., Connolly, J. J. T., & Langemeyer, J. (2019). Digital co-construction of relational values: Understanding the role of social media for sustainability. *Sustainability Science, 14*(5), 1309–1321.

Castillo-Abdul, B., Ortega Fernandez, E., & Romero-Rodriguez, L. M. (2022a). *Corporate social responsibility communication of male luxury fashion brands: Analysis on Instagram, Facebook and TikTok*. Management Decision.

Castillo-Abdul, B., Pérez-Escoda, A., & Núñez-Barriopedro, E. (2022b). Promoting social media engagement via branded content communication: A fashion brands study on Instagram. *Media and Communication, 10*(1), 185–197.

Colasante, A., & D'Adamo, I. (2021). The circular economy and bioeconomy in the fashion sector: Emergence of a "sustainability bias". *Journal of Cleaner Production, 329*, 129774.

D'Adamo, I., & Lupi, G. (2021). Sustainability and resilience after COVID-19: A circular premium in the fashion industry. *Sustainability, 13*(4), 1861.

Dissanayake, G., & Sinha, P. (2015). An examination of the product development process for fashion remanufacturing. *Resources, Conservation and Recycling, 104*, 94–102.

Fletcher, K., & Grose, L. (2012). Fashion & sustainability: Design for change. Hachette UK.

Greco, S., & De Cock, B. (2021). Argumentative misalignments in the controversy surrounding fashion sustainability. *Journal of Pragmatics, 174*, 55–67.

Henninger, C. E., Alevizou, P. J., & Oates, C. J. (2016). What is sustainable fashion? *Journal of Fashion Marketing and Management: An International Journal, 20*(4), 400–416.

Jiménez-Marín, G., Galiano-Coronil, A., & Tobar-Pesántez, L. B. (2021). Organizational communication and social marketing strategies targeting Spanish consumers of fashion. Sustainability as a form of happiness management. *Corporate Governance: The International Journal of Business in Society*.

Lopes, P., Rodrigues, R., & Varela, M. (2022, April). Social networks clothes shopping and the influence of brand image and perceived benefits on purchase intention. *European Conference on Social Media, 9*(1), 95–102.

Orminski, J., Tandoc, E. C., Jr., & Detenber, B. H. (2021). # sustainablefashion–a conceptual framework for sustainable fashion discourse on twitter. *Environmental Communication, 15*(1), 115–132.

Park, H. J., & Lin, L. M. (2020). Exploring attitude–behavior gap in sustainable consumption: Comparison of recycled and upcycled fashion products. *Journal of Business Research, 117*, 623–628.

Tighe, D. (2022). *Global apparel market – statistics & facts*. The Statistics Portal.

Xie, X., Hong, Y., Zeng, X., Dai, X., & Wagner, M. (2021). A systematic literature review for the recycling and reuse of wasted clothing. *Sustainability, 13*(24), 13732.

The Role of Ecolabels in Sustainable Fashion Consumption

Mariana Pereira Silva

Introduction

The offer for sustainable fashion has tremendously increased (Fletcher, 2008). Nonetheless, consumption has not been keeping up with this growth (Andrews et al., 2013). This phenomenon, known as the attitude-behavior gap, has already been demonstrated in literature (see, e.g., Beard, 2008; D'Souza, 2015; Freise et al., 2015), together with its underlying motives; however, the industry has yet to find viable solutions to counteract its consequences. While reading about the attitude-behavior gap in the fashion industry, it was clear that most issues were connected to the same problem: consumers lack information on sustainable fashion (McNeill & Moore, 2015) or when that information is provided to them, they lack the tools to understand it (Lindersson, 2017). McNeill and Moore (2015) demonstrated that consumers educated on sustainability increase their sustainable fashion consumption. Nowadays, most of fashion's sustainability information and education is transmitted through ecolabels (Ferrari et al., 2013). Ecolabels are a communication tool which highlights a product's ethical and sustainable features (McShane et al., 2015), thereby guiding the consumers' purchasing decisions, since specialized entities interpret the information for them and make it more accessible and easily understood (Ferrari et al., 2013).

M. P. Silva (✉)
IADE, Lisbon, Portugal

A. Cruchinho, J. C. Neves (eds.), *Sustainability in Fashion: Creativity, Education, and Innovation*, https://doi.org/10.1007/978-3-031-80991-0_16

Development

This work follows a mixed methods research design, based on two stages: the exploratory and the confirmatory stage. The exploratory stage consists of qualitative methods, such as descriptive literature review, four exploratory interviews with fashion and sustainability professionals, and a focus group with consumers. The confirmatory stage follows a quantitative approach based on a consumer survey with 134 participants. The final data was analyzed considering the research question: How can ecolabels help increase sustainable fashion consumption?

Can Communication Help Overcome the Attitude-Behavior Gap?

The negative impacts of the fast fashion industry are becoming increasingly more visible, enabling a feeling of rising awareness among consumers (Bajtelsmit et al., 2019). Because of that, consumption habits are nowadays based not only on price and quality but also on society-centered values (Langguth & Schnee, 2018), like sustainability. Nonetheless, while increasing awareness allows for a growing number of sustainable products to be available for consumers, these rarely appear to leave the stores' shelves (Andrews et al., 2013). In reality, what society says is important, especially in matters of ethical and sustainable consumption, diverges from what is actually practiced (d'Astous & Legendre, 2009). This phenomenon has been referred to as the attitude-behavior gap (Beard, 2008; D'Souza, 2015; Freise et al., 2015), which means that consumer intentions do not always translate into their actual behavior (Lin & Park, 2020).

There are still few solutions regarding how to overcome the consequences of the attitude-behavior gap; however, most of them seem to be based on communication practices. In the case of sustainable fashion, the role of communication can be even increased, because consumers cannot simply see sustainability—they cannot look at an organic cotton t-shirt and immediately identify it as organic, products need to be perceived as sustainable (McNeill & Moore, 2015). Perceptions can greatly influence purchasing habits, and they mainly rely on two factors: information and education. First, consumers rely greatly on brands to gather the information they need to perceive a product as sustainable (Phau & Ong, 2007). Only through "clear, transparent and easy-in-use communication" (Borboni, 2019, p. 195) can consumers be aware of all the characteristics that define a sustainable fashion product. Second, it has been demonstrated that the more knowledgeable consumers are on sustainability, the more likely they are to engage in sustainable fashion consumption (McNeill & Moore, 2015; Lindersson, 2017). Providing more information and education to consumers will increase their awareness regarding the impact of their shopping habits (Auger et al., 2003). The hardship, however, remains in finding how to communicate this (Kenning et al., 2014), and this is where ecolabels have a role to play.

The Different Types of Ecolabels

McShane, Noseworthy, and Pancer (2015, p. 162) define ecolabels as a "wide range of certification labels that are used to signal various ethical qualities about products (e.g., ethical sourcing, fairtrade, energy efficiency, labor practices, animal rights, environmental orientation, etc.)." Companies adopt this tool to identify and promote sustainable products (Lenk & Rotkirch, 2021) by communicating its eco-friendly features (Ferrari et al., 2013). The ISO 14020:2000 standard—the most common standard used by researchers (see, for example, McShane et al., 2015; Lindersson, 2017; Gonçalves & Silva, 2021)—divides ecolabels into three groups (International Organization for Standardization, 2000). The first type corresponds to environmental claims published by a third-party organization, for instance, when an independent entity awards a specific certification based on the fulfilment of a set of environmental criteria (International Organization for Standardization, 2000), like the EU Ecolabel. The second type of ecolabeling relates to self-declared environmental claims (International Organization for Standardization, 2000). It occurs when manufacturers or retailers make statements about the environmental performance of their own products (Blengini & Shields, 2010). This ecolabeling type is really common in the fashion industry, for example, in the case of H&M and their "Conscious" label. Lastly, the third type consists of a "quantified declaration of the environmental performance of products throughout their life cycle" (Blengini & Shields, 2010, p. 485), also known as Environmental Product Declaration (EPD) (Gonçalves & Silva, 2021). Nonetheless, it is not widely known by consumers and will not be discussed in this investigation.

While the second type of ecolabeling, also known as self-ecolabeling, is commonly associated with green skepticism (Peattie & Crane, 2005; Ferrari et al., 2013) and inconsistency (Lindersson, 2017; Halme & Turunen, 2021), the first type is usually considered more reliable and easier to understand by consumers (Gershoff & Frels, 2015; Lindersson, 2017). Since they belong to an independent entity, Type I ecolabels use the same indicators for evaluating products from different brands (Blengini & Shields, 2010), and these indicators will most likely be more objective and trustworthy than the ones established by brands to evaluate their own products (Curran, 2001). When referring to independent entities, these can both be private businesses, like GOTS or Fair Trade, or legal authorities, such as the European Union, which have the power to establish mandatory guidelines for communicating environmental attributes. For Ferrari et al. (2013, p. 263), these norms "could prevent biases or misleading messages being conveyed to consumers and could enable a more consistent and synergetic action by other influent stakeholders (…) to support or boycott the purchasing of certain products on the market."

Some methods are already being taken into action, like the European Product Environmental Footprint (PEF), which is creating an A to E certification scheme (Gonçalves & Silva, 2021), similar to the method adopted in the EU Energy Label. The PEF guide was launched in 2013 aiming to provide a cohesive directive for measuring and informing consumers and companies on the impact of a product during its lifecycle (European Commission, 2013; Bach et al., 2015; Gonçalves & Silva, 2021). However, the European Commission has yet to explain how this measuring tool will coexist with other labeling schemes defined by the ISO 14020:2000 standard, namely Type I ecolabels (Bach et al., 2015) where the European Union already has its own tool, the EU Ecolabel. The latest information regarding the fashion industry details that the European Commission has been studying ways "to integrate PEF method in EU Ecolabel criteria through the development of PEF studies to identify hotspots when needed, and through the 'pilot project' on the revision of EU Ecolabel criteria for textiles and footwear apparel" (European Commission, 2020, p. 7).

Results of the Exploratory Stage

After an extensive literature review, the exploratory stage continued with a round of interviews, where the author talked with fashion and sustainability experts to further understand the impact of ecolabels on sustainable fashion consumption. The interviewees were Silvia Ferratini, coordinator of the EU Ecolabel team at the European Commission, Mafalda Afonso, co-founder of the sustainable Portuguese fashion brand Sunkissed, Sara Andrade, journalist at Vogue Portugal, and Anabela Gonçalves, researcher at Faculdade de Ciências da Universidade de Lisboa. First, all interviewees defined ecolabels in accordance with the Type I category of the ISO 14024 standard, thereby neglecting self-ecolabeling strategies. It was also implied that even though certain social aspects may be guaranteed, these labels are mostly used to communicate the environmental performance of a product. All experts agreed that ecolabels provide truthful and reliable information which empowers consumers. For that reason, these certifications can act in accordance with current legislation to further confirm and validate a product's level of sustainability. But ecolabels also come with problems. In many cases, brands are presented with a complex set of criteria that needs to be fulfilled to obtain the organization's seal of approval. For Silvia Ferratini, this complexity is precisely one of the factors that the EU Ecolabel needs to work on. The other problem raised was the lack of visibility around ecolabels. When Mafalda Afonso presents Type I ecolabels to consumers, she claims to see no reaction because most consumers do not understand what these organizations actually do. All interviewees agreed that there is a lot that needs to be done by legislators since ecolabels are still a voluntary tool presented by brands to inform consumers about their sustainable practices.

For the second part of the exploratory stage, the author conducted a focus group composed of ten participants, characterized in Annex 1, to discuss the shopping habits of average consumers and how ecolabels could impact their fashion consumption. When asked to define ecolabels, participants showed not to be familiar with the concept. Although Zara and GOTS' ecolabels looked familiar to 30% of the participants, no one could tell where that familiarity came from. Regarding regular clothing labels, the price was the number one information they looked for, followed by materials and the place of production, where "Made in Portugal" was the most sought-after. Overall participants were impacted by the presence of independent certifications, even though they did not know the entities that were behind them. They were, however, familiar with other certification systems from other industries. Both the EU Energy Label and the traffic light food label had a significant impact on consumers' purchases. All participants, with no exception, knew the systems very well and were aware not only of their purpose but also of how they function, since it was, in their words, "easy to understand." Two focus group members even said: "It would be great if we had that on clothing."

Results of the Confirmatory Stage

After gathering information on the topics of sustainable fashion and consumer behavior, the author conducted a survey targeted at general consumers as the confirmatory stage of this study. In total, there were 134 responses to the online questionnaire, from Portuguese consumers demographically characterized in Annex 2. Furthermore, four hypotheses and two sub-hypotheses were established to analyze the quantitative data obtained from the questionnaire. The following paragraphs will analyze the results of the questionnaire in relation to each hypothesis, with a summary of the results in Annex 3.

H1: Consumers consider Type I ecolabels more reliable than Type II.

The first hypothesis touched on the different types of ecolabels defined by the ISO 14024 standard. On the questionnaire, Type I ecolabels were portrayed by the GOTS label and the EU Ecolabel, whereas Type II ecolabels took the form of a Zara "Join Life" label. The Zara ecolabel was the one most recognized by the participants—67.9% said they knew what it was, while only 32.1% and 47% declared to recognize the EU Ecolabel and the GOTS label, respectively. However, when consumers had to choose which ecolabel they preferred, 84.3% of the participants chose a GOTS certification label, and only 5.2% selected the Zara "Join Life" label. Therefore, consumers consider Type I labels more reliable than their Type II counterparts, which confirms H1.

H2: The environmental dimension (materials) has more impact than the social one (place of production) for fashion consumers.

Place of production and garment composition are two components regularly detailed in current clothing labels. During the exploratory stage of this research, materials seemed to be of the highest importance to the general audience. And during the survey, when asked to choose between a garment made in Portugal composed of 100% polyester and a garment made in Bangladesh of 100% organic cotton, 42.5% selected the first option and 48.5% selected the second one. Since Bangladesh has a negative connotation as a place of production for fashion, this leads us to assume that materials were guiding consumers which chose the second option. A second comparison, however, refuted that idea because 64.9% of the participants selected a 100% recycled polyester garment made in Portugal, while 28.4% voted for one of 100% cotton made in Bangladesh. This means that we cannot assume either that the place of production or the garment's composition is always the most relevant, since the value of each component changes according to the situation which is placed in front of the consumer.

H3: Label systems like the energy efficiency certification and the nutritional traffic light are effective tools to increase sustainable fashion consumption.

Besides the color green, label systems were another visual cue analyzed in the consumer survey. Both the energy efficiency certification and the nutritional traffic light label were recognized by a big majority of respondents (98.5% and 96.3%, respectively) and scored really high regarding their efficacy. This was mostly proven by the results of the comparisons presented in the following sections of the questionnaire. When choosing between a garment made in Portugal, composed of 100% cotton, priced at €40, and a garment made in Bangladesh, composed of 100% polyester, priced at €10, 50.7% of respondents selected the first option and 40.3% the second one. Later in the survey, the same comparison but with the presence of an energy efficiency certification system (which evaluated the first garment with an A and the second one with an F), 79.1% of respondents selected the first option, which means that 29 people changed their choice because of the added label system. A similar thing happened with the second label system, the nutritional traffic light label. When this label was added to the regular clothing label, 76.6% chose the first option, meaning that 27 participants changed their choice compared to when there was no labeling system at all. This means that both tools positively impacted consumers' choices, by making more people choose the most sustainable option.

H4: Ecolabels help increase sustainable fashion consumption in all consumers.

The overall results of the consumer survey showed that the more information is provided regarding the sustainability of a garment, the more consumers will prefer that garment over its not-so-sustainable counterpart. The results of previous hypotheses confirm that ecolabels help increase sustainable fashion consumption in all consumers.

H4.1: Ecolabels are more effective in already eco-conscious consumers.

For analyzing the level of education of consumers, the author looked for subtle clues—like the means through which consumers obtained their knowledge on this topic—that would help define the participant's knowledge of sustainable fashion. First, the author analyzed the group of participants who had changed their decision on which garment they would buy after introducing label systems. Out of the 29 people who shifted their buying behavior with the introduction of the energy efficiency certification system, only 20.7% met the criteria to be considered as having a high level of education on sustainable fashion. After that, the author analyzed the answers of the people who claimed to always get their sustainable fashion knowledge from books, magazines, or academic resources. Among those nine answers, seven kept choosing the most sustainable garment throughout all comparisons, meaning that the presence of different visual cues or even prices did not impact their choice. Overall, we may say that ecolabels are more effective in consumers who know less about sustainable fashion since consumers that are already characterized as eco-conscious do not need an added incentive to make the most sustainable choice.

H4.2: Ecolabels are less effective in price-sensitive consumers.

According to the survey's results, price ranked second in terms of importance immediately after size (606 points and 608 points, respectively, out of 670). We already saw that there were 29 answers where consumers shifted their choice from the lowest-priced garment (or from seeing the choice as indifferent) to the highest-priced and most sustainable garment thanks to the presence of ecolabeling systems. However, there were still 26 people who kept choosing the lowest-priced garment, regardless of how the clothing label was portrayed. Price sensitivity can also be seen in the cases where the most sustainable garment's price was changed from €40 to €30. When there were no systems or visual cues, the price drop saw an increase of 11.2 percentage points in people choosing the most sustainable garment. In the presence of certification systems, the increase was only 6.1 or 8.5 percentage points, depending on the type of system. This confirms the hypothesis that ecolabels are still effective in price-sensitive consumers but are less effective in this group compared to the general audience.

Conclusion

"Can ecolabels help increase sustainable fashion consumption?" At this stage, we can say that there is enough evidence to respond affirmatively. However, there are some factors that need to be considered to maximize the ecolabels' efficacy. First, there needs to be a unified definition for the concept of ecolabels. The results of both the exploratory and the confirmatory stage showed that this term should be only used for describing independent certifications provided by third-party entities, which would contribute to minimizing greenwashing attempts. Second, changing how current clothing labels portray information to consumers is essential. During this research, we have proven the efficacy of visual schemes and systems, and the author believes that this should be the future of clothing labels. Some brands have been trying to fix the lack of information issue by adding QR codes to labels with more data about the garment. Nonetheless, we know that most consumers will not take the time to read information that cannot be easily and quickly accessed. Therefore, visual systems could fill this gap: they are not only quick to read and understand, but they are also easy to read and understand. Lastly, one of the best tools that can be used to increase sustainable fashion consumption is education. Without educated consumers, there will always be barriers to how much ecolabels can do to increase sustainable shopping habits. Education would be the key to empowering consumers in their choices, as it also would be the key to pressuring brands into making more sustainable products. Overall, it is through education that we enable a long-lasting change in the fashion industry.

Annexes

Annex 1: Focus Group Sample Characterization

	Gender	Age	Occupation
Participant 1	Female	51	Kindergarten teacher
Participant 2	Female	53	Entrepreneur/clothing store owner
Participant 3	Female	50	Dressmaker
Participant 4	Male	57	Accountant
Participant 5	Male	43	Business operations manager
Participant 6	Female	21	Call-center operator
Participant 7	Female	48	Warehouse manager for a fashion brand
Participant 8	Male	25	Student
Participant 9	Male	17	Student
Participant 10	Male	55	Graphic designer

Annex 2: Consumer Survey Sample Characterization

Gender

	Frequency	Percent	Valid percent	Cumulative percent
Male	18	13.4	13.4	13.4
Female	112	83.6	83.6	97.0
Nonbinary	4	3.0	3.0	100.0
Total	134	100.0	100.0	

Age Group

	Frequency	Percent	Valid percent	Cumulative percent
<18	4	3.0	3.0	3.0
18–25	82	61.2	61.2	64.2

	Frequency	Percent	Valid percent	Cumulative percent
26–41	25	18.6	18.6	82.8
42–57	23	17.2	17.2	100.0
Total	134	100.0	100.0	

Employment Situation

	Frequency	Percent	Valid percent	Cumulative percent
Student	41	30.6	30.6	30.6
Student-worker	22	16.4	16.4	47.0
Full-time employee	59	44.0	44.0	91.0
Part-time employee	4	3.0	3.0	94.0
Unemployed	6	4.5	4.5	98.5
Other	2	1.5	1.5	100.0
Total	134	100.0	100.0	

Academic Level

	Frequency	Percent	Valid percent	Cumulative percent
Middle school	4	3.0	3.0	3.0
High school	33	24.6	24.6	27.6
Bachelor's degree	76	56.7	56.7	84.3
Master's degree	21	15.7	15.7	100.0
Total	134	100.0	100.0	

Annex 3: Summary of the Results

Hypothesis	Was it confirmed?	Explanation
H1	Yes	84.3% of consumers trusted a Type I label, like the GOTS ecolabel, against 5.2% who preferred a Type II label, represented by Zara
H2	No	While materials were more important in some cases, the place of production seemed more relevant in others, so we cannot assume that there is always one that has more value than the other. It varies according to the situation

Hypothesis	Was it confirmed?	Explanation
H3	Yes	For both labels, the number of consumers who chose the most sustainable garment went up at least 25 percentage points when these new systems were added to a regular clothing label
H4	Yes	An overview of the results and previous hypotheses confirms that ecolabels help increase sustainable fashion consumption in all consumers
H4.1	No	Only 20.7% of people who changed their choice to a more sustainable one thanks to ecolabeling systems were perceived as eco-conscious. The participants with higher knowledge were always choosing the most sustainable option, regardless of how the label was presented
H4.2	Yes	Even though 29 people shifted their buying decisions thanks to ecolabels, 26 kept choosing the lowest-priced option

Bibliography

Andrews, D., Cronin, J., Gleim, M., & Smith, J. (2013). Against the green: A multi-method examination of the barriers to green consumption. *Journal of Retailing., 89*(1), 44–61. https://doi.org/10.1016/j.jretai.2012.10.001

Auger, P., Burke, P., Devinney, T., & Louviere, J. (2003). What Will consumers pay for social product features? *Journal of Business Ethics., 42*(3), 281–304. https://doi.org/10.1023/A:1022212816261

Bach, V., Eggers, D., Finkbeiner, M., Gül, S., Lehmann, A., & Spielmann, M. (2015). Benchmarking and environmental performance classes in life cycle assessment—Development of a procedure for non-leather shoes in the context of the product environmental footprint. *The International Journal of Life Cycle Assessment., 20*(12), 1640–1648. https://doi.org/10.1007/s11367-015-0975-7

Bajtelsmit, V., Bloodhart, B., Diddi, S., McShane, K., & Yan, R. (2019). Exploring young adult consumers' sustainable clothing consumption intention-behavior gap: A behavioral reasoning theory perspective. *Sustainable Production and Consumption., 18*(2019), 200–209. https://doi.org/10.1016/j.spc.2019.02.009

Beard, N. (2008). The branding of ethical fashion and the consumer: A luxury niche or mass market reality? *Fashion Theory., 12*, 447–468. https://doi.org/10.2752/175174108X346931

Blengini, G. A., & Shields, D. J. (2010). Green labels and sustainability reporting: Overview of the building products supply chain in Italy. *Management of Environmental Quality: An International Journal., 21*(4), 477–493. https://doi.org/10.1080/13527266.2010.490420

Borboni, E. (2019). Communication of sustainability and ethical issues in fashion: The Lombardy's experience. In N. Kalbaska, T. Sádaba, F. Cominelli, & L. Cantoni (Eds.), *Fashion communication in the digital age* (pp. 181–187). Springer.

Curran, M. A. (2001). Developing a tool for environmentally preferable purchasing. *Environmental Management and Health., 12*(3), 244–253. Available from https://www.emerald.com/insight/content/doi/10.1108/09566160110392308/full/html

European Commission. (2013). Commission recommendation (2013/179/EU) of 9 April 2013 on the use of common methods to measure and communicate the life cycle environmental performance of products and organisations—Annex II: Product environmental footprint (PEF) guide.

Official Journal of the European Union. Available from https://eur-lex.europa.eu/legal-content/EN/TXT/?uri=CELEX%3A32013H0179

European Commission. (2020). *Strategic EU Ecolabel Work Plan 2020–2024*. European Commission. Available from https://ec.europa.eu/environment/ecolabel/documents/EU%20Ecolabel%20Work%20plan%202020-2024%20Dec%202020.pdf

d'Astous, A., & Legendre, A. (2009). Understanding consumers' ethical justifications: A scale for appraising consumers' reasons for not behaving ethically. *Journal of Business Ethics., 87*, 255–268. https://doi.org/10.1007/s10551-008-9883-0

D'Souza, C. (2015). Marketing challenges for an eco-fashion brand: A case study. *Fashion Theory., 19*(1), 67–82. https://doi.org/10.2752/175174115X14113933306824

Ferrari, E., Iraldo, F., Testa, F., & Vaccari, A. (2013). Why eco-labels can be effective marketing tools: Evidence from a study on Italian consumers. *Business Strategy and the Environment., 24*(4), 252–265. https://doi.org/10.1002/bse.1821

Fletcher, K. (2008). *Sustainable fashion and textiles: Design journeys*. Earthscan.

Freise, M., Will, C., & Strähle, J. (2015). Communication of sustainability at European fashion shops online. *International Journal of Economics, Commerce and Management., 3*(7), 71–86. Available from https://publikationen.reutlingen-university.de/frontdoor/deliver/index/docId/224/file/224.pdf

Gershoff, A. D., & Frels, J. K. (2015). What makes it green? The role of centrality of green attributes in evaluations of the greenness of products. *Journal of Marketing Research, 79*(1), 97–110. https://doi.org/10.1509/jm.13.0303

Gonçalves, A., & Silva, C. (2021). Looking for sustainability scoring in apparel: A review on environmental footprint, social impacts and transparency. *Energies, 14*(11), 1–33. https://doi.org/10.3390/en14113032

Halme, M., & Turunen, L. L. M. (2021). Communicating actionable sustainability information to consumers: The shades of green instrument for fashion. *Journal of Cleaner Production., 297*(15), 1–10. https://doi.org/10.1016/j.jclepro.2021.126605

International Organization for Standardization. (2000). *ISO 14020:2000 Environmental labels and declarations — General principles*. International Organization for Standardization. Accessed 28 Nov 2021. Available from https://www.iso.org/standard/34425.html

Kenning, P., Meise, J. N., Phillips, D. M., & Rudolph, T. (2014). Feed them facts: Value perceptions and consumer use of sustainability-related product information. *Journal of Retailing and Consumer Services., 21*(4), 510–519. https://doi.org/10.1016/j.jretconser.2014.03.013

Langguth, J. & Schnee, S. (2018). Undressing sustainability communication: An exploratory study of sustainable fashion brands' online message. Master's thesis, Lund University. Available from https://lup.lub.lu.se/student-papers/search/publication/8945609

Lenk, L. T & Rotkirch, I. (2021). Green millennials? A qualitative study on the impact of green marketing on European female millennials purchase behavior in the fashion industry. Master's Thesis, Jönköping University. Available from https://www.diva-portal.org/smash/get/diva2:1561872/FULLTEXT01.pdf

Lin, L., & Park, H. (2020). Exploring attitude–behavior gap in sustainable consumption: Comparison of T recycled and upcycled fashion products. *Journal of Business Research., 117*(2020), 623–628. Retrieved from https://doi.org/10.1016/j.jbusres.2018.08.025

Lindersson, M. (2017). *Applying a lens of practice theory to understand the role of sustainability-labels in sustainable fashion shopping*. Master's Thesis, Lund University. Available from https://lup.lub.lu.se/luur/download?func=downloadFile&recordOId=8911692&fileOId=8911696

McNeill, L., & Moore, R. (2015). Sustainable fashion consumption and the fast fashion conundrum: Fashionable consumers and attitudes to sustainability in clothing choice. *International Journal of Consumer Studies., 39*(3), 212–222. https://doi.org/10.1111/ijcs.12169

McShane, L., Noseworthy, T. J., & Pancer, E. (2015). Isolated environmental cues and product efficacy penalties: The color green and eco-labels. *Journal of Business Ethics., 143*(1), 159–177. https://doi.org/10.1007/s10551-015-2764-4

Peattie, K., & Crane, A. (2005). Green marketing: Legend, myth, farce or prophesy? *Qualitative market. Research, 8*(4), 357–370. https://doi.org/10.1108/13522750510619733

Phau, I., & Ong, D. (2007). An investigation of the effects of environmental claims in promotional messages for clothing brands. *Marketing Intelligence & Planning., 25*(7), 772–788. https://doi.org/10.1108/02634500710834214

WeAr Design: Developing Co-design Frameworks for Sustainable Fashion Futures

Ruxandra Lupu, Anca Rosca, and Dana Tanase

Context

Fashion sustainability is complex and urgent, yet hard to tackle through the dominant ecological model that focuses on the rationalization of resources rather than on enabling wearers to reconnect on a personal level to clothes. Finding new ways to engage consumers with clothes through strategies such as co-creation, where designers and customers establish a deeper connection than through the traditional fashion production model, are fundamental for restoring the bond with our clothes and rethinking fashion sustainability.

The WeAr Design project addresses the rather new, yet promising approach to co-creation in fashion through the lens of action research. It was conducted between May 2020 and March 2021 and was funded by the WORTH Partnership call. The core aim of the project is to push the boundaries of co-creation in fashion, understood as those practical approaches through which consumers are involved in the process of fashion creation, instead of buying already-made clothes. The result was the creation of a new, participatory process, through which customers can be involved in the design process, and thus to rekindle the relationship between people and their clothes (relationship which is getting lost through digitalization). Led by three partners from Italy, Romania and the United Kingdom, with expertise in the arts, digital humanities, user-centric design and fashion, the project resulted in the

R. Lupu (✉)
Marie Curie Postdoctoral Fellow, University of Cyprus, Nicosia, Cyprus
e-mail: lupu.ruxandra@ucy.ac.cy

A. Rosca
Product Lead, Signat Health, Philadelphia, PA, USA

D. Tanase
Owner & fashion designer at Dana Tanase, Brașov, Romania
e-mail: atelier@danatanase.com

A. Cruchinho, J. C. Neves (eds.), *Sustainability in Fashion: Creativity, Education, and Innovation*, https://doi.org/10.1007/978-3-031-80991-0_17

generation of six experimental design items, created following a collaborative process that used an experimental co-design framework.

The Methodology

Action research framed the three parts of the project: the first phase mapped common interests of designers and consumers through co-creation experiments (e.g. storytelling); during the second phase, partners developed a tailor-made co-creation process based on the concept of sharing personal stories and tested this process by conducting six collaborative experiments; the last phase included an evaluation of the project outcome and the dissemination of the six created design items at Milan Design Week 2021. The project was divided into three parts: a preparatory phase, an exploratory phase and the final evaluation & dissemination of results.

Preparation
First, partners researched existing possibilities offered by co-creation in fashion: how designers and clients can meet half-way during the design process. The research revealed that attempts to involve clients in the design phase of clothes is almost inexistent, with few isolated exceptions in Asia (Japan) and Northern Europe. Brands such as Mon Purse, Shoes of Prey, The Mighty Company, Dresshirt and others that offer personalized designs were not included as authentic co-creation but as personalization, insofar as the choice is always limited to an existing array of colours/shapes/models, and the only hands-on activity that clients are involved in consists in clicking and/or dragging elements on the screen. Based on this initial research, partners adopted a definition of co-creation as a personal and hands-on involvement of the client in the design process, which goes beyond the personalization of products and offers meaningful experiences for both designers and clients. With this definition in mind, and trying to find out how such a process could take place, partners organized an experimental event in the frame of the Romanian Design Week in 2020. During the event, five selected designers/artists met visitors and engaged in discussions, some of them creating quick product prototypes together. Project partners observed interactions between designers and visitors during the event and extracted patterns of interaction and common models of behaviour. The analysis showed that visitors engaged with designers mostly through stories (about their own life, hobbies and passions), based on which designers could imagine new products for them.

The second phase used the results of the first part to create a framework that could support the co-creation process between designer and client. Informed by the results of the Design Week, partners set up a co-creation process that included the following steps:

(a) Client thinks about a story they care deeply about (from their own lives or something they like) and shares it with the designer (through email/over phone/in person).

(b) The designer uses the story as inspiration to imagine a product that follows the storyline.
(c) Designer and client meet (virtually or physically) to discuss the story.
(d) Designer makes first sketches of the product based on the story and initial talk and shares them with the client.
(e) Designer refines the sketches and meets the client again.
(f) Designer selects the best sketch and works further on it (designer can meet the client again for final review or not).
(g) Designer finalizes the product and sends it to the client.

As a result of this workflow, project partners issued a call and selected six designers (four fashion designers, one jewelry maker and one paper artist). Some designers suggested potential clients, others were matched with people in the partners' network who expressed the desire to take part in the experiment. Each pair had 1 month time to undergo the entire process at the end of which a new product was created. Two teams worked fully remotely, using video conferencing methods. The other four teams used a mixed approach of physical and digital interaction.

The last phase of the project was dedicated to conducting an evaluation of the process and disseminating the results. As part of the evaluation, individual interviews were conducted with both designers and matched clients. The results of the evaluation highlighted how this process represented an exciting and deeply meaningful experience, with many surprising elements for both sides. Nevertheless, it also highlighted how clients need to be appreciative of and open to artistic processes, to fully engage without prejudices in such a process and without being too controlling about the design process. Moreover, the means through which client and designer communicated varied a lot, indicating that in order for this process to work, a mix of communication channels should be available, and its structure should be jointly decided by designer and client. Some of the unveiled challenges on the other hand included the fear of the designer to let the client get too involved in the process, risking for clients to become patronizing, as well as the high costs associated with this process compared to traditional design processes. As part of the dissemination of results, the six objects were virtually showcased at Milan Design Week 2021.

Experiments provided a deeper insight into how we can open the design process towards a more participatory direction, where clothes are produced according to a meaningful process that is rooted in personal stories, identities and relationship with our clothes. The process not only expanded the horizons of both co-creators, but we also observed the braking down of invisible barriers and fears currently existing between the fashion designer and the consumer. Storytelling proved fundamental in this process, highlighting how personal stories as well as their sharing can bring people together to shape a more sustainable future for the fashion sector.

Co-created Pieces

The Corset: Designer—Oana Parcalab

The corset (Fig. 1) was chosen as an 'iconic' accessory because of its symbolism and its history. Designer and client decided to rethink together the corset from a contemporary perspective, as a 'statement' of identity and as a historical piece at the same time. The collaboration stemmed from the profound interest in the corset as a support for the body. Through redesign, the corset becomes a contemporary affirmation of female emancipation. The traditional use of the corset underneath clothes is also turned around; worn on top of clothes it becomes a statement of a woke personality and self-trust. Adornments like visual elements and spikes underline its gender-neutrality. The printed design on the stitched leather surface recalls tattooed skin, alluding to intimacy.

Fig. 1 The Corset

The Harness: Designer—Alexandra Abraham

Born from a desire to grow wings and fly (expressed by the client each year for Orthodox Easter), the harness embodies the deepest childhood wishes of the client. Made from leather, the harness is the highest artistic expression of the artist's and client's ideas and concepts distilled as geometrical shapes that encapsulate the body. The green element connecting parts of the design, flows vertically like a lifeline. Each curve, each pattern and colour possess a specific symbolism that is intimately connected to the client's personality as her biggest wish: to fly (Fig. 2).

The Anorak: Designer—Dana Tanase

An enchanted alliance was born between two friends (artist and client) who share the passion for clothes and art (both like to paint on clothes). The idea is based on giving a new life to clothes by adding an artistic touch as a sign of the wearer's personality (Fig. 3). Repurposing and slow fashion stand at the core of the creation process of this garment. The design method weaves in perfectly aspects of intimacy

Fig. 2 The harness

Fig. 3 The Anorak

and friendship that unites designer and client. The anatomy of the heart breathing a rainbow of colours is a literal depiction of the essence of the design. The fusion between different materials builds the character of the garment, underlining the upcycling process, but also gender-neutrality of universal elements such as friendship.

Trench Coat: Designers—Emilia Bosoi and Larisa Croitoru

The trench coat (Fig. 4) as a partnership between the designers and the client (owner of memories), blends in the colours and the texture of the tree trunk. Both colours (brown) and patterns (cross section of tree trunks) stand for the wearer's deep connection to the times spent at their grandmother's house as a young child. The deep roots of a tree echo the childhood memories imprinted in early life. Just like your family surrounds one with safety, security and love, the coat wraps around the body igniting the emotion of essential bonds to the origins.

Fig. 4 The trench coat

Brooch: Designer—Vika Tonu

'Organized Serendipity' is a collaborative project consisting in seven jewelry pieces, for which the input was the personal and corporate stories of the client (Fig. 5). The series illustrates the key ingredients for successful collisions: passion, trust, patience, persistence, level playing field, the right agenda and energy as a unit of measurement. The brooches are made from six neodymium disc magnets, set in a handcrafted silver structure. Solely by magnetic forces, a variety of elements are put on top of this structure to create this jewelry as artwork. Only to be given and received, never to be sold or bought, the pieces will have their own journey and links in order to connect, collide and surprise. There is no matter, only energy!

Paper Crane: Designer—Cristian Mariancıuc

Based on a Japanese legend, a person who folds one thousand origami cranes will be granted a wish by the gods. This paper crane (Fig. 6) was ideated as a gift from a grandmother to her grandchild and blends in memories and colours into the fine

Fig. 5 The brooch

craftsmanship of paper folding. Its colours reflect ideas of non-violence and strength—the essence of martial arts—and are transmitted as wishes to be bestowed upon the child as a sign of good luck and abundance.

Conclusion

This experimental project provided more insight into how we can open the design process and steer it towards a more participatory direction, where clothes are produced according to a meaningful process that is rooted in and reflects our identity and strengthens our relationship with our clothes. Storytelling proved fundamental in this process, highlighting how personal stories and the process of sharing them can bring people together, to create a more sustainable future for the fashion sector.

Ruxandra Lupu is a visual artist and digital humanities researcher currently working as a Postdoctoral Researcher and creative entrepreneur. Her work is positioned within the broader field of sustainable innovation and focuses on the role of co-creative and design practices to explore the

Fig. 6 Paper crane

intersection of film, fashion and new technologies. Ruxandra was trained as a traditional printmaker in Italy and obtained her practice-led PhD from the University of Leeds with a project that combined her knowledge of crafts with film studies. Her research interests cover the areas of responsible innovation, haptic visuality and imaginative ethnography.

Anca Rosca is an experienced Technical Product Consultant, a keen Agile Thinker, a Data Science enthusiast and a world traveller. She enjoys working cross industries and exploring the intersections between tech, art and health science. Anca has more than 15 years of experience leading teams in building products in various industries such a logistics, finance, health and small business providers. Her passion is building products that people love by employing lean thinking models and user-centric design methods and techniques that focus on the needs of the users in each phase of the design process. Anca is fascinated by how technology has changed and is continuously affecting the relationship between consumer and fashion brands and loves working in interdisciplinary groups to explore new levels of thinking and collaboration.

Dana Tanase is a Romanian fashion designer working under her own brand name 'Dana Tanase'. She is a graduate of the Arts School in Constanta and holds a Master in Fashion Design (2011) from the European Institute of Design in Milan. Dana has been designing clothes, casual and also evening wear with her visual signature, a hand-painted ballerina, for more than 15 years, partnering with important names such as Nadia Comaneci and local TV shows. Dana is always experimenting with approaches such as co-creation, which enable her to establish a unique and closer connection with her clients.

Circular Fashion as a Business Model

Andréa dos Anjos Moreiras, Isabel Cristina Scafuto, and Milton Terumitsu Sogabe

Introduction

The current business model of the fashion industry is fast fashion, it translates that the main attributes are sales and generation of economic growth. (Grose & Fletcher, 2011) The result is easily discarded clothes, driven by low quality, with a focus on short-lived trends, causing clothes to become obsolete because they are "out of Fashion" (Ferronato & Franzato, 2015).

With the economic growth of the fashion industry promoted by fast fashion, the focus has turned to the production processes, the difficult working conditions, and the environmental deterioration caused by the production stages. This has awakened the attention of public authorities and institutions. The alignment of the fashion industry's productive chain to the labor laws and the environment has gained prominence and supporters for the cause; after all, the glamour of Fashion is not present in the whole chain or in the way the earth's natural resources are being consumed.

The fashion industry needs to separate increased sales volume from real value creation. People and the planet need to be valued above business profit. See in Fig. 1 how much a garment produced by the fast fashion model can cost.

With a great impact on climate, the fashion industry is responsible for 2–8% of global carbon emissions, and if the current pace is maintained, the sector may not meet the goals of reducing emissions by half by 2030 (ONU Brasil 2021b).

Pressure from society made a difference to the results at the 26th United Nations Climate Change Conference (COP26), the main UN summit to debate climate issues and create strategies for a common good, which took place in Glasgow,

A. dos Anjos Moreiras (✉) · M. T. Sogabe
Universidade Anhembi Morumbi (UAM), São Paulo, Brazil

I. C. Scafuto
Universidade Nove de Julho (Uninove), São Paulo, Brazil

A. Cruchinho, J. C. Neves (eds.), *Sustainability in Fashion: Creativity, Education, and Innovation*, https://doi.org/10.1007/978-3-031-80991-0_18

Fig. 1 How much does fast fashion cost?. (Source: Fashion Revolution Brasil)

Scotland, from October 31 to November 12, 2021, and gathered representatives from almost 200 UN countries and more than 100,000 people.

People are threatened with global warming above safe levels for humanity. Globally, there are initiatives to stimulate public commitment from major Fashion companies, such as the UN Fashion Industry Charter for Climate Action. Recognizing their huge role in promoting climate change, in an integrated message, a number of Fashion and sustainability organizations have signed an open letter that highlights the key role of the industry and says leaders must recognize this important in order to act.

The Charter's mission is to lead this industry to reduce greenhouse gas (GHG) emissions by 2050 and still keep global warming below 1.5 °C. At its launch in 2018, it had 130 companies and 41 supporting organizations committed to climate action. The document provides a framework for stakeholder dialogue and engagement on climate action (ONU Brasil 2021b).

At COP26, the fashion industry renewed its commitment to actions to combat climate change, the new Charter was presented in a session exploring how to move the fashion industry toward a zero net emissions pathway and the requirement for the creation of incentive mechanisms for supplier engagement on the decarbonization pathway, and existing solutions that can be scaled up and accelerated.

During the event, a call was issued for policies to encourage the use of traceable and environmentally friendly materials. "At a time when the climate crisis is accelerating to unprecedented levels, we need the real economy to be a leader in climate action. Strengthening the commitments of the Charter signatories is an excellent example of this leadership," said Niclas Svenningsen, UN Climate Change Global Climate Action Manager (ONU Brasil 2021a).

The request to leaders is to commit to the following recommendations: collective action to achieve zero emissions by 2050; efforts and creativity to solve the waste problem; increase corporate responsibility for their global supply chains; support skills development in education to encourage children to learn to make, mend and reuse their clothes; and frame any solutions to the climate emergency around business models that shift the focus from profit to a welfare economy (Artuso, 2021).

Therefore, sustainability in Fashion is necessary and fundamental to slow down the climate crisis.

Sustainable Fashion

Sustainable Fashion provides for environmental improvement of flows along the production chain and as a social improvement by offering proper working standards and conditions for workers. According to Niinimäki (2015), Sustainable Fashion can be done through incremental innovation with small improvements to existing products and systems and design innovation for a sustainable society (Niinimäki, 2015).

Fashion, to be considered sustainable, needs to fulfill some criteria:

- In the management model: (1) a Circular Economy with "cradle to cradle" philosophy changing processes; (2) a collaborative economy, clothing library, or exchange bazaars, which advocate that access is more important than ownership; (3) system B, new market identity with high social, environmental, and transparency standards.
- Technology: (1) fabric recycling; (2) innovation; (3) blockchain[1]
- Social: (1) social inclusion, such as resocialization of ex-convicts; (2) valuing minorities, such as women overcoming domestic violence
- Upclycling:[2] Resigning (1) clothes; (2) accessories; (3) shoes

[1] Internet-based technology that records transactions in a decentralized, public, anonymous, and traceable way.

[2] Transforming a garment that already exists or a residue for a product with higher added value.

- Colors: (1) natural dyeing; (2) colored latex; (3) organic cotton that is born colored
- Alternative fibers: (1) hemp; (2) pineapple; and (3) orange, among others (Moreiras, 2019)

The slow Fashion means working in search of the best productivity, knowing the before and after of what is generated, and correcting and exceeding quality with creativity. This unhurried attitude makes the process environmentally friendly, contemplating the quality of the final product (Livni and Escuder, 2022).

The zero waste means the maximum use of the fabric by modeling. The technique advocates not generating waste. "Zero Waste also presents itself as a design-oriented practice, but subordinated to a culture of aesthetics and value, with a discourse focused on sustainability and proposing the reduction of raw material waste" (Fraga, 2020, p.115).

Inclusive Fashion, which produces clothes for people with visual impairment, dwarfism, mobility problems, and amputated organs, among others, is a segment of people forgotten by brands and with little representation in events and Fashion editorials, but with great importance.

It is increasingly necessary to rethink production, so the more criteria the company meets, the more sustainable it will be.

Design and Economia Circular

The current economic model is the linear economy and may be under threat due to the limited availability of natural resources and the limits of the planet in terms of its capacity to assimilate the pollution generated. This model is known as "cradle to grave" and is based on the extraction of natural resources to be raw materials for products that will be used until they are discarded as waste.

To counter this model that is harmful to the environment, researchers Michael Braungart and William McDonough have created a "cradle to cradle" philosophy, or Circular Economy, in which waste is used as input for the production of new products. It is a new model that seeks to unlink the consumption of finite resources to global economic development. This response to challenges regarding the issue of resources for companies could attribute growth, generate more jobs, and reduce environmental impacts (Ellen Macarthur Foundation 2017). One definition of circularity is to limit the consumption of finite resources, reuse products at their end of life, and eliminate waste from the system (Global Fashion Agenda).

The two economic models can be analyzed in Fig. 2.

Creativity has never been more important, and design has become an essential tool for building more sustainable futures through a circular and innovative approach with a fresh look at materials and processes. Designers and the business sector may have a more comprehensive view of the relevance of developing sustainable products and be encouraged to promote strategic design with the aim of improving the efficiency of production systems and products, but for this to happen, they need to

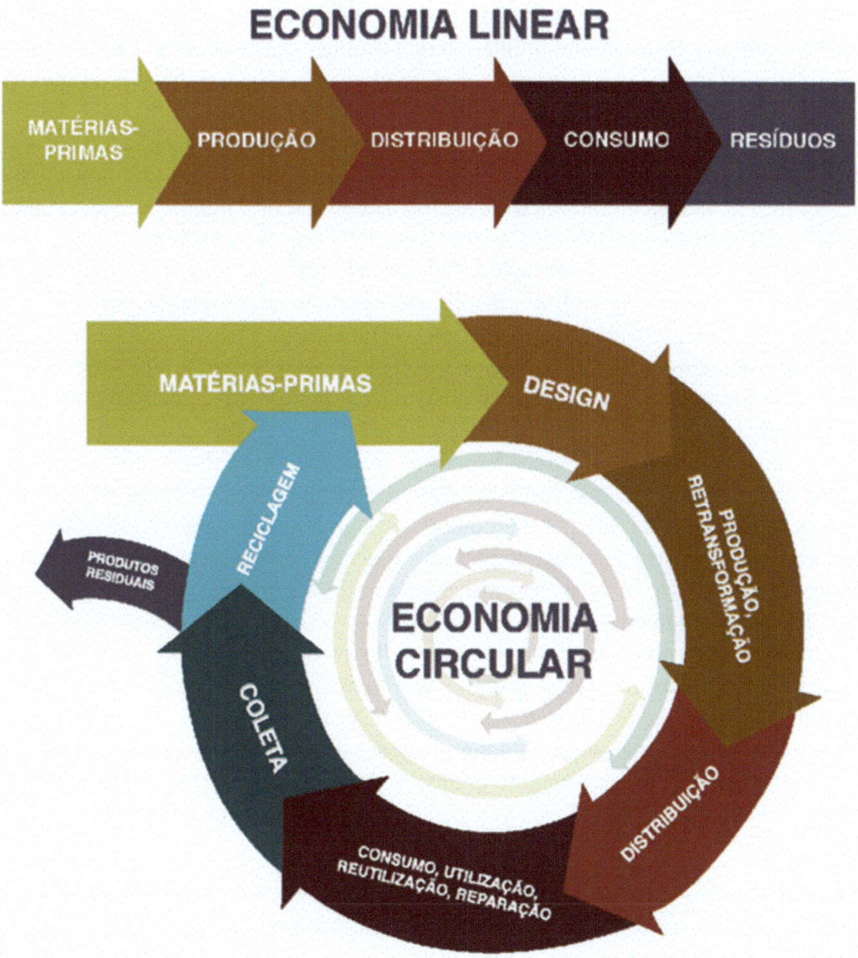

Fig. 2 Linear and Circular Economy. (Source: European Commission (2014, p. 2–3). Adaptation: Avila et al. (2018))

have knowledge and engagement on some of the social, environmental, economic, and political aspects relevant to the productive sector (Braga, 2012).

In the paradigm shift toward sustainable development,[3] the design of sustainable products emerges as a response to the advancement of business competitiveness and the transformation process toward a more responsible economy, it can collaborate with the sustainability of production channels, distribution, consumption, and disposal of products inorder to reduce pressures from production activities on the environment (Braga, 2012).

[3] It is a model that seeks to conserve and restore the environment that the economy uses to meet the needs of current generations without compromising the future of new generations.

New products need to be designed with ecological thinking, guided by the regenerative economy, in which durability and recyclability are as important as functionality and aesthetics, and to support this thinking, the Sustainable Development Goals (SDGs) can serve to guide the design of these products.

The SDGs were built on a global negotiation process that began in 2013 and included the participation of several countries in their discussions and definitions regarding a global agenda adopted during the United Nations Summit on Sustainable Development in September 2015, consisting of 17 goals and 169 targets to be achieved by 2030. This agenda includes global actions in the areas of poverty eradication, food security, agriculture, health, education, gender equality, reduction of inequalities, energy, water and sanitation, sustainable patterns of production and consumption, climate change, sustainable cities, protection and sustainable use of oceans and terrestrial ecosystems, inclusive economic growth, infrastructure, and industrialization.

Sustainability and the Circular Economy are at the top of the textile industry's agenda, with challenges and solutions that promise to radically change this sector. And design plays a key role. "[...] definitely, design for sustainability can be recognized as a kind of strategic design, that is, the design of strategies applied by companies that have seriously imposed themselves to the prospect of environmental sustainability" (Manzini & Vezzoli, 2002, p. 23).

One of the key points in discussions about the practice of Circular Economy has been its recognition in product design. Representatives of the design community state that *"waste is a design mistake."* And 80% of the factors causing the negative life cycle environmental impacts of a product are defined during design.

Thus, it becomes fundamental to gradually design the elimination of waste from the economy, with all the advantages associated with this action (Royal Society of Arts, 2013).

The Circular Fashion

The future of Fashion is Circular. Circular Fashion is a holistic design approach that refers to products that are designed and produced to be used and reused several times for as long as possible, and when they can no longer be reused, they are regenerated into raw material for the production of new products. This Circular model aims to break the linear model, which is one of the criteria of Sustainable Fashion.

Circular Fashion, which follows the Circular Economy, is an economic model that proposes to increase the life of materials and products from various stages of use. In Fig. 3, circularity in Fashion can be observed, "Circular Fashion rethinks textiles, processes, forms of consumption, design, and even business models that preserve natural capital, extend the life span of products, capture maximum value from existing products and structures, and eliminate the concept of waste" (Schuch, 2020).

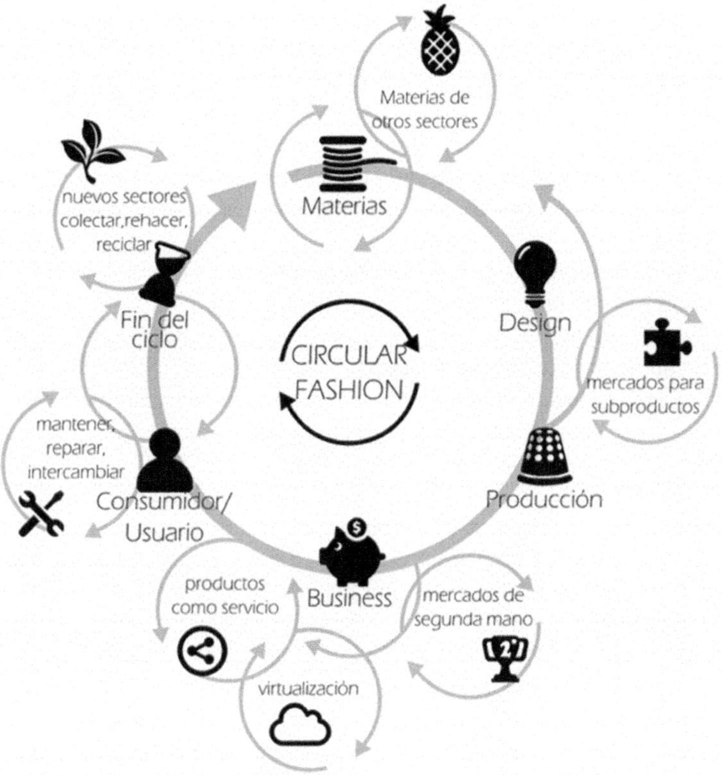

Fig. 3 Circular Fashion. (Source: Schuch, 2020)

The 16 principles of Circular Fashion were defined by Anna Brismar, owner of the Swedish consulting firm Green Strategy, who sought in the principles of Circular Economy and sustainable development the circularity of raw material with a view to its infinite cycle, without it having a loss of value. Learn about the principles:

1. Design with purpose: Develop products and services guided by the regenerative and restorative concepts of the Circular Economy, aiming to maintain products, components, and materials at their high level of utility and value at all times.
2. Design for longevity: Develop products with durable materials, extending product life, avoiding continuous consumption and, consequently, waste accumulation.
3. Design for resource efficiency: Develop products that consume fewer human resources, time, capital, energy, water, and fuel, among others.
4. Design for biodegradability: Develop products foreseeing decomposition without harming the environment.

5. Design for recycling: Develop products contemplating recycling as an opportunity for new products and new concepts, promoting the recyclability of materials and avoiding the generation and accumulation of waste.
6. Source and produce locally: Develop products using local raw materials and local labor, region, or country, stimulating the development, autonomy, and the maintenance and income generation of producer groups, cooperatives, and artisans through the valorization of their knowledge, techniques, and traditional know-how.
7. Toxic-free sourcing and production: Develop products using low-impact materials, minimizing or even eliminating toxic dispersion.
8. Source and produce efficiently: Develop products using appropriate strategies and technologies for clean production.
9. Source and produce with renewable sources: Use renewable resources as much as possible in production processes, making use of solar, wind, as well as wave and tidal energy and biomass.
10. Source and produce with good ethics: Produce products ensuring decent and safe working conditions in all its processes, from the cotton plantation to the sewing workshops. Compensate women, ethnic groups, and artisans with fair wage transparency of the actors and their working conditions, and the origin and type of materials used.
11. Providing services to support the long life of products: Implement strategies inserted in the company's performance, evaluating the product's life cycle and the consumers' post-sales demands, such as repairs, reuse, repair, and recycling.
12. Reuse, recycle, or compost all waste: Use strategies that are part of the companies business model for the reuse of textile fibers, textile waste and upcycling, reuse of garments to design new products.
13. Encourage broad collaboration: Stimulate fair trade, awareness, and information for various audiences such as consumers, entrepreneurs, and public institutions for the creation of collection points and disposal of clothing and textile waste, shared knowledge for more conscious consumption, the approach of all actors involved to stimulate the exchange of sustainable practices. Creation of a connection between brands, producers, and consumers, for the valorization of traditional knowledge.
14. Wear, wash, and repair with care: Caring for your clothes is an investment, as it is directly linked to the durability of the garment. Thus, the consumer has the right to information on how to care for his or her clothes.
15. Renting, borrowing, trading secondhand pieces, or customizing the piece instead of buying a new one: Collaborative economies like clothing libraries;
16. Buy quality instead of quantity: It is the practice of conscious consumption; always remind the consumer about the responsibility of buying clothes; he should always ask questions: Can I buy that outfit? Do I need these clothes? Who made these clothes?

The time between becoming aware and applying the principles of Circular Fashion in the development of products and services cannot be long. Preserving the

Planet is a complex attitude, and the Fashion system cannot act with delay. It is a matter of respecting ecological limits in which humanity and the Planet are threatened (Factum & Quaresma, 2017).

Conclusion

Sustainability in Fashion is necessary and fundamental to the industry that is undergoing profound transformations. The more sustainable the production, the more it will contribute to the maintenance of the environment for future generations. Sustainable Fashion is a differentiated proposal to make Fashion, for it demands respect for the environment, for people, and for biodiversity. The entire chain must be observed and the processes valued, origin evaluated, choice of raw materials with less impact, and a fair relationship with the producer, product, and consumer. From the plantation of cotton, which can be organic, besides protecting the land and the farmers, it requires less water for the plant to develop, to the Fashion shows that must present diversity in the professionals involved and an ecological structure and production in the Fashion shows generating the least possible residue.

In Fashion, initiatives have emerged that question and challenge the linear economy standards, enabling adaptations or even the appearance of opportunities guided by creativity and innovation based on a design using Circular Economy guidelines.

Product design is a great ally in this process, because the conception phase is very relevant. It is worth emphasizing the importance of thinking about the purpose and function of each piece that will be developed. Fashion products must contribute to the positive development and well-being of ecosystems and societies, and their life cycle must not cause any negative socio-environmental impact.

Some strategies are needed to enable long life and maximize the circularity of products and materials in society, such as designing products that can be easily disassembled or refurbished in order to facilitate the repair, redesign, reuse, and recycling of materials.

Another strategy is to create timeless products using high-quality materials and durable construction to promote durability, longevity, and attractive products to a wide range of users.

To avoid unnecessary inventory, design for order reduces waste by creating products that maximize resource efficiencies, such as zero-waste patterning, minimal waste, modular design, and limited material combinations. Choosing organic, recycled, or biodegradable materials that can be safely processed, used, reused, and recycled or composted at the end of their useful life. Using renewable energy for selecting, producing, transporting, and distributing, high resource efficiency in order to decrease the negative environmental impacts caused by fossil fuels.

Repair, reuse, recover, and recycle all materials along the chain and deliver consumer benefits that allow products to be kept longer, such as secondhand clothes, exchange or loan, and clothing library. Another technique widely used within

Circular Fashion is upcycling, which extends the useful life of products, promoting a decrease in the extraction of virgin resources.

And with a dedicated eye to social issues, fair labor relations, appreciation of women, minorities, plurality, and diversity, among other actions that can positively impact lives.

Circular Fashion promotes solutions that can radically change this sector; it is a Fashion with more socio-environmental responsibility, with a careful look to a responsible production chain, and with less GHG emissions, effluents, and waste, slowing down the climate urgency, but it requires reflection and planning throughout the process, starting from conception to consumption, the way to go is long, but fortunately, there are companies already surfing this wave. Through collaboration, united by transparency and a vision for regenerative and circular economies, Fashion can be a transformative agent and has the opportunity to use its influence for the good of the environment in the face of the greatest threat that humanity faces, the climate crisis, but for Fashion's potential to be recognized as part of the solutions, efforts must be collective and aimed at the common good.

References

Artuso, E., (2021). Cop-26: Os compromissos, as emissões, as comunidades afetadas e a indústria da moda [on-line]. *Carta Capital / Fashion Revolution*. [Consultado em 22 novembro 2021]. Disponível: https://www.cartacapital.com.br/blogs/fashion-revolution/cop-26-os-compromissos-as-emissoes-as-comunidades-afetadas-e-a-industria-da-moda/

Avila, A. P. S., Maciel, D. M. H., Silveira, I., & Rech, S. R. (2018). Os Resíduos Têxteis Sólidos no Contexto de Abordagens Sustentáveis: Ciclo de Vida, Economia Circular e Upcycling. *MIX Sustentável., 4*(3) ISSN 24473073.

Braga, J. C. (2012). *O design como ferramenta estratégica para a produção sustentável no Brasil: estudo de caso do ventilador Spirit e da cadeira ICZERO1*. Mestrado em Ciências Humanas, Universidade Federal de Uberlândia.

Ellen Macarthur Foundation. Economia Circular [on-line]. *Circular Economy Learning Hub*. [Consultado 12 dezembro 2021]. Disponível: https://archive.ellenmacarthurfoundation.org/pt/economia-circular/conceito

Factum, A. B. S., & Quaresma, U. (2017). Articulação para sustentabilidade social e ambiental no sistema de moda via artesanato. Em: *13° Colóquio de moda UNESP, 11 de outubro de 2017, Bauru, Brasil*. Universidade Estadual Paulista UNESP.

Ferronato, P. B., & Franzato, C. (2015). Open Design e Slow Fashion para a Sustentabilidade do Sistema Moda. *Moda Palavra E-periódico., 9*(Especial), 104–115.

Fraga, D. (2020). *O zero waste frente à pragmática do consumo no setor de corte da confecção do vestuário: a falácia do aproveitamento na modelagem com foco na redução do resíduo*. Centro de Ciências Exatas, Naturais e Tecnologias Programa de Pós-Graduação em Tecnologia Ambiental, Universidade de Ribeirão Preto.

Global Fashion Agenda [on-line]. *Global Fashion Agenda*. [Consultado 20 agosto 2022]. Disponível: https://www.globalfashionagenda.org

Grose, L., & Fletcher, K. (2011). *Moda & sustentabilidade: design para mudança*. SENAC.

Livni, A., & Escuder, F. A 100 años del movimiento FUTURISTA en Italia reciclamos su manifiesto en Uruguay. [on-line]. *Ana Livni*. [Consultado em 26 de agosto de 2022]. Disponível: https://www.analivni.com/filosofia

Manzini, E., & Vezzoli, C. (2002). *O desenvolvimento de produtos sustentáveis: os requisitos ambientais dos produtos industriais*. EDUSP.

Moreiras, A. (2019). A Moda na Agenda 2030 | I4SSS Instituto [on-line]. *YouTube*. [Consultado 25 agosto 2022]. Disponível: https://www.youtube.com/watch?v=gYJkpXl1Hzk

Niinimäki, K., (2015). Ethical foundations in sustainable fashion. *Textiles and Clothing Sustainability* [on-line]. *1*(1). [Consultado 20 agosto 2022]. Disponível: https://doi.org/10.1186/s40689-015-0002-1.

ONU Brasil. (2021a). Indústria da moda intensifica ambição climática durante COP26 [on-line]. *Brasil*. [Consultado 20 agosto 2022]. Disponível: https://brasil.un.org/pt-br/157504-industria-da-moda-intensifica-ambicao-climatica-durante-cop26

ONU Brasil. (2021b). Na COP26, mais de 130 empresas de moda prometem reduzir emissões de CO2 [on-line]. (2021). *ONU News*. [Consultado 22 novembro 2021]. Disponível: https://news.un.org/pt/story/2021/11/1769992

Royal Society of Arts. (2013). Investigating the role of design in the circular economy [on-line]. [Consultado 20 agosto 2022]. Disponível: https://www.thersa.org/globalassets/pdfs/reports/great-recovery-executive-summary-june-2013.pdf

Schuch, A. B. (2020). Moda Circular antes que seja tarde demais! Brasil Eco Fashion [on-line]. *Brasil Eco Fashion*. [Consultado 19 agosto 2022]. Disponível: https://brasilecofashion.com.br/moda-circular-antes-que-seja-tarde-demais/

Knitting Sustainability: Applying a Life Cycle Design Approach to Knitwear

Giulia Lo Scocco and Martina Motta

Introduction[1]

The question is often raised as to what the most appropriate description of design is. One possible definition is to frame it as understanding the existing situation and creating new situations (Simon, 1988; Simonsen et al., 2014). This approach anchors designers in the present moment and requires them to observe and understand the existing carefully.

In current times, this definition immediately recalls the needs of our planet and people: being a designer today, in any field, cannot be separated from the continuous spread and growing need for sustainability.

If for other fields like architecture and product design concepts like longer-lasting commodities, less wastage, recycled materials have gradually become part of every new project and are widely accepted by the public, the fashion industry is struggling to adopt them, possibly due to its vital cultural component, theorized by Roland Barthes back in mid-1900: clothes constitute themselves as semantic structures whose meaning is incessantly formed and re-formed, generating the constant need of renewing and change (Barthes, 2006). The two words *sustainable* and *fashion* become almost an oxymoron, searching for balance between the necessity of renewal and the need to intervene in a global industry that affects the planet and people. To overcome the difficulty of integrating sustainability into the fashion

[1] Giulia Lo Scocco wrote the *Abstract* and the paragraph *Life Cycle Design in the field of Knitwear: complex design decisions as opportunities*. Martina Motta wrote the paragraphs *From Design to Life Cycle Design, Methodology* and *The impact of LCD on the stakeholders*. Both the authors wrote the *Introduction* and *Conclusion*.

G. Lo Scocco (✉) · M. Motta
Politecnico di Milano, Milan, Italy
e-mail: giulia.loscocco@polimi.it; martina.motta@polimi.it

A. Cruchinho, J. C. Neves (eds.), *Sustainability in Fashion: Creativity, Education, and Innovation*, https://doi.org/10.1007/978-3-031-80991-0_19

industry, it is essential to understand the different processes that govern it, moving from the general level to a detailed analysis of the specific branches of the fashion system.

Both authors of this chapter have research and professional background in the Italian knitting industry. Knitwear is a unique case to study because of its strong link to all stages of raw material processing. These production techniques are inherently sustainability and zero-waste oriented, right down to the recent upgrading of software and technologies that help predict prototyping and optimize resources.

To frame the concept of sustainability within the field of knitwear design, the authors have chosen to relate it to the product lifecycle analysis (Giannone et al., 2019) to understand whether any particular reason might make it more complex than in other sectors, such as those mentioned above, architecture, and product design.

From Design to Life Cycle Design

Life cycle analysis is a systematic analysis of the potential environmental impacts of products or services during their entire life cycle (Piontek & Müller, 2018). This assessment examines the possible environmental effects during the whole life cycle of a product (production, distribution, use, and end-of-life phases) or service. The evaluation also includes upstream processes (e.g., suppliers) and downstream processes (e.g., waste management) associated with production (e.g., production of raw, auxiliary, and operational materials) and the use phase and disposal (e.g., waste incineration) (Amicarelli et al., 2022).

Stemming from the Life Cycle Analysis methodology, Life Cycle Design operatively supports the fashion industry in designing low-environmental impact cloth. Research in design for sustainability (Ceschin & Gaziulusoy, 2016; Vezzoli, 2018; Vezzoli et al., 2021) is working on transferring the Life Cycle Design approach (LCD), widely studied for product design and other fields, to fashion.

By following this methodological strategy, designers consider the environmental impact generated at each stage of a product's life: raw materials, production, transport, distribution, use, and end of life. The final aim of adopting such a methodology is to optimize resources and continuously improve a product or service regarding its environmental impact.

This approach to fashion requires designers to abandon the design model based solely on a garment's aesthetic and functional design in favor of an attitude of analysis of the product's life cycle. In other words, designers are called upon to make a paradigm shift from the mere creation of garments to the design of garments based on the consideration of a fair and less environmentally impactful life cycle (Vezzoli et al., 2022). With this approach, design becomes every decision taken throughout the entire process, from the original concept to how we wear clothes. Its final goal finally evolves to minimize the overall environmental impact of all life cycle phases with a systematic approach defined by researchers with strategies, guidelines, methods, and tools.

Suppose the role of designers is evolving in this direction. It becomes fundamental to observe their work not just as merely creative and imaginative people but as decision-makers whose resolutions impact all the phases of products' life cycle and thus on all the stages and stakeholders on the supply chain. To Conti, creative people can "recognize new opportunities by looking at things from a new approach" (ibid., p. 31). They consequently have the opportunity and the responsibility to navigate and innovate a consolidated system from the inside, in a moment when all the practices and the processes of the fashion industries require innovation and evolution.

But what is the actual role that designers play in this scenario, and what choices can they make so that their influence can be measured in the path toward the sustainability of textile and knitwear products in particular? In which phases of the process is the designer able to have the most effective influence?

According to the European Commission (2014), 80% of the environmental impact of a product is defined at the time of its design. Upstream activities in the supply chain, such as raw material sourcing and garment processing, are responsible for about 70% of the CO2 emissions of a clothing product, while the remaining 30% of emissions are associated with the downstream part of the supply chain, namely the sale, use, and disposal of the garment (Rulli, 2022). This means that at the design stage it is possible to significantly influence the actual sustainability content by considering raw materials, manufacturing processes, and the product's fate once its original function has been fulfilled.

Through the academic investigation they carry out as a research group, the authors intend to probe the role of knit designers along the production process to outline the best skill set they should possess. In addition, it is helpful to identify what relationships knit designers intersect with the multitude of stakeholders active along the supply chain, which is complex and fragmented as it is composed of breeders and farmers, spinners, knitters, certifying bodies, fashion brands, knitting machine manufacturers, and others who contribute knowledge and creativity to the entire process.

Against this variegated scenario, the chapter aims to outline what an LCD approach means for knitwear designers, who the stakeholders are during the development process, and what role they play in the decision-making process.

Methodology

The research reported in the chapter is the result of the analysis of data and information collected during the last five years (2018–2022) of academic investigation in the field of knitwear, addressing it as a design discipline, as a cultural and creative industry, and as a productive sector. Data have been collected by combining direct observation inside companies along the supply chain (including the above-cited typologies, i.e., spinners, knitters, certifying bodies, fashion brands, and knitting machine manufacturers) and qualitative interviews.

Direct observation has been made in 22 companies to analyze the ongoing practices and their evolution over the last five years, considering the changes under the shift toward sustainability.

The interviews have been divided into:

- Ten interviews with professionals working along the supply chain investigating the dynamics of the knitwear industry
- Seven interviews with academics and designers investigating the practices of knitwear design
- Eleven interviews with professionals and designers that, in light of the previous findings, deepen the topic of sustainability

The research allowed the authors to compare the knowledge about sustainable practices in the fashion industry reported in the literature with a direct and more specific knowledge rooted in the knitwear field, to detect similarities and unique features, and consequently to identify opportunities for future research and actions.

Life Cycle Design in the Field of Knitwear: Complex Design Decisions as Opportunities

So far, we have discussed the importance of having a holistic attitude toward fashion sustainability, identifying the LCD methodology (Giannone et al., 2019) as a concrete supporting approach in this direction. As we now turn our attention to the knitwear industry, it is essential to recall how, while being a component of the fashion system, knitwear employs its practices, materials, equipment, and technologies and, consequently, adopts specific solutions for sustainability.

As published in previous works (Motta & Conti, 2020; Lo Scocco, 2023), being a knitwear designer means combining creative sensitivity with design culture and specialized knowledge of knitwear technology. Against this background, knit designers adopting an LCD approach are called upon to act on two fronts. On the one hand, they must perform a complex and fragmented set of tasks requiring a very high level of knowledge of the product, workflow, and techniques. On the other, they have the unique opportunity to harness the inherent nature of knitting techniques and technologies to finalize a more responsible way of creating and producing.

So, what does it mean to integrate an LCD approach into knitwear design practice? And what practical decisions can designers make to increase the sustainability metrics of their products? Declined to the knitwear sector, the assessment would range from the extraction and processing of raw materials, such as cotton, wool, viscose, and polyester (the so-called cradle), through the product's design, production, distribution, and utilization phases, to the recycling or final disposal of its constituent materials (the so-called grave).

In the following diagram, a typical knitwear life cycle is schematized in ten different phases, each belonging to one of the five categories of LCD outlined by Vezzoli (2018).

1. Preproduction: raw materials extraction
2. Production: spinning and dying of yarns, knitting, finishing (industrial washing), linking and sewing
3. Distribution: packaging and transport
4. Use: use
5. End of life: disposal or recycle

Designers play a central role in this assessment as they can intervene at all stages of the life cycle (Fig. 1), devising targeted solutions and taking responsibility for the overall performance of the finished product in terms of environmental impact. They can rethink each stage of the process from an LCD perspective, from the research phase, through production, marketing, maintenance, and end-of-life management.

So, while consumers can only intervene downstream in the process, upstream, it is the designers in team with product managers who have the power to control much of the product's environmental impact. If they are the first to think about sustainable design, far fewer aspects can be mitigated at later stages.

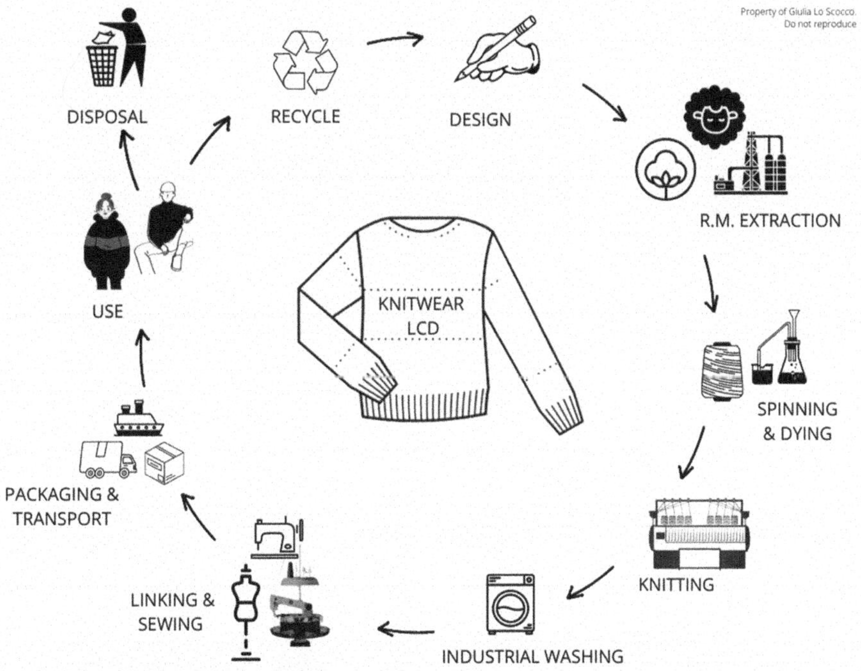

Fig. 1 Stages of life cycle

Table 1 Designers' influence in the sustainability shift

ACTION	ENVIRONMENTAL BENEFIT	LCD STAGE
Take advantage of knitting simulation software	Simulating the knitting stitches during the prototyping phase up to the anticipation of the 3D garment allows only the necessary material to be purchased, avoiding waste	Production
Favor fully fashion garments	To avoid fabric waste generated by the cutting process (not necessary in knitwear)	Production
Devise strategies to extend product life	To avoid the end consumer getting rid of the garment before its end of life	Use
Design knitwear that can be machine-washed instead of dry-cleaned	To avoid the use of perchloroethylene, a substance that is highly dangerous to humans and the environment	Use
Create products that can be repaired or reconditioned	To move the moment of the end of life away in time	Use; End of life
Increase the emotional connection between object and consumer	To encourage less fast and more lasting purchases. For example, by studying a customization service on the garment itself	Use
Defining practices that allow the reuse of products once they have reached the end of their life cycle	To foster a textile recycling culture	Use; End of life
Design low impact/reusable packaging	To minimize the release of plastics into the environment and the use of disposable materials	Distribution
Design mono-material articles	It facilitates recycling. Even a small percentage of elastane in addition to the base material or a thread of a different material used for sewing or embroidery can undermine the possibility of proper disposal or recycling of the product	End of life
Make sure that the jumper is easy to disassemble	To facilitate proper recycling	Recycle (End of life)

Table 1 analyzes the actions designers can put in place at different stages of the LCD.

The Impact of LCD on the Stakeholders

Working with such a complex subject matter makes it essential for designers to establish a close dialogue between all the different professional figures they encounter, inside and outside a company: from the product developer to the spinner who

Table 2 Outline of the knitwear design process (Author, 2019)

Activity	Stakeholders
Color and trend research	Spinners / Trend Forecasters + **Designers**
Yarns development and production	Spinners + **Designers**
Yarns trade fairs (Pitti Filati, Première Vision)	Spinners / Knit Factories / **Designers** / Brands / Students and Fashion Schools / Machine Producers / Certificatory Bodies
Research and creative design process	Brands + **Designers**
Design of the collection	Brands + **Designers**
Study of prototypes with design and pattern amendments	Knit Factories + **Designers**
Pre- production, sample development, selling, and consolidation of orders	Knit Factories + Brands
Production and quality control	Knit Factories + Third Contractors
Press, PR, and distribution	Brands / Communication Agencies / PR + **Designers**

produces the yarn, from the technician who works on the knitting machine to make the garment, to the software programmer who controls the coding on the computers. Whether colleagues, suppliers, or customers, the sustainable shift is profoundly changing how all these figures perform their tasks and relate to each other.

Combining the evidence reported in previously published research (Author, 2019) with the findings that emerged with the direct observation and the interviews, this paragraph retraces the steps of the creative process of knitwear design in light of the changes that new sustainable practices are bringing. The creative process, as described in Designing Knit Designers (Motta, 2019), is a succession of phases where designers constantly meet the expertise of the stakeholders involved (Table 2). The following paragraph aims to update the existing outline by understanding how the traditional modus operandi is evolving under new sustainable requirements, introducing new players, generating emerging assets between them, and new opportunities for change.

1. Color and Trend Research on Yarns

The knitwear design process does not begin with the idea inside the fashion brands headquarters but some steps earlier, with the research on trends and colors made by spinning mills to design their yarn collections (Motta, 2019). Now that sustainability is recognized as a widespread requirement, the trend research that used to anticipate the development of yarns has partially lost its centrality as the main driver for renovation in favor of more concrete actions that today guide a large part of the new collections toward the development of sustainable yarns.

Designers interact with:

Spinners
Trend forecasters

New interactions/influences:

Sustainability consultants

2. Yarns Development and Production

Spinners are in the spotlight as the ones who process the raw matter, namely one of the most critical steps of the process due to chemical release, soil consumption, harmful dyes, and unethical animal treatment. If traditionally, innovation for yarns was intended in color, texture, or technical performances, today designers and brands increasingly demand "sustainable" yarns that almost always coincide with "certified" yarns.

To gain certifications, spinners need to trace the entire process from the farm or the field to the finished yarn, monitoring all the steps with 100% transparency. This means they must be aware of the animal welfare or the impact of growing vegetable fibers, the resources consumed, and the substances released during transportation, washing, dyeing, twisting, and finishing. The effort is huge if we think that only the biggest companies manage to keep the entire process inside. In contrast, smaller ones rely on external suppliers and third contractors, making the monitoring activity intricate and expensive.

The amount of information and actions that sustainability requires confirms the importance for designers to be aware of materials and their features, to make knowledgeable choices by knowing the meaning, boundaries, and limits of each yarn, and to build a reliable and professional relationship with their suppliers.

Designers interact with:

Spinners

New interactions/influences:

Farmers and fibers growers
Certification bodies
Sustainability consultants
Chemical formulators

3. Yarns Trade Fairs

The developed yarns are presented at yarn trade fairs (e.g., Pitti Filati in Florence, Première Vision Yarns in Paris). Fairs are still an essential window for spinners, while recently, the standard formats suffered the pandemic closure and the travel bans, triggering discussion on the traditional appointments that are becoming more fluid, versatile, and open to a broader public.

The interest in sustainability is changing how spinners and knitwear factories present their products, exactly as it is happening with fashion shows and sampling of garment collections. Spinners like Lanificio dell'Olivo and Filpucci offer digital presentations, so that designers can receive virtual samples, visualize yarns, consult and finalize orders remotely. Nevertheless, touch is still considered the primary sense for the insiders to evaluate the feeling of a yarn and to envision the aspect it will get in the shape of a garment; haptic technology is promising even in this direction, as engineers are developing devices that mimic the experience of touch by stimulating areas of the skin with vibration (Tripicchio et al., 2022; Lee & Choi, 2022).

Designers interact with:

Spinners
Knit factories
Students and fashion schools
Brands
Machine producers
Certificatory bodies

New interactions/influences:

Sustainability consultants

4. Research and Creative Design Process
If the more inspirational phase of this stage is almost unchanged, sustainability is changing the more commercial side and the guidelines for the architecture of the collection. The data on previous sales, market trends, and the success of items are evolving as the public sets its preferences on more sustainable choices.

Designers interact with:

Brands (Commercial department and style department)

New interactions/influences:

CSR department
Public

5. Design of the Collection
This phase can be kept inside the brand style departments, demanded to external studios, or entirely delegated to manufacturers who transform sketches into a finished product. In recent years, the highly technical characteristics of knitwear products reduced the operational range of designers and privileged the decisions of more technical figures. Previous research reports the need for specialized knit designers expressed by knitwear factories and product developers. Today this need is even more urgent, as the higher specialized designers are, the more they can take back their role of decision-makers and introduce sustainable decisions to avoid waste of resources, materials, time, and in light of the raised attention to raw materials properties and features. As emerged in many of the qualitative interviews collected in Talks on a sustainable fashion and textile system: a qualitative analysis on stakeholders (Motta, 2022), the work of design and product development is a shared process based on longstanding relations with suppliers, made not by controlling them with audits and inspections but relying on their decades-long efforts in making sustainable choices.

Designers interact with:

Brands (style department)

New interactions/influences:

Spinners
Knitwear factories

6 and 7. Study of Prototypes with Design and Pattern Amendments + Preproduction Samples Development, Selling, and Orders

At this stage, designs are transformed into actual garments through studying pattern and paper patterns, further definition of technical solutions and technologies, and programming garment pieces on computers.

Design decisions (e.g., control of garment construction, feasibility, priority of fully fashioned, or seam-free knitting) involve pattern-makers and machine programmers, who are entrusted with finding the best technical solutions and machine technologies to be used to achieve the desired result, but also can provide virtual anticipation of the final result with the software, facilitating the workflow and avoiding waste. By interacting with these figures, specialized designers can exploit all the most updated tools and techniques to act in the most sustainable way possible, to minimize the impact of a costly phase in terms of human resources, competencies, materials, and time.

Designers interact with:
Knit factories (pattern-makers, programmers and machine operators)
Brands (commercial department)

New interactions/influences:
Updated software experts
Machine producers

8. Production and Quality Control

Given its influence on prototyping and sampling, sustainability also significantly impacts production. Due to economic reasons and the need for specialized laboratories that own diverse typologies of machines, production in knitwear has always been fragmented and subcontracted. Although this organization is still one of the strengths that ensure the high-quality products of Made in Italy, it becomes a criticality when it comes to the sustainable issues of transparency and traceability of the supply chain, as the higher the number of suppliers, the higher—complex and time demanding—is the process of collecting information and data from them. This theme is one of the main levers that triggered the debate around reshoring as a strategy, managing sampling and production back together (e.g., Benetton TV3100 project in 2016) (United Colors of Benetton, 2016), thanks to the most updated knitting machinery. However, some initial findings (Guagnini et al., 2022) suggest that the phenomenon is far from becoming a widespread practice, as it implies high costs, both for relocation of production and production inputs.

Designers interact with:

Knit factories
Brands (commercial department)
Machine producers

9. Press, PR, and Distribution

Commitment to the environment is increasingly important in communication campaigns and marketing operations. It is revolutionizing retail uses, as it requires the systematic control of all distribution stages and is bringing radical innovations in how products are sold to the public.

If these are changes common to the entire fashion system, with respect to knitwear, some advantages must be pointed out. Thanks to its traditional production modalities, knitting facilitates the adoption of personalization, made-to-order, collaborative design with the consumer (see Kniterate[2] project and Unmade[3] project), repairing solutions, and upcycling. All these new practices have sustainable consequences as they reduce waste, avoid leftovers, enforce the emotional bond with the knitted garment, and improve the length and use of products.

Designers interact with:

Brands (communication department, retail department)
Retailers

New interaction/influences:

App developers
Service providers
Final customers

Conclusion

In the course of the chapter, it repeatedly emerges how in a scenario of profound change for the fashion system, it fosters the unique opportunity for designers to play a central role.

When focusing on the knitwear field, the last fifteen years have witnessed a progressive impoverishment of design content in favor of a new centrality of knitwear technicians, due to the worsening of the technical difficulties underlying its textile technology and the scarcity and low effectiveness of design support software (Lo Scocco et al., 2022). Thus, while knitwear designers experience a residual status in the manufacturing phase, they become essential players in the sustainability challenge as their design decisions still present a tight bond with the materials and the production techniques, and thus a mutual relation and impact with the practices of suppliers.

Hence, we have seen how, upstream of the process, the designers and product managers have the power to control much of the product's environmental impact. If they are the first to think about sustainable design, there are fewer aspects to mitigate in the subsequent stages. In this respect, Life Cycle Design could represent a unique opportunity to bring the role of the knitwear designer back to center stage.

[2] https://www.kniterate.com/. Accessed September 14th, 2022.

[3] https://www.unmade.com/. Accessed September 14th, 2022.

It is also interesting to observe how this change can be grounded in—and reinforced by—some of the most typical and unique features of knitwear we have seen along the chapter: the centrality of raw materials, the zero-waste manufacturing techniques, the chance to reuse unraveled yarns, the personalization and made-to-order supported by evolving digital manufacturing tools.

All these elements, which were already present in the most traditional past, are today the elements to be exploited for the future with an LCD approach, which makes the designer's role reemerge as fundamental in the sustainability process of a knitwear product.

Bibliography

Amicarelli, V., Bux, C., Spinelli, M. P., & Lagioia, G. (2022). Life cycle assessment to tackle the take-make-waste paradigm in the textiles production. *Waste Management, 151*, 10–27.

Barthes, R. (2006). *Il senso della Moda. Forme e significati dell'abbigliamento*. Giulio Einaudi Editore.

Ceschin, F., & Gaziulusoy, I. (2016). Evolution of design for sustainability: From product design to design for system innovations and transitions. *Design Studies, 47*, 118–163.

European Commission, Directorate-General for Energy, Directorate-General for Enterprise and Industry. (2014). *Ecodesign your future*: *how eco-design can help the environment by making products smarter*. Available at: https://data.europa.eu/doi/10.2769/38512. Accessed 13th August 2022.

Giannone, D., Vezzoli, C., & Conti, G. M. (2019). Life cycle design (LCD) guidelines for environmentally sustainable clothing care systems: An open and operative tool for designers. In *3rd LeNS world distributed conference* (pp. 854–859).

Guagnini, C., Corti, F., & Benedetti, M. (2022). *Il fashion tornerà di moda*? SACE report. Available at: https://www.sace.it/studi/dettaglio/focus-on-il-fashion-tornera-di-moda. Accessed 10 Sep 2022).

Lee, H. K., & Choi, D. (2022). Can I touch the clothes on the screen? The mental simulation for touch in online fashion shopping. *Journal of Fashion Marketing and Management*, 1–18.

Lo Scocco, G. (2023). Technology epiphany and knitwear design. Critical perspectives and possible applications. In *Meeting of research in music, arts and design* (pp. 209–219). Springer.

Lo Scocco, G., Maria Conti, G., & Motta, M. (2022). The digitalization of knitwear: Towards a redefinition of the conventional design boundaries. In *Intelligent human systems integration (IHSI 2022): Integrating people and intelligent systems* (Vol. 22, pp. 1–10). AHFE INTERNATIONAL.

Motta, M. (2019). *Designing Knit Designers*. FrancoAngeli.

Motta, M. (2022). Talks on a sustainable fashion and textile system: A qualitative analysis on stakeholders. *Modapalavra EPeriòdico*, 15(35), 97–156.

Motta, M., & Conti, G. M. (2020). Focused practices for future changes. *DIID - Disegno Industriale Industrial Design Journal, 72*(20), 70–79. ISSN:1594-8528.

Piontek, F. M., & Müller, M. (2018). Literature reviews: Life cycle assessment in the context of product-service systems and the textile industry. *Procedia CIRP, 69*, 758–763.

Rulli, F. (2022). *Fashionisti consapevoli-Vademecum della moda sostenibile*. Dario Flaccovio Editore.

Simonsen, J., Dvabo, C., Strandvad, S. M., Samson, K., Hertzum, M., & Hansen, O. E. (2014). *Situated design methods*. The MIT Press.

Simon, H. A. (1988). The Science of Design: Creating the Artificial. Design Issues, 4(1/2), 67–82. https://doi.org/10.2307/1511391

Tripicchio, P., D'Avella, S., & Ruffaldi, E. (2022). Real-time numerical simulation for accurate soft tissues modeling during haptic interaction. *Actuators, 11*(1), 17.

United Colors of Benetton. (2016). TV31100. Press release. Available at: https://www.benettongroup.com/it/mediapress/comunicati-statements/tv31100/. Accessed 15 Sep 2022.

Vezzoli, C., García Parra, B., & Kohtala, C. (Eds.). (2021). *Designing sustainability for all: The Design of Sustainable Product-Service Systems Applied to distributed economies*. Lecture Notes in Mechanical Engineering.

Vezzoli, C. (2018). *Design for Environmental Sustainability: Life cycle Design of Products* (2nd ed.). Springer.

Vezzoli, C., Conti, G. M., Macrì, L., & Motta, M. (Eds.). (2022). *Designing sustainable clothing systems. The design for environmentally sustainable textile clothes and its product-service system*. FrancoAngeli.

Rhizomatic Mapping Model: A Framework for Representing Circular Practices Towards Transformative Textile Scenarios

Tamames-Sobrino María

Introduction

The research begins by highlighting insights into systemic changes and transformative areas that the industry is currently experiencing. The condition of the fashion value chain as an agent of reticular reconstruction, capable of redesign and regenerate the actual context (Ellen McArthur Foundation, 2017 and Amed et al., 2020). The fashion industry tends towards a circular systematization which presents the opportunity to reimagine our systems. To do so, businesses need to consider, address and rethink the governance of cultural, social and environmental impacts of their current models and corresponding supply chains (Amed, et al., 2020). The textile supply chain has completely transformed its practices, modifying the notions, parameters, criteria and linear processes. It has given rise to a new genre whose central axis is circularity to ensure prosperity (Lehmann et al., 2018).

The new challenges go through generating resilient regenerative systems capable to integrate those ecological, technological and social impacts. During many decades, sustainability has been occupying a subaltern position of the "great policies" as a beautification tool in actions generally isolated and disjointed. Compliance with corporate due diligence obligations and trusting dialogues are essential in order to configurate active participation, engagement and reconnection (Steensen et al., 2018a, b).

Will then discuss the impact that production hierarchies and their complexity have on traceability and transparency attempts. Organizations have realized how necessary it is to give sustainable strategy a core new value: a catalyst and energizer role. A transformative structure that fertilizes and vivifies new profitable

T.-S. María (✉)
Fashion Department Researcher, UDIT. Universidad de Diseño, Innovación y Tecnología, Madrid, Spain
e-mail: maria.tamames@udit.es

A. Cruchinho, J. C. Neves (eds.), *Sustainability in Fashion: Creativity, Education, and Innovation*, https://doi.org/10.1007/978-3-031-80991-0_20

dimensions. Obsolete and already ineffective models cannot be perpetuated. The research discusses how the conduct criteria have become a core factor to be incorporated extensively into the sustainable culture to ensure maximum value retention, justice, fairness and equality, where ecosystems have a fundamental presence.

Action that broadens the autonomy and decision-making capacity of textile industry players is needed, establishing direct dialogue and agile interactions. Thus, textiles players must address strategic questions in order to give answers to complex processes. This analysis questions how most attempts to raise and spread sustainable awareness occur at the textile-making stage and its design processes. Sustainable is often primarily used by the industry and the brands as a greenwashing marketing tool. In the 2017 report by Global Fashion Agenda (GFA) industry workers identified several barriers to sustainability such as short-term thinking, resistance to change, lack of company resources, undefined roles and lack of skills (Kerr & Landry, 2017). In this context, transparency not only provides a link between workers and customers, but it provides workers and their unions the needed information as it allows for interactions to be settle in an efficient manner (Ditty et al., 2018). Given the findings discussed in Markard et al. (2012), it is imperative to develop transformative scenarios that are system-oriented to effectively guide and inform both policy and business practices concerning the circular transition within the textile supply chain. This approach is essential for enhancing the consistency of understanding regarding circular practices. Accordingly, this chapter identifies the circular dimensions and key enabling actions, examines their systemic relationships, and delineates strategies that promote effective interactions. The conceptual foundation of this study is reinforced by the work of Niinimäki and Moreira (2020), who provide a detailed exploration of sustainable textiles and garments within the context of a circular economy. Our methodology involves a comprehensive review of academic literature, supported by examples from industry practices that demonstrate the application of these concepts. A reflective practice approach is employed throughout this study to deepen the understanding of key circular strategy concepts. This method allows for a nuanced exploration of various definitions and aids in critically outlining the main interactions that drive the evolution of transformative processes within the industry. The chapter begins with an introduction to the concepts of traceability and transparency, crucial for the circular transformation of the textile sector. This is followed by an overview of the dimensions for circular transformation, supported by an analysis of the most recurrent circular concepts observed in the textile industry.

Conceptual Background: Key Concepts

Circularity, Traceability and Transparency Concepts

Circularity A sustainable development approach, perhaps the only one in the long term, is articulated through a circular economy that mimics the closed cycles of nature itself. Circular transitions aimed at creating a new way of designing for a circular future where materials are designed, produced, used and disposed of in radically new ways. The circular model forces you to rethink everything from start to finish. It is a new way of understanding the processes of production and management of resources, a continuous cycle that feeds and regenerates itself.

Circular transitions (London, 2016) was the first global event to bring together academic and industry research related to the fashion industry for a circular economy. The topics explored new perspectives around various areas such as the design of new materials, the implementation of emerging technology and social innovation to the design of systems and tools.

Sustainable textile development can contribute to circularity, as it could potentially boost resources consumption, enhance creative economies, reinforce social interactions and preserve local grassroots heritage. For this, a deep understanding of all the interfering systems and the interconnective flows of their elements is necessary. The textile industry is currently not sustainable. The depletion of natural resources and social behaviour has a great impact on profitability and business opportunities in the short, medium and long term.

The linear system leaves untapped economic opportunities, puts pressure on resources, the natural environment and its ecosystems. Creates significant negative social impacts at local, regional and global scales. The economic value of these externalities is difficult to quantify, although the report Pulse of the fashion industry (2018) estimated that the global benefit to the world economy could be around 160 billion euros (USD 192 billion) in 2030, if the fashion industry faced all these environmental and social consequences of the current status quo (Lehmann et al., 2018).

In this context, some of the values and principles on which this new scenario is based are determining factors, such as cooperation, collaboration, multidisciplinary, empathy, conviviality, interdependence, tolerance, accessibility and transparency.

Traceability Traceability is understood as the set of those pre-established and self-sufficient procedures that allow knowing the history, location and trajectory of a product along the supply chain at any given time, through certain tools (Safety Committee AECOC). An extensive and detailed look around own operations can also help to identify problems and ultimately strengthen the efficiency and integrity of the supply chain.

Traceable system components may include identification devices, operators who generate databases, administrators who audit the system and bodies that certify the system. Internal traceability is based on obtaining data by recording the signs that the product leaves as it moves through the entire chain. However, the external one obtains evidence from other independent agents through different types of certifying bodies or auditors. A sustainable traceability tool can be anything that simplifies sustainability work: checklists, software, networks, environmental assessment methods, eco-labels, legislation, etc. The development of methodologies to transmit these indications in a standardized way among the different agents of the chain is essential.

Transparency It is defined as the ability to make information (e.g. on product specifications, chemical inputs, materials used and production practices) available to all stakeholders of the supply chain, allowing common understanding, accessibility, comparability and clarity (Ellen McArthur Foundation, 2020). Major brands and retailers publicly communicate the most information about their policies, commitments and objectives on human rights and environmental issues and significantly less about the results and outcomes of their efforts. Despite progress made on transparency, stakeholders remain opaque, and legislation is required to hold to account measurement on impacts (Fashion Revolution, 2021). To enhance decision-making capabilities of managers, consumers, and investors, we need to leverage data that yields meaningful transparency regarding sustainability impacts. In this context, traceability is regarded as a key requirement for transparency, as it facilitates the acquisition of crucial information that can later be disclosed with relevant stakeholders (Kumar et al., 2017).

As companies still face the challenge of making sure the data they get is accurate, they are setting higher standards for supply chain transparency and traceability. Pressure to establish supply chain transparency and adopt circular business models will fuel demand and help justify scaling pilot projects to mainstream applications (Business of Fashion, 2022). This shift is essential to navigate the ongoing transformations within the textile industry effectively. Moreover, it underscores the importance of cultivating and fortifying relationships among various stakeholders to advance towards a robust circular supply chain. This strategic alignment not only facilitates sustainability but also strengthens industry resilience and profitability in the long term.

Defining a Sustainable Strategy Towards Traceability and Transparency

Some literature is emerging about the applicability of sustainable strategy to different company areas such as suggestions to use practices that determined the process of improving the traceability and transparency. Similarly, Benson and Ross (1998) found the importance of the incorporation of a culture of ethical behaviour. Implementation of sustainability in the corporate culture determines the character (formal and informal) and its degree of scope concerning two different concepts such as commitment and innovation.

Organizations have realized how necessary it is to give sustainable strategy a core new value: a catalyst and energizer role. A transformative structure that fertilizes and vivifies new profitable dimensions. Obsolete and already ineffective models cannot be perpetuated.

Dimensions for Circular Systematic Transformation

Thinking, Models and Resources

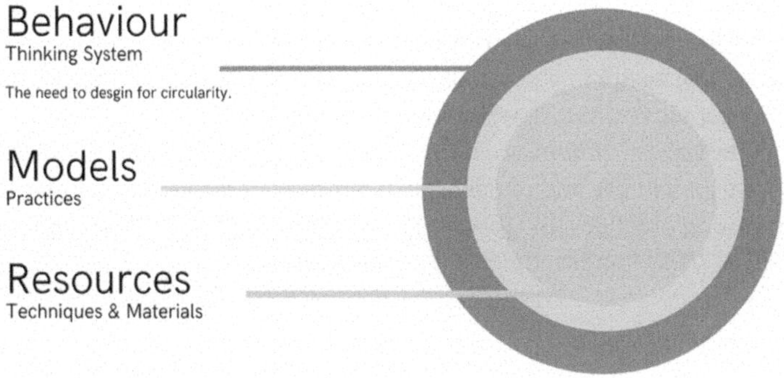

Dimensions for circular design systematic transformation

New practices arise in pursuit of more circularity to reduce specific impacts throughout textile lifecycle stages such as producing, designing, manufacturing, use and end-of life. A classification of initiatives to this approach is proposed in Table 1 grouping textile behaviour, practices and resources. The parameters defined in each dimension help in setting clear actions for achieving these transformations.

This classification indicates a panel of possibilities for diversifying the involvement in circular implementation. Some initiatives are hybridizing different concepts by combining their methods, which broadens the scope of strategies that can be effectively applied. Parameterizing information involves the meticulous organization and structuring of data to ensure it is measurable, analyzable, and comprehensible. Transmitting and disseminating this information in a standardized manner

Table 1 Parameters definition of dimensions for circular transformation. Classification of actions for circular-based approach

Concept	Actions	Objectives
Behaviour/textile knowledge	Shared values and principles	Simple and easy to understand: Common language and codes Aspirational: replicable
	Parametrize, transmit and disseminate information in a standardized way among the different stakeholders of the supply chain	
	Generate effective and efficient training and communicative flows	
Practices/textile models	Generate traceable, transparent, dynamic, flexible, innovative, cooperative, resilient and regenerative models	Applicable and accessible: adaptable to different sizes
Resources/textile techniques and materials	Map and evaluate impact	Focused on the impact: measurable
	Create rigorous and comprehensive techniques	
	Promote the identification of resources impact reduction practices	

across different supply chain agents guarantees consistency and uniformity, critical for maintaining clarity and preventing misunderstandings. Furthermore, creating effective and efficient communicative flows is vital for the accurate delivery of messages and optimal resource utilization. These models should enable easy monitoring, maintain openness and clarity, adapt to changes, encourage creativity and innovation, withstand and recover from disruptions, and restore and renew resources. Additionally, mapping and evaluating the impact involves pinpointing and assessing the effects of various activities on the environment, economy, and society. Creating rigorous and comprehensive techniques entails developing thorough and detailed methods to enhance sustainability and efficiency. By promoting the identification of resource impact reduction practices, the discovery and implementation of strategies that minimize negative effects on resources are encouraged.

Model

This section describes the Rhizomatic Mapping Model (RMM), and their fostering and hindering factors. The model provides a concept for mapping and understanding how systematic changes can grow circularly, giving rise to nodes, working as connective nets. Proving how indefinite growth can foster evolving vigorous and healthy structures. This could assist agents to maximize value across the system and outlines a new scenario in which practices occur. This theoretical model is zooming a framework consisting of points, nets, directions and interactions. The RMM explores the widest economic, social, cultural and environmental contexts.

A significant part of the study is focused on the formulation of impact areas. Seven categories of action have been defined: coding, integration, assignment, training, motivation, evaluation and verification. Within the context of such scenario, consistency visions arise based on data gathered with Design-driven Material Innovation (DDMI) methodology (Tanttu et al., 2016). The various textile strategies were compiled into a framework to highlight concerns about welfare, economic viability and regenerative nets towards circular schemes.

Rhizome Model

From Deleuze and Guattari (1977), we learned to think of the rhizome as a form of knowledge organization that overcomes hierarchical structures. That conception is used as a device to conceive thinking and making sustainable textile culture. The conceptual roots and junctions are linked to a kind of aphorisms that accompany these knots.

The map consists of a tool that can help to discloser circular textile concepts in a holistic narrative by deeply exploring how practices and people could interconnected. The framework designs thinking and action systems and weaves radical interactions (Fig. 1). This approach merges experience and emotional-affective learning. As a constellation, it helps to provide a synchronous understanding from the reality of the macrosystem.

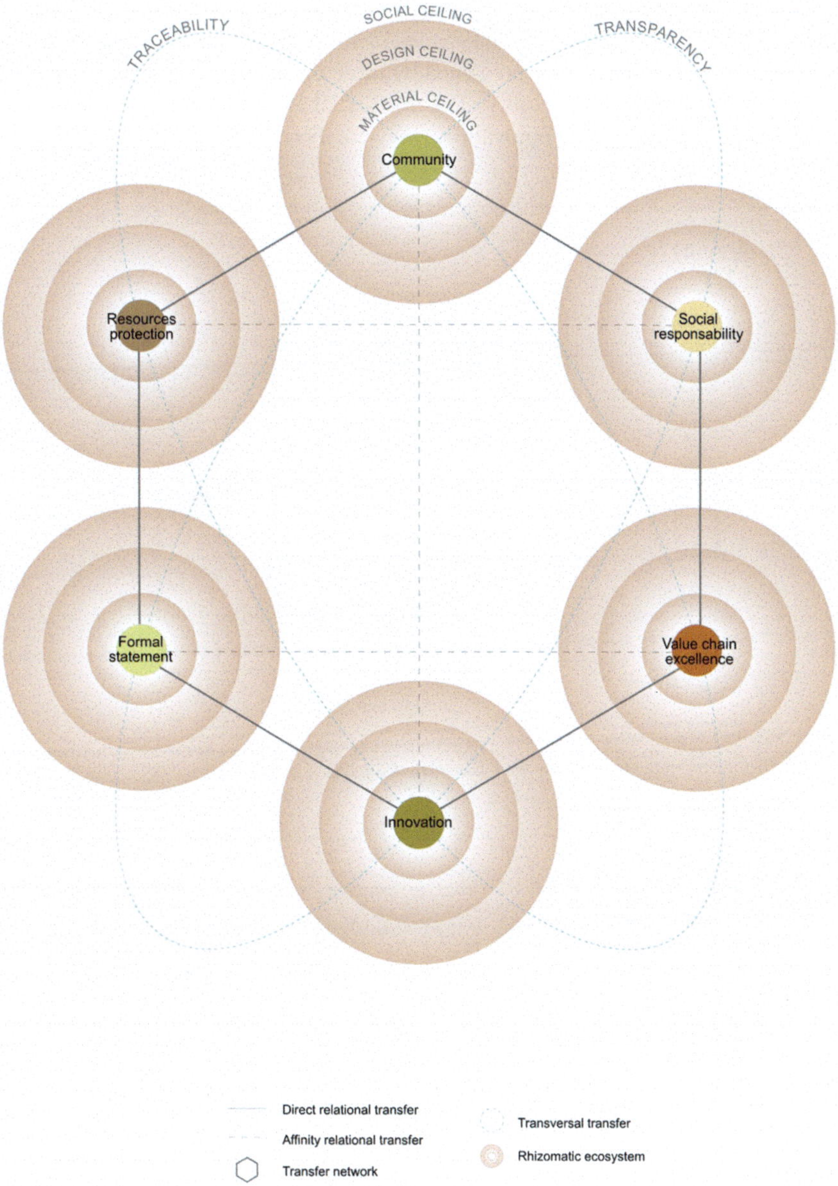

Fig. 1 Rhizomatic Mapping Model (RMM). Framework for representing circular textile design strategy

Transfer Networks and Rhizomatic Ecosystems

The pentagon represents the interdependences between concepts. Any point can and should be connected to another. Under this principle of connection, transfers designate transverse and reciprocal movements. A multiplicity factor occurs and changes in nature increasingly affecting semiotic, material and social flows (Deleuze & Guattari, 2002), cultivating new landscapes of thought and intervention. Material, design and social ceilings are represented by three concentric rings which sets each rhizomatic ecosystem: Community (Fig. 2), Social Responsibility (Fig. 3), Value Chain Excellence (Fig. 4), Innovation (Fig. 5), Formal Statement (Fig. 6) and

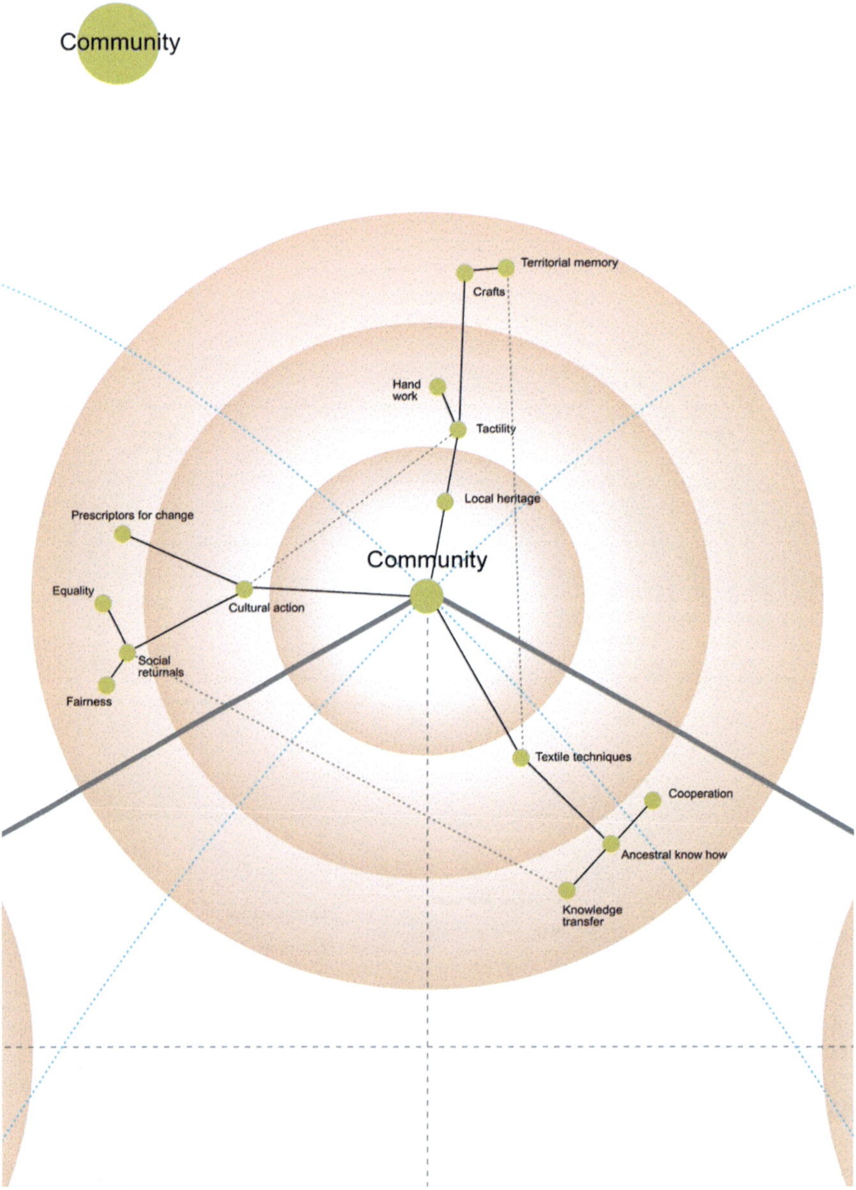

Fig. 2 Rhizomatic Mapping Model (RMM). Community Ecosystem

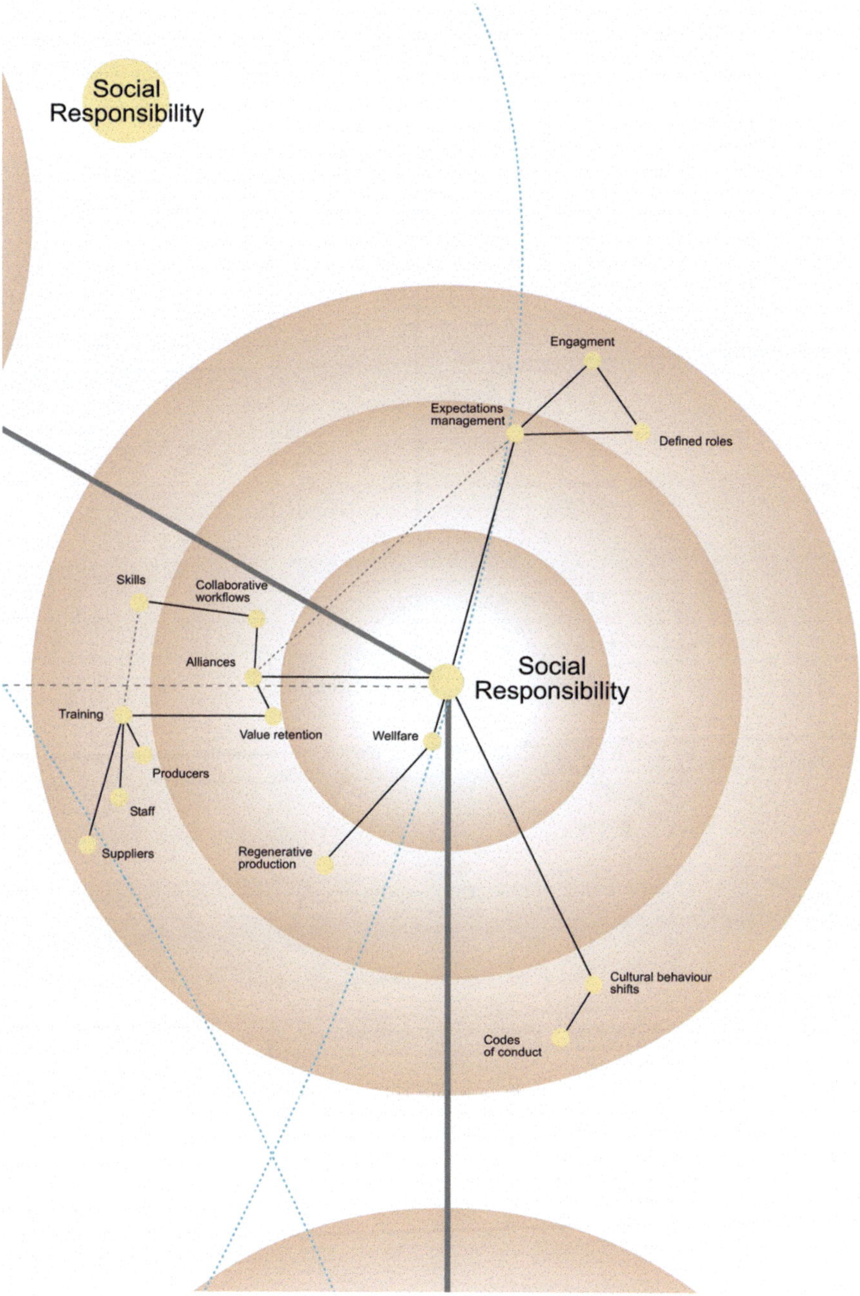

Fig. 3 Rhizomatic Mapping Model (RMM). Social Responsibility Ecosystem

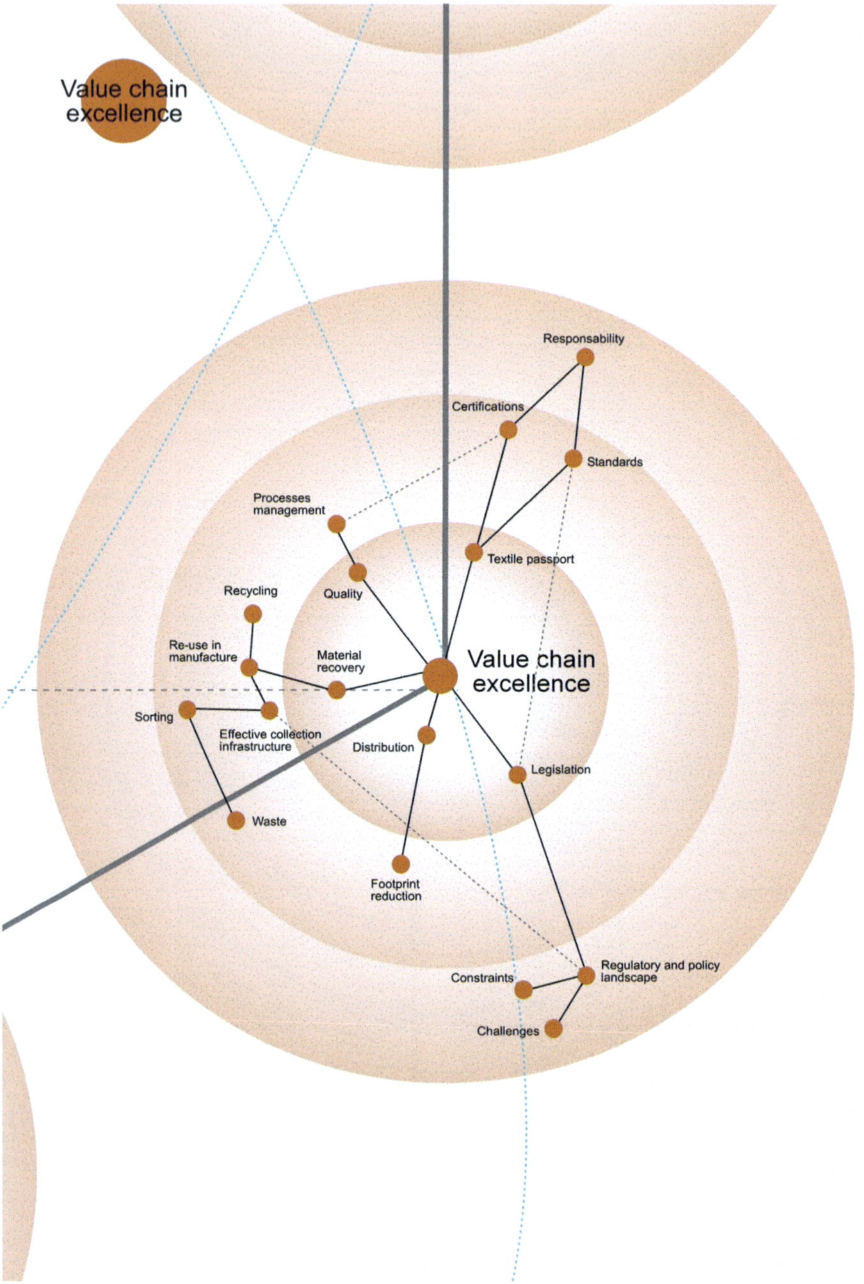

Fig. 4 Rhizomatic Mapping Model (RMM). Value Chain Excellence Ecosystem

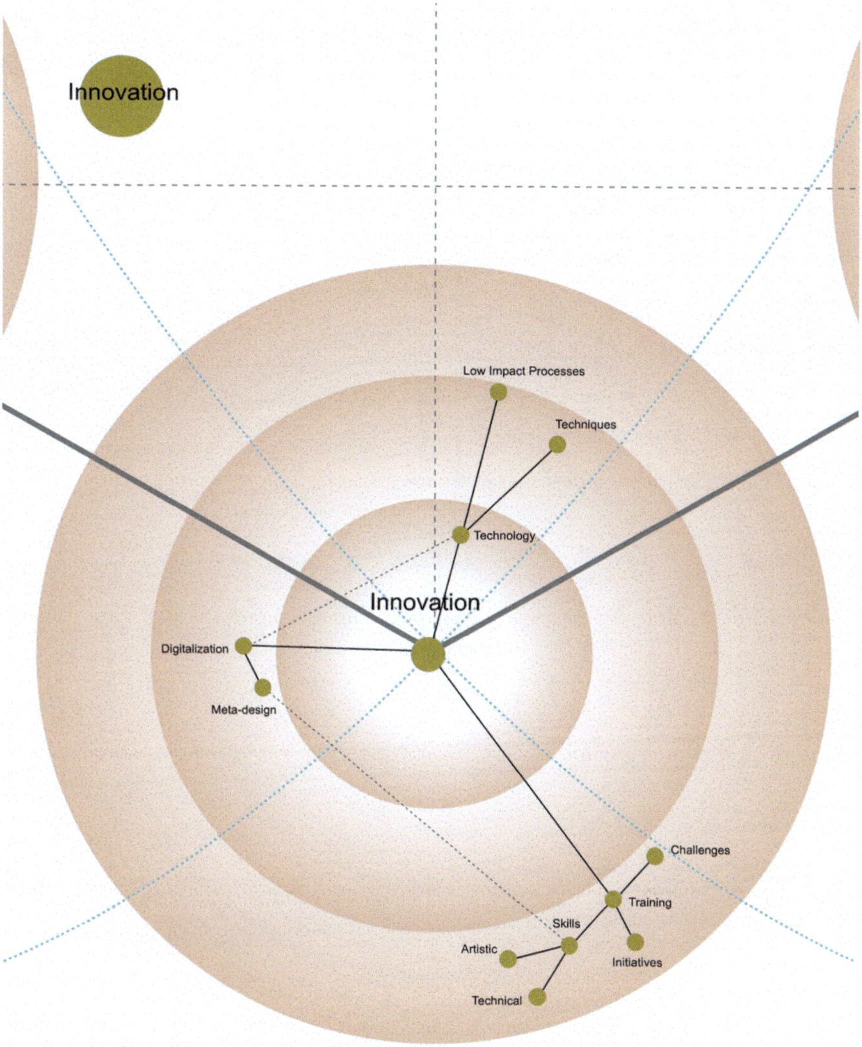

Fig. 5 Rhizomatic Mapping Model (RMM). Innovation Ecosystem

Resources Protection (Fig. 7). Both transparency and traceability are pointing to the dotted area that gathers the variables. This emphasis on transversality suggests the potential cross-sectional effect setting behavioural change streams.

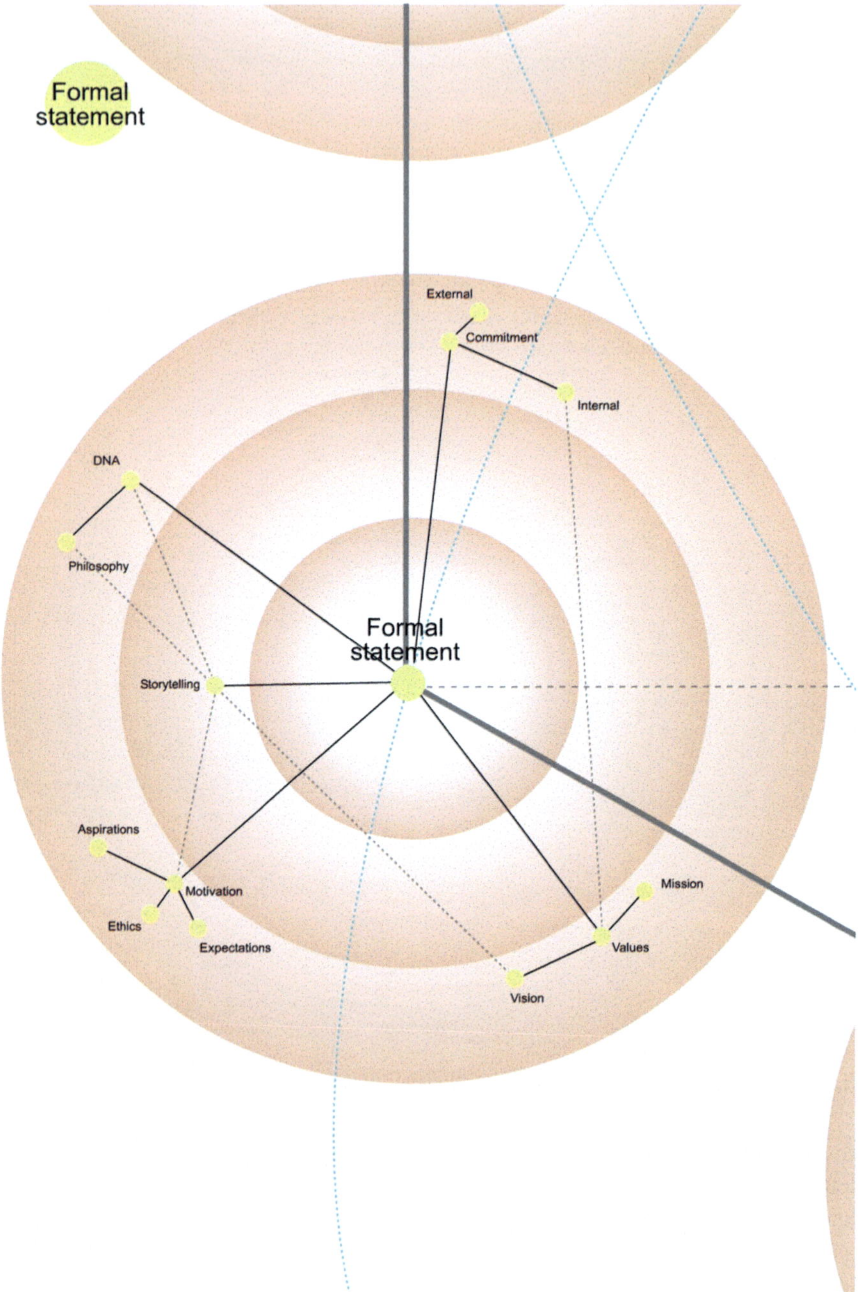

Fig. 6 Rhizomatic Mapping Model (RMM). Formal Statement Ecosystem

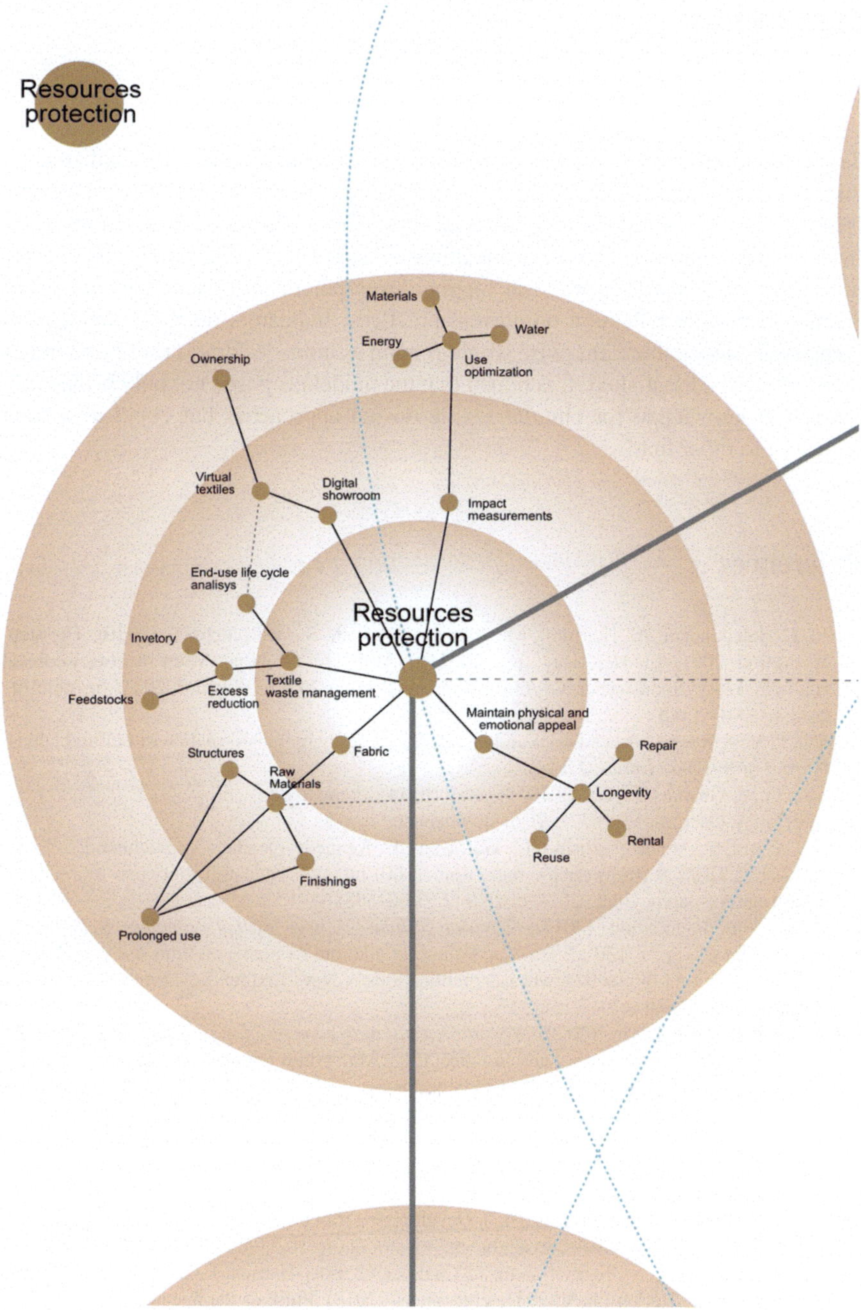

Fig. 7 Rhizomatic Mapping Model (RMM). Resources Protection Ecosystem

Conclusions

In conclusion, the requirement of redefining the meaning and connotations of concepts, systems, processes and materials is an inescapable condition of sustainable development, therefore, it forces us to configure these new codes. This holistic and disruptive reshaping of understanding has the potential to lead us towards a new thriving plot. This research undertakes an exploration which aims to deliver holistic overview that empowers to make meaningful improvements. Overall, this chapter has presented a research proposal for the development of a framework model to promote a critical reflection on current practices. Indicators must be operational, comparable and viable, otherwise would remain as mere declarations of good intentions or wishes dissatisfied. It is hoped that the model proposed not only brings into relief new knowledge for circular textile design approaches but could also have relevance to other fields.

References

Amed, I., Balchandani, A., Beltrami, M., Berg, A., Hedrich, S., & Rölkens, F. (2020). The state of fashion 2021. In *McKinsey* (pp. 10–116) Available at: https://www.mckinsey.com/~/media/mckinsey/industries/retail/our%20insights/state%20of%20fashion/2021/the-state-of-fashion-2021-vf.pdf

Benson, J. A., & Ross, D. L. (1998). Sundstrand: A case study in transformation of cultural ethics. *Journal of Business Ethics, 17*(14).

Deleuze, G., & Guattari, F. (2002). *Mil mesetas* (p. 14). Pretextos.

Deleuze, G., & Guattari, F. (1977). *Rizoma*. Valencia.

Ditty, S., Carry, S., Aidan, S., Ilishio, L., Walravens, F., Kennedy-Fletcher, A., Iracheta, E., & Lai, M. (2018). *Fashion transparency index* (pp. 1–60) https://issuu.com/fashionrevolution/docs/fr_fashiontransparencyindex2018?e=25766662/60458846

Ellen McArthur Foundation. (2017). *Circular fashion—A new textiles economy: Redesigning fashion's future* (pp. 1–150). Ellen MacArthur Foundation. Available at: https://www.ellenmacarthurfoundation.org/assets/downloads/publications/A-New-Textiles-Economy_Full-Report_Updated_1-12-17.pdf

Ellen McArthur Foundation. (2020). *Vision of a circular economy for fashion*. Sustainable by Design. Explorations in Theory and Practice. Ellen MacArthur Foundation.

Fashion Revolution. (2021). Fashion transparency index 2021. In *Fashion revolution* (pp. 6–14) Available at: https://issuu.com/fashionrevolution/docs/fashiontransparencyindex_2021

Kerr, J., & Landry, J. (2017). *Pulse of the fashion industry, 2017*. Global Fashion Agenda. Available at: https://globalfashionagenda.com/wp-content/uploads/2017/05/Pulse-of-the-Fashion-Industry_2017.pdf

Kumar, V., Hallqvist, C., Ekwall, D. (2017). Developing a framework for traceability implementation in the textile supply chain. Available at: https://doi.org/10.3390/systems5020033.

Lehmann, M., Tärnberg, S., Tochtermann, T., Chalmer, C., Eder-Hansen, J., Seara, D. J. F., Boger, S., Hase, C., von Berlepsch, V., & Deichmann, S. (2018). Pulse of the fashion industry 2018. In *Boston consulting group global fashion agenda*. Available at: https://www.globalfashionagenda.com/pulse-of-the-fashion-industry-2018-report-released/

Markard, J., Raven, R., & Truffer, B. (2012). Sustainability transitions: An emerging field of research and its prospects. Research Policy, 41(6), 955–967.

Niinimäki, K., & Moreira, N. (2020). Sustainable Textiles and Garments in the Context of a Circular Economy.

Steensen Nielsen, K., Gwozdz, W. (2018a). Report on geographic differences in acceptance of alternative business models. http://mistrafuturefashion.com/wp-content/uploads/2018/05/Mistra-Future-Fashion-Report-3.1.2.1.pdf.

Steensen Nielsen, K., & Gwozdz, W. (2018b). Report on geographic differences in acceptance of alternative business models. Available at: http://mistrafuturefashion.com/wp-content/uploads/2018/05/Mistra-Future-Fashion-Report-3.1.2.1.pdf.

Tanttu, M., Kohtala, K., & Niinimäki, K. (2016). Can design-driven material innovation also drive circularity? In *Circular transition: A Mistra future fashion conference on textile design and the circular economy 23–24 November 2016*. Chelsea College of Arts & Tate Britain.

Queer Is In: Non-binary Representation in the Portuguese Fashion

Sofia Batista, Graça Guedes, and Daniel Cardoso

Introduction

In recent years, the Western binary notion of gender has been challenged to favour more comprising terms. Non-binarity is one of them. Under the notion of trans identity, this umbrella term encompasses anyone who does not define their gender as either "masculine" or "feminine". Therefore allowing for identities like gender fluid, demi gender, agender, or polygender. Thus, despite these concepts only gaining popularity at the beginning of the 2010s decade (McNabb, 2018), what was once called the "Gender Revolution" by National Geographic (Gutierrez-Morfin, 2016) is now estimated to correspond to approximately 1% of the world population's gender identity (Ipsos, 2021). In the United States (Wilson & Meyer, 2021), those who self-identify outside of the binary are predominantly individuals between 18 and 29, of white ethnicity, living in an urban environment, and with relatively high economic instability.

The expansion of non-normative gender identities, like non-binarity, also has the potential to challenge society's binary views on topics such as clothing. Realising one's identity does not have to be contained in polarised views of one's existence might allow for more creative uses of fashion as an element for self-expression.

S. Batista
2C2T, Minho University, Guimarães, Portugal

G. Guedes (✉)
2C2T, Minho University, Guimarães, Portugal

Lisboa, Portugal
e-mail: mgg@det.uminho.pt

D. Cardoso
CICANT—Lusófona University, Lisbon, Portugal

FCSH—Nova University of Lisbon, Lisbon, Portugal

A. Cruchinho, J. C. Neves (eds.), *Sustainability in Fashion: Creativity, Education, and Innovation*, https://doi.org/10.1007/978-3-031-80991-0_21

Judith Butler's work on gender performativity goes along with this notion. To Butler, gender "is real only to the extent that it is performed" (2010, p. 426), and its institution is made "through the stylisation of the body"(2010, p. 419). Therefore, clothing—a non-negotiable mandatory aspect of everyday life in most situations of contemporary Western urban existence—is essential in perpetuating (or challenging) oppressive gender norms and roles.

Fashion, as an industry, also has a vital role. The decision to incorporate or exclude these issues depends on each brand's corporate side. Nonetheless, in a world where especially younger generations are making their purchasing decisions depending on a company's values (McKinsey & Co, 2018), brands are increasingly starting to feel the need to actively participate in social justice and sustainability (Llamas, 2022). Gender identity and LGBTQI+ rights are no different. However, as a result, cases of tokenism and pinkwashing are also increasing. That is the practice of making only a performative effort to appear inclusive—in this case, by trying to appeal to the LGBTQI+ community. An approach that only tends to continue happening since brands are now tapping into the economic potential of queer individuals (those who are not heterosexual or cisgender) and community allies. "Pink money" is the term used to refer to the purchasing power of these individuals, and estimates rounding the yearly 3.7 trillion dollars are driving many businesses to start catering to these customers' needs (Baron, 2022). Ergo, issues of representation arise.

As consumers' demands for more robust practices of social sustainability increase, themes of diversity, inclusion, and well-being are starting to enter the concerns of multiple fashion brands. A process that is not always seamless since it challenges what has come to be the industry's modus operandi for many years (Elkington, 1997; Gwilt & Rissanen, 2011; von Busch, 2018). Western cultural dominance is present in many aspects of everyday life, and fashion does not escape it (Jansen, 2019). An indisputable and unspeakable ordinance seems to be in place regarding the types of identity accepted, celebrated, and as a result, seen as universal or desirable: white, thin, young, heterosexual, and cis bodies—as was heavily publicised by brands and fashion media outlets (Gaugele & Titton, 2019). Now that the paradigm is shifting, it is interesting to observe if (and if yes, how) brands adapt.

Therefore, this chapter explores how these processes ensue in a Portuguese context—a country whose history is strongly marked by colonialism, fascism, and Judeo-Christianity (Santos, 2018). A series of interviews with non-mainstream Portuguese queer fashion designers explore how they use fashion as a mechanism for self-expression, view recent attempts at fashion representation by larger companies, and perceive the impact of these marketing actions on the community members.

Methodology

Despite considerable efforts in incorporating queer issues into the fashion industry, such measures have only started to enter the Portuguese agenda in the past few years—especially when considering non-binary people. Hence, this chapter's central objective was to analyse the queer, non-normative gender identities' representation in the country's current fashion scene. A qualitative narrative research method was, thus, chosen to explore the viewpoints, given meanings, and experiences of designers and creators that are part of the LGBTQI+ community and have shown an interest in working on issues related to gender normativity and further contributing to defy the status quo.

The selected participants are the faces behind the fashion brands: Amor de la Calle [literally, Love of the Street] (Martim Alvarez), Francisco Félix (eponymous), Ivan Hunga Garcia (eponymous), Mystirgo (Raquel Borges), and Tiago Bessa (eponymous). Living in Portugal grants the five of them a particular understanding of the industry's relationship with matters such as gender nonconformity. Additionally, all of them are young creatives who started their careers less than 10 years ago. As a result, not only are they a part of the generation(s) who place special significance on social and environmental values of fashion production and consumption, but also they are not heavily intertwined with the consolidated ways of an industry that is now facing major transformations.

Martim is a cisgender man designer who uses masculine pronouns (he/him). His work with Amor de la Calle aims to bring queer, non-normative paths to Portuguese

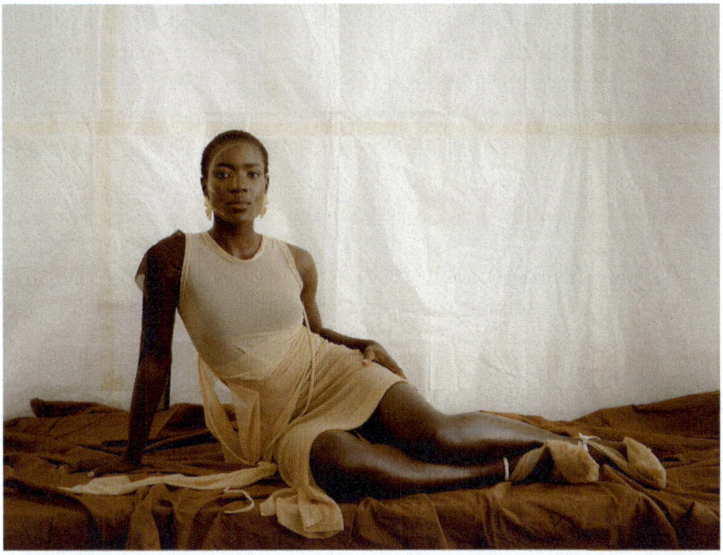

Fig. 1 Amor de la Calle's collection outfit

Fig. 2 Francisco Félix's collection outfit

and Iberian traditions (Fig. 1). The brand inaugurated its website and started sales at the beginning of 2022.

Francisco is a cisgender man designer who, too, uses masculine pronouns (he/him). Since 2015, Francisco Félix, as a brand, has focused on empowering people to express themselves whatever, however, and whenever they want (Fig. 2).

Ivan is non-binary and uses gender-neutral pronouns (such as they/them). Through the Ivan Hunga Garcia brand, they use fashion as an instrument of exploring new possibilities for humanity, no longer constrained by people's categorisation (Fig. 3). Their collections have been shown in Lisbon's Fashion Week since 2022 and are, so far, unavailable for acquisition.

Raquel is a cisgender woman who uses feminine pronouns (she/her). Her brand, Mystirgo, uses patchwork to create and assemble all-unique pieces that look past binary views of gender (Fig. 4). Through the brand's Instagram page (@mystirgo), customers can buy Raquel's creations since October 2021.

Tiago is non-binary and goes by all pronouns (feminine, masculine, and gender-neutral). Their work focuses on gender, mainly acting as an output of personal experiences with the topic. Their latest collection, "Hermaphrodite", reached the international limelight and granted Tiago the winning title of the European competition, ModaPortugal (2021 edition). The pieces are not purchasable (Fig. 5).

Fig. 3 Ivan Hunga Garcia's collection outfit

Fig. 4 Mystirgo's collection outfit

Fig. 5 Tiago Bessa's collection outfit

The research design consisted of a semi-structured interview, carried out digitally (using the video conferencing platform Zoom or in person). The conversations ranged from 16 to 86 minutes—all were recorded and later transcribed for further examination. Additionally, as the interviews were in Portuguese, supplemental translation was necessary for divulgation purposes.

Various questions were developed surrounding four types of relationships (all regarding the designer): their interest and work in the fashion field, their work and the dimension of gender; their perception of the industry's usage of gender; and their audiences. Afterwards, a thematic analysis method of coding the data from the interviews' transcriptions allowed for identifying and analysing other broader groups of themes and topics (Braun & Clarke, 2012).

The following dissemination of findings is done with the use of gender-neutral language. Thus, resorting to pronouns other than those traditionally linked to maleness (he/him) and femaleness (she/her) in order to respect and honour the ways each designer chooses to express their identity (Hord, 2016). Accordingly, this sometimes implies the use of the pronouns they and them, typically used in the plural, to refer to a singular designer—in particular, Ivan Hunga Garcia and Tiago Bessa.

Results

The themes of authenticity, polarisation, and a sense of community were the most prevalent throughout all of the interviews. However, they revealed themselves to be heavily influenced and conditioned by one another, seldom occurring as independent phenomena.

First, it became apparent that all five designers use fashion not only as a career but also as a creative output for self-expression. As Raquel describes it, “fashion is experimenting, fashion is putting our essence into what we are doing”. Likewise, Francisco corroborates, “the collections always revolve very much around what is happening in my life during that period”. A practice that sometimes means entirely dismissing the commercial aspect of fashion creation, as happens with Ivan and Tiago’s work. For Ivan, their “purpose is still the performative context of the show itself. The clothing is the highlight itself and not so much a response to a specific client”, and as for Tiago “my concern is never to sell (…) it was to be able to express myself”. Thus, it might exist solely as a form of art and artistic expression.

Moreover, these views denote a level of representing authenticity (as a product of social practices) that cannot be disassociated from who the designers are as individuals. So it is inevitable that their work also conveys elements of personal values, modes of being, and overall identity. Consequently, as members of the LGBTQI+ community, a bridge is forged surrounding the affinities between them and other members, whose lives might (also) defy the commonly social-imposed norms. As a result, a dual relationship is created. Not only are the specific experiences of a queer person being represented on a public platform, but this process is carried out by someone who—most likely—is aware of the many specificities because they, too, have lived through similar episodes. Francisco explains: “Basically, my personal experiences were the same personal experiences that almost everybody else had, in one way or another, (…); with me working, and exploring my work, based on my social experiences, it is almost inevitable that the community does not see that”. Besides, the possibilities for social representation are accentuated. Martim elucidates how being queer influenced his brand’s decisions, “I do not know, trivial stuff, but enjoying watching RuPaul’s Drag Race, obviously I am going to want to dress up drag queens at some point, or admiring trans people who have immense power and immense strength, obviously I am going to want to see those people wearing my stuff”. Hence, even if it comes from an everyday life aspect, such as watching a famous drag queen’s TV show, it still allows for queer and trans visibility produced by a community member who understands and respects its people.

Overall, social activism—especially regarding human and climate rights—is becoming more prevalent among younger generations due to digital resources like social media helping in public mobilisation (Campos et al., 2016). As a consequence, brands are expected to voice their stances. However, these actions do not always come across as the most genuine. Designers like Tiago quickly point out: “everything I do is about my queerness and all of my sources of inspiration come from queer people, queer activists, the community itself, queer history. I think it is lovely and there is so much to explore, it is a universe in itself, and sometimes it falls a little bit short because people only see the surface of things—like, a pride collection, just with a rainbow printed on it, for god's sake". As a consequence, questions of intent are accordingly raised. Tiago further explains: "I think we as creators have a little bit of a tendency to create—it's not that it has no purpose—but sometimes you create because you have to create", and now, since the onus is on topics such as gender identity, Francisco points out how "every brand there is not a brand that does not put in a press release, in a website description, that the brand serves to empower people or to empower women". Yet, he continues, "but obviously there are brands that we know their purpose five years ago was not that". Therefore, a division is seemingly drawn between brands' motifs for comprising social issues in their practices.

The designers interviewed, both cisgender and non-binary alike, shared their vision of how gender played a central role in their work for a while. They perceived gender as a reflection of who they are as individuals. All of them expressed how, even when dressing for everyday life, they did not make decisions based on the typical binary notions of clothing. Moreover, at some point or another in their lives, they actively went against what was expected by society—in this case, by purchasing items that others assumed would not be for them. Not necessarily to provoke a reaction from others but, as Francisco explains, "I stopped buying most of my clothes in the men's section many years ago; simply because I found it more interesting". Nonetheless, this trouble with gender norms was sometimes met with a reaction, resulting in questions such as "are you Martim male or a Martim female?" or puzzling proclamations like "that shirt has the buttons on the right" (something standardised in "men's" clothing).

Now, with their brands, the idea of gender is used in numerous ways. For Tiago, it is the base for their inspiration, "while my classmates had other stories to tell, I realised that what I really had to talk about was my relationship with gender, which was not always good (…)". Raquel tells us how it was no longer about defining fashion by the gender binary. "Why do I have to buy 'men's' clothes in a specific part of the store when we can simply have a shop with no 'women's' clothes or 'men's' clothes, just clothes? This idea is the origin of the concept of my brand". For Ivan, it is about representing their personal views on non-binarity and ultimately entirely dismissing the need for a gender distinction "because deep down it is something you do not need" more, "it ends up being a natural manifesto, I think. There is not really that much concern for diversity if you are already within it".

Despite Ivan's use of fashion "as a form of expression or clothing, fashion, as an industry, is quite binarised". Some understand the more modern notions within the same industry, while others do not. The problem arises when despite this, they continue to appropriate these topics for personal gain. Martim, for example, argues that some brands do it in an uninspiring way that "sometimes it even sounds a bit… fake". There is the case of fashion shows and how "(…) you do not see many trans people walking the runway, I mean internationally you do, but on a national level it only started happening 2/3 years ago". The Amor de la Calle's designer argues that it might have to do with an attempt to reach certain brands' goals and specific target audiences. "(…) you also see that maybe your audience will welcome these people when they see them in a campaign, on the catwalk, on Instagram, whatever; and I think more brands are targeting trans people, queer people, on catwalks also for that very reason (…). But I think that from this also comes the ambition to put yourself alongside international brands". Ivan directly calls this practice "tokenism", describing it as

> the appropriation of cultural movements that you are not exactly present, but want the appropriation of that audience, and for them to feel represented. These brands end up giving little hints that they are either inclusive or sustainable, only to try to appeal to a certain type of people.

However, in the designer's opinion, "sometimes that shows that you are not into something since you force yourself to prove it too hard".

In general, some of the examples given by the designers express a fashion industry composed of two very different approaches. The brands often marketed to the general public show no translation to the more-corporate side of the business. Ivan illustrates this dichotomy through what happened with the Black Lives Matter movement.

> They started publishing the teams that are behind the brands. There were some that you saw that it was genuine, others that you realised (…) this dichotomy between the image of the brand itself and the communication it adopts depended on the stakeholders themselves.

Those who grant the power positions to make those decisions chose to offer the market brand concepts that represented the new approaches to fashion only for commercial goals.

Regarding gender, despite the moments of queer public representation, such as the catwalks, Francisco alludes to the fact that "at the end of the day you go to the website, and you are going to have the two categories: men and women". Moreover, the same thing happens with physical stores. Ivan brings up how "it ends up being completely unnecessary existing a genderless collection if you have a store with a women's zone, a men's zone, and then a section that is non-binary. Why not just put everything together?" A clear example is the non-binary product line sorted on the Regenbogen brand's website as having separate women's and men's versions (see

Fig. 6 Regenbogen's Non Binary collection sorted by "women" and "men" categories, as highlighted in brand shop online

Fig. 6). For Ivan, this emphasises how "you see that is not genuine. Someone decided they wanted to have a non-binary t-shirt but they did not think about the whole social context that it has, and non-binary is a social movement it is not (…) a connotation, it is not just a note on people's identity". Here, the concept of genuineness is employed regarding Ivan's notions of what is a non-binary gender identity. However, as with any concept that covers many individuals, one must acknowledge that each person will have a different experience and may not adhere to the same definitions or identity concepts.

Ultimately, the conversation falls on the impacts of the fashion industry's utilisation of queer experiences, topics, and aesthetics. While Ivan states, "if your goal is to make products, make products, do not try to be socially active", Francisco argues that "it is fine that everyone wants to make money at the end of the day, that is great, it is just that I think for a brand to have longevity, to really have a big impact (…) there can not be two separate things if you do not align your values to the commercial part (…) I do not think it makes sense, and it exists a lot, unfortunately". For both designers, the issue with using gender and queer topics has to do with non-genuine ways to increase a brand's visibility and profits. As a combative reaction, Francisco decided to change the approach to how he had been addressing these topics. "A little bit to break that we are a trend, that our community is a trend and we have to be constantly using it for commercial purposes, I think that is very unfair and very ungrateful". "Now it is starting to be more normalised, but this is not a trend, it is not something that is going to pass, (…) we have always been here, we have always existed. The community has always existed, and I think people need to start normalising that", Francisco adds.

Fortunately, despite some money-driven attempts at queer representation, Tiago acknowledges a silver lining. They explain that "when I see another collection that is based on Pride, of course, I roll my eyes because there comes a time when you are saturated. But I think the important thing is to remember that it does not stop being important (…). If for us it may be repetitive, for other people it may be as important as it was years ago". Ivan, too, mentions how "at least it is a universal conversation. Even if things like that non-binary t-shirt for men and women come up, it is already something that exists, that hits the mainstream, that then gets inserted into people's vocabulary". This a reminder that when working for an audience, the project often transcends one's initial intention. In this case, that might mean positively impacting the queer community through the increased representation of their realities in the fashion industry—despite original motifs.

Conclusion

This research aimed to break down binary and polarised worldviews. Besides working with non-traditional ideas of gender, the intent was to avoid setting dichotomies between dimensions such as small designers/big corporations and their relations to genuine/non-genuine practices or positive/adverse effects on the LGBTQI+

community. This practice is crucial, especially, when working with queer theory and topics that actively seek to question, challenge and resist essential categorisations of identities or individuals (Barker & Scheele, 2016).

Overall, whether as clothing or as an industry, fashion can shape people's perceptions about what is or is not acceptable. One example of this is trendiness and what is, at a certain point in time, considered to be trendy. Trends have the ability to transform something previously unpopular into something widely known and worn by most people. Thus, this chapter explored the recent concern about incorporating LGBTQI+ issues into Portuguese fashion. Through the views and opinions of five young non-mainstream queer national designers—and by concentrating on their lived experiences—the research aimed to expand the work on the intersection between gender and queer studies and fashion scholarship.

However, despite this chapter's focus on gender-diverse representation, there is still a lack of a racial lens in this work. As queer visibility increases in the fashion world, we must account for how "queer visibility in the mainstream media depends on discourses of whiteness" (Kohnen, 2016, p. 3). Additional research is, therefore, needed to explore how to transpose this movement into the fashion industry regarding race and size, age, and able-bodies.

Finally, this chapter also explored the importance placed on money in capitalist societies. In this case, how the financial power of a subsection of the queer community granted them a front stage where fashion marketing campaigns were concerned, rendering their potential as customers more relevant than the potential backlash from conservative sections of the society while still objectifying that selfsame social group. This reflection might allow for more conscious consumption practices, as one realises their money dictates the values and the type of society where they live. On the other hand, it also allowed for sharing practices where money was not a primary factor; art was. Hence, the perspectives shared by the designers could also promote diverse approaches to work, livelihood, and the importance that professional activities retain in our lives. The way this unfolds will need continued scholarly attention, a more intersectional approach, as well as interdisciplinary work between fashion studies and the field of social sciences.

Acknowledgements This work is financed by Project UID/CTM/00264/2019 of 2C2T—Centro de Ciência e Tecnologia Têxtil, funded by National Funds through FCT/MCTES.

References

Barker, M.-J., & Scheele, J. (2016). *Queer: A graphic history*. Icon Books Ltd.

Baron, K. (2022). Prepping for pride 2022 & beyond: Engaging a booming LGBTQ+ consumer landscape [WWW document]. *Forbes*. URL https://www.forbes.com/sites/katiebaron/2022/04/11/prepping-for-pride-2022%2D%2Dbeyond-engaging-a-booming-lgbtq-consumer-landscape/. Accessed 19 Sept 2022.

Braun, V., & Clarke, V. (2012). Thematic analysis. In *APA handbook of research methods in psychology, Vol 2: Research designs: Quantitative, qualitative, neuropsychological, and biological*

(APA handbooks in psychology®) (pp. 57–71). American Psychological Association. https://doi.org/10.1037/13620-004
Butler, J. (2010). Performative acts and gender constitution: An essay in phenomenology and feminist theory. In *Feminist theory reader: Local and global perspectives* (pp. 419–430). Routledge.
Campos, R., Pereira, I., & Simões, J. A. (2016). Ativismo digital em Portugal: um estudo exploratório. *Sociologia, Problemas e Práticas, 82*, 27–47.
Elkington, J. (1997). *Cannibals with forks: The triple bottom line of 21st century business*. Capstone.
Gaugele, E., & Titton, M. (Eds.). (2019). *Fashion and postcolonial critique*. Sternberg Press; Akademie der bildenden Künste. (Publication series of the Academy of Fine Arts Vienna, volume 22).
Gutierrez-Morfin, N. (2016). National geographic explores gender through a new issue, a documentary. Available via NBC News. https://www.nbcnews.com/feature/nbc-out/national-geographic-explores-gender-through-new-issue-documentary-n695966. Accessed Oct 2020.
Gwilt, A., & Rissanen, T. (2011). *Shaping sustainable fashion: Changing the way we make and use clothes*. Earthscan.
Hord, L. C. R. (2016). Bucking the linguistic binary: Gender neutral language in English, Swedish, French, and German. *Western Papers in Linguistics/Cahiers linguistiques de Western, 3*(1) Available at: https://ojs.lib.uwo.ca/index.php/wpl_clw/article/view/966. Accessed: 26 Nov 2022.
Ipsos. (2021). LGBT+ Pride 2021 Global Survey.
Jansen, A. (2019) Decolonising fashion: Defying the "white man's gaze". *Vestoj: The Platform for Critical Thinking on Fashion*. Available at: http://vestoj.com/decolonialising-fashion/. Accessed: 25 Nov 2022.
Kohnen, M. E. S. (2016). *Queer representation, visibility, and race in American film and television: Screening the closet* (Routledge research in cultural and media studies). Routledge.
Llamas, V. G. (2022). Between transgression and tokenism: The simultaneous inclusion and exclusion of trans and gender non-conforming bodies in fashion. *Fashion Style & Popular Culture*, 1–22. https://doi.org/10.1386/fspc_00128_1
McKinsey & Co. (2018). *'True gen': Generation Z and its implications for companies*. Available at: https://www.mckinsey.com/industries/consumer-packaged-goods/our-insights/true-gen-generation-z-and-its-implications-for-companies. Accessed: 23 Nov 2022.
McNabb, C. (2018). *Nonbinary gender identities: History, culture, resources*. Rowman & Littlefield.
Santos, A. C. (2018). Luta LGBTQ em Portugal: duas décadas de histórias, memórias e resistências. *Revista TransVersos, 14*, 36–51. Available at: https://doi.org/10.12957/transversos.2018.38654
von Busch, O. (2018). *Vital vogue: A biosocial perspective on fashion*. Amazon Digital Services LLC – KDP Print US.
Wilson, B. D. M., & Meyer, I. H. (2021). *Nonbinary LGBTQ adults in the United States*. William Institute.

Sustainability in Fashion Design Courses: Strategies and Tools for Teaching

Bruna Lummertz Lima and Suzana Barreto Martins

Introduction

Undergraduate courses in Fashion Design in Brazil date back to the end of the 1980s (Pires et al., 2008). In their origin, most of these courses appeared as an adaptation of foreign courses, without considering the specificities, often, of the Brazilian identity (de Lima, 2018). Undergraduate Fashion Design courses, slow fashion, a term conceived to designate sustainable fashion, has been approached in the last 12 years to name the development and production processes of products that consider sustainable issues (Fletcher and Williams, 2013).

Today in Brazil, there are approximately 150 undergraduate Fashion Design courses (de Lima, 2018). With the increased debate on sustainability and its noticeable insertion in companies and fashion projects, it is important to reflect on the articulation between the teaching of fashion and sustainability, so that these courses train professionals who are better prepared for the market challenges.

Thus, teaching sustainability in Fashion Design courses has been a major challenge for teachers around the world. As an example, there are several data that place the fashion industry as a polluting source of the environment (Gwilt, 2014), which is a challenge in classroom and teaching practice and requires teachers to constantly rethink from pedagogical practice to sustainability (de Lima, 2018).

Given this scenario, this study proposes to investigate, from publications in the area of teaching sustainability in fashion, strategies that have been used in teaching in the last 10 years, with regard to the incorporation of content related to Fashion

B. L. Lima (✉)
Federal Institute of Santa Catarina, Florianópolis, Brazil
e-mail: bruna.lummertz@ifsc.edu.br

S. B. Martins
State University of Londrina, Londrina, Brazil
e-mail: suzana.barreto@uel.br

A. Cruchinho, J. C. Neves (eds.), *Sustainability in Fashion: Creativity, Education, and Innovation*, https://doi.org/10.1007/978-3-031-80991-0_22

and sustainability, with a focus on undergraduate courses in Design and Fashion Design.

This study seeks to carry out a bibliographic review containing the main strategies, seeking to help researchers and teachers in the area of design and fashion for sustainability to enhance their classroom practices.

Therefore, this work, regarding its development, was organized in four stages: The first evidences the literature review on the subject in question; second one presents the work research process and the collected data; the third one brings the data analysis and the fourth one presents the partial conclusions of the study.

Development

Literature Review

Sustainability has been addressed in several areas of knowledge with greater concern in recent decades. In Fashion Design area, issues related to environmental and social impacts throughout the fashion product project have motivated alternative methodological propositions to linear design (Gwilt, 2014).

With intense challenges today, the fashion industry and their workers need to seek new paths and the professional who works in this market is challenged, given their responsibility to shape the way products and services are built (Moreno et al., 2016). The designer, being identified by this strategic role, must coordinate the entire process of incorporating sustainability and build relationships with many people involved, such as seamstresses, model makers, cutters, assistants, and the company's top management to make the same possible (Gwilt, 2014).

Moreno et al. (2016) highlights the importance of universities working on broader and more integrated concepts of sustainability, from the beginning of the Design and Fashion Design course, evidencing the incorporation of the sustainability tripod to the professional's performance (van der Velden, 2016). According to Gwilt (2014), designer needs to understand sustainability as part of the design process.

In this sense, the inclusion of sustainability in the development of new products encompasses the concept of Design for Sustainability (DfS) which addresses the integration of environmental, social and economic criteria into the product development process, along with elements that serve to minimize adverse impacts on the product. Throughout its life cycle (Crul & Carel Diehl, 2009).

To this end, universities and Fashion Design courses need to dedicate themselves to professional training, so that during the course, they acquire skills, a critical and reflective sense to act in companies in an assertive and strategic way, in terms of sustainability and especially, the design of products and or system-product-service (PSS).

With regard to the professional's performance regarding sustainability, Fletcher and Groose (2011) point out four possible actions for the sustainable fashion designer, aiming at the incorporation of sustainability in fashion:

(a) The designer can act as an educator-communicator, disseminating information to industries and consumers, helping these actors to better understand materials, processes, and sustainability practices.
(b) As an activist, operating in different fields, promoting sustainable fashion and articulating partnerships between different institutions that benefit industries, the community, and the environment.
(c) As a facilitator in which it acts seeking to implement sustainable design strategies in a viable way.
(d) Entrepreneur who designs new business models centered on sustainability, in which success is measured in social, environmental, and economic terms.

Therefore, based on these notes, the next section presents the research process carried out.

Research Process

Research is a rational and systematic process, which aims to provide answers to the problems that are proposed, through the use of scientific procedures. This research is defined as applied, as it aims to generate knowledge for practical application, aimed at solving specific problems and local interests (Marconi & Lakatos, 2010).

With regard to nature, this study will be an exploratory research, since its purpose is to develop and clarify concepts and ideas related to teaching fashion for sustainability, with a view to formulating new concepts, as well as questions that may guide further studies (Marconi & Lakatos, 2010).

A bibliographic survey was carried out between May and June 2022, on the Brazilian platform Periódicos Capes and also on Google Scholar, based on the keywords "teaching," "sustainable fashion," and "fashion teaching."

Based on the findings of "Periódicos Capes," the research was taken to the Google Scholar platform and from the readings of abstracts, the 12 works presented below were selected (Table 1).

The results raised were separated into two columns, called Strategy (procedure) and Tools (instruments used for teaching).

Based on the research carried out, three strategies for teaching sustainability in undergraduate courses in Fashion Design were identified. They are extension projects, research projects, and teaching.

(a) *Extension project:* With regard to extension projects, it was identified in the analysis of the works studied that they happen, at first, through partnerships between educational institutions and cooperatives or organizations. Then, partnerships make room for teaching to be carried out in these environments, favor-

Table 1 Strategies and tools for teaching sustainability

Autores	Estratégia	Ferramentas
Martins et al. (2012)	Extension project	Workshop with students, partner companies and the artisan community Exchange of knowledge Generation of seamless parts Use of textile waste
Fletcher and Williams, 2013	Extension project	Partnerships with organizations and cooperatives Immersion work with students Cooperation Reflexivity
Perez et al. (2015)	Search group	Creation of research group Contributions of design thinking Workshops Teamwork Creation of products and new business models
Williams (2016)	Research project *Shared talent*	People-centered education Practical educational model of sustainability Collaboration
Camargo and Rüthschilling (2016)	Teaching	Partnership with cooperatives Collection project with disused clothes or textile waste Theoretical reflection Seminar Technical visits Use of sustainable design strategies
dos Santos and Perez (2017)	Teaching	Elective subject Digital fabrication Open design Life cycle planning Clothing construction without need for machine sewing (THE POST-COUTURE) Validation with experts

(continued)

Table 1 (continued)

Autores	Estratégia	Ferramentas
de Lima (2018)	Teaching	Design disciplines with a sustainability approach Sustainability discipline at the beginning of the course Modeling and sewing disciplines with sustainability techniques Disciplines of textile technology with study of fibers and processes
	Research/extension project	Research topic in course conclusion papers Research and extension projects
Mazzarella et al. (2019)	Research project	Co-creation workshop Interviews with researchers Case study Theoretical study (understanding of sustainability in a systemic way)
Silveira and da Silva (2019)	Teaching/extension project	Partnership with organizations Co-creation; Theoretical reflection Generation of product alternatives Textile experimentation with waste Oral presentation
Perez et al. (2020)	Systematic analysis of work-teaching sustainability in design courses	Cross-cutting theme Active teaching-learning methodologies
Troiani et al. (2022)	Asystematic analysis of work-teaching sustainability in design courses	Promotion of good practices Technical visits Case studies Speeches Parts development Approaching concepts indirectly in PPCs

ing exchange, collaborative knowledge, and manual knowledge, such as craft techniques and local materials, for example.

(b) *Project and research group:* As a second strategy, projects and research groups appear. In this space, it is possible to highlight the course completion works that adopt the theme of sustainability or even the creation of projects and research groups within universities, often motivated by external demands, arising from companies close to educational institutions.

(c) *Teaching:* As a third strategy, reports of teaching experiences appear, carried out through interdisciplinary projects in curricular units of undergraduate courses in Fashion Design, sustainability disciplines at the beginning of graduation or even the approach to sustainability in a different way, in disciplines of a technical nature, such as pattern making, sewing, and textile technology.

Given the organization of the data carried out, seven tools were created for teaching sustainability in undergraduate courses in Fashion Design, based on the similarities between words (concepts) and repetition, common to at least three studies studied. Accordingly, the categories are presented below:

(a) *Theoretical reflection:* Present in seven of the works consulted, the theoretical study of the theme of sustainability appears as an important tool for teaching in undergraduate courses. According to the studied authors, this tool can appear in the form of seminars, readings, course conclusion works, or even, in the creation of research groups, in which students are participants.
(b) *Co-creation/collaboration:* As the second most cited strategy in the works, the process of co-creation and collaboration appears. Here, sustainability projects are developed by different actors, such as teachers, students, partner organizations, companies, and/or cooperatives. The main idea is that exchanges enable the development of more assertive products.
(c) *Disciplines:* Specially mentioned in Lima's thesis (2018), disciplines focusing on sustainability are an important tool for teaching the subject in undergraduate courses in Fashion Design. As presented, the disciplines can be theoretical, practical, projective, or for textile studies. In all cases, the theme of sustainability may appear occasionally, with the use of one or more sustainable design techniques. It should be noted that in some universities, the theme of sustainability is approached in an interdisciplinary way, especially in project or theoretical disciplines, with the supervision of more professors and with two or more participating disciplines.
(d) *Workshops:* Mentioned in three works studied, workshops appear as a tool for teaching sustainability, especially when partnerships occur with different organizations and are organized within course disciplines or from extension projects. The idea is to encourage the exchange of knowledge and bring University teaching closer to the knowledge of artisans, for example.
(e) *Partnerships with cooperatives/organizations:* Presented in three works consulted, partnerships appear as a resource for teaching sustainability, precisely because they provide students with experience, research in loco and extension in other spaces beyond the university. From this, students can experience techniques and learn about realities, working beyond the environmental dimension of sustainability, by trying to also include the economic and social dimensions.
(f) *Use of textile waste:* Present in three of the works studied, the use of textile waste appears as a tool for the development of fashion products, such as accessories and clothing, through experimentation with these materials. It is interesting to point out that this tool normally appears in studies related to the co-creation strategy and workshops, where professors and universities interact with cooperatives and other organizations.
(g) *Development of seamless garments:* Mentioned in two works, the manufacture of seamless garments appears as a strategy used in environments where sewing machines are not mastered. In some cases, the seamless designed part is devised

using digital fabrication technologies. In others, it is designed from the free experimentation of the material, using creative techniques.

Results Analysis

As mentioned in the data collection, it is possible to notice that the teaching of sustainability already has some strategies and tools that have been used in the last 10 years by teachers in Brazil and also in the United Kingdom.

However, it should be noted that the tools in most cases are used in the field of technique, related to the practice in the development of new products, especially in the construction of garments or fashion accessories. In addition, it should be noted that these techniques used are often related to the environmental dimension of sustainability. As examples it is possible to mention: the use of textile residues for experimentation and product development.

In the social scope of sustainability, for example, it can be observed that the workshops or partnerships with cooperatives carried out by universities, work the context, providing experience and knowledge for students, especially collaborative learning (Martins et al., 2012; Fletcher and Williams, 2013; Silveira & da Silva, 2019).

In contrast, the economic sphere of sustainability is not mentioned in the practical tools used by professors in the studied courses. Thus, a gap for teaching is highlighted here, since it belongs to the triad of sustainability (environmental, social, and economic). Another issue that deserves attention and presents itself as a gap is the approach to teaching with a focus on PSS (system-product-service), since it was not found in the bibliographic survey. Such issues contradict the concept of Design for Sustainability (Crul & Carel Diehl, 2009), and the works of authors Gwilt (2014), Moreno et al. (2016), and Van der Velden (2016) cited in the theoretical foundation of this research.

With regard to theoretical tools (disciplines and research projects), a more transdisciplinary work is perceived, which promotes greater reflection on the sustainability theme, involving analysis of the product's life cycle, study of the environmental, social, and economic spheres, readings, and reflections. However, this tool has appeared in two works consulted and deserves attention from professors who work in higher education courses in fashion (de Lima, 2018; Perez et al., 2015; Fletcher and Williams, 2013).

Conclusion

This work presented a compilation of strategies and tools for teaching Sustainability, used by teachers and documented in scientific articles, published in periodicals and congresses, in the last 10 years. It is important to emphasize that the research that originated this work is in progress and willing to discuss and elucidate the mechanisms that can help this teaching, so necessary and relevant for the training of professionals who will work in the fashion market.

With regard to the findings, it is believed that it was evident that most of the tools are in line with project practice, as well as experimentation using waste or alternative materials. Here, it is understood that there is the creation of scenarios from gaps and that students design something, seeking to solve this initial problem.

On the other hand, two gaps in teaching were identified. The first deals with the theoretical study of the sustainability macro-theme, which appeared in smaller numbers, which stands out as a gap that needs to advance, both in research and in the classroom experience. This question is the basis for several choices during the design of a fashion product or service, since, for the student or professional to be more assertive in favor of sustainability, they need a foundation.

The second points to the absence of tools that work with the system-product-service in fashion, especially with regard to services. In the works consulted, the authors identified that the teaching of sustainability, for the most part, focuses on the development of products, limiting the activity of designing in fashion to products.

It is believed that the findings of this work are a starting point to expand the discussions on this theme and that they open space for new works. As researchers and teachers, the authors wish to instigate concern with the qualification of teaching, the preparation of teachers to work with such a complex theme that is sustainability and, consequently, of students and new professionals for the job market.

References

Camargo, C. W., & Rüthschilling, E. A. (2016). Procedimentos metodológicos para projeto de moda sustentável em ambiente acadêmico. *Modapalavra e-periódico, Florianópolis, 9*(17), 299–312. https://doi.org/10.5965/1982615x09172016299. Disponível em: https://revistas.udesc.br/index.php/modapalavra/article/view/1982615x09172016299. Acesso em: 1 jul. 2022.

Crul, M. R. M., & Carel Diehl, J. C. (2009). *Design for sustainability*. Paris. Disponível em: Acesso em: 5 fev. 2017. https://www.researchgate.net/publication/258278093_Design_for_Sustainability_D4S_Manual_and_Tools_for_Developing_Countries.

de Lima, V. F. T. (2018). *Ensino Superior em Design de Moda no Brasil: Práxis e (In) Sustentabilidade*. São Paulo. Tese (Doutorado) – Faculdade de Arquitetura e Urbanismo Da Universidade De São Paulo. Área De Concentração: Design e Arquitetura, 291p. Disponível em: https://teses.usp.br/teses/disponiveis/16/16134/tde-19122018-154908/pt-br.php. Acesso em: 25 abr. 2022.

dos Santos, A., & Perez, I. U. (2017). Ensino de moda para atuação em novos contextos de design e de produção: sustentabilidade, open design e fabricação digital. *Revista de Ensino em Artes, Moda e Design, Florianópolis, 1*(1), 149–174. https://doi.org/10.5965/2594463012017149. Disponível em: https://www.revistas.udesc.br/index.php/ensinarmode/article/view/10162. Acesso em: 1 jul. 2022.

Fletcher, K., & Groose, L. (2011). *Moda e sustentabilidade: design para mudança* (Tradução de Janaína Marcoantonio). Senac.

Fletcher, K., & Williams, D. (2013, maio 1). Fashion education in sustainability in practice. *Research Journal of Textile and Apparel, 17*(2), 81–88. https://doi.org/10.1108/rjta-17-02-2013-b011.

Gwilt, A. (2014). *Moda sustentável: um guia prático* (1st ed., Tradução de Márcia Logarço). Gustavo Gili.

Marconi, M. A., & Lakatos, E. M. (2010). *Fundamentos da metodologia científica* (7th ed.). Atlas.

Martins, S. B. , Daher, M. Z., & Pinheiro, N. (2012). *Moda, sustentabilidade e inclusão: retraços que tecem histórias*. Resultado de uma Oficina. VIII Colóquio de Moda.

Mazzarella, F., Storey, H., & Williams, D. (2019, abr. 1). Counter-narratives towards sustainability in fashion. Scoping an academic discourse on fashion activism through a case study on the Centre for Sustainable Fashion. *The Design Journal, 22*(1), 821–833. https://doi.org/10.1080/14606925.2019.1595402.

Moreno, M., de los Rios, C., Rowe, Z., & Charnley, F. (2016). A conceptual framework for circular design. *Sustainability, 8*, 1–15. Disponível em: http://www.mdpi.com/2071-1050/8/9/937. Acesso em: 30 dez. 2016

Perez, I. U., Martins, S. B., & de Sampaio, C. P. (2015). Pesquisa científica em moda e sustentabilidade sob a perspectiva do design: relato de experiência. In *Colóquio de moda, 11., 2015, Curitiba. Anais [...]*. Abepem.

Perez, I. U., Moura, M., & da Cruz Landim, P., Martins, S. B. (2020, ago. 18). Abordagem da sustentabilidade no ensino em design no brasil: ênfases e lacunas. *Revista Gestão & Sustentabilidade Ambiental, 9*, 184. https://doi.org/10.19177/rgsa.v9e0i2020184-203

Pires, D. B., et al. (2008). *Design de moda: olhares diversos* (Vol. 423). Estação das Letras e Cores.

Silveira, M. d. S., & da Silva, A. R. (2019). Projeto educativo moda sustentável. In *Congresso nacional universidade, ead e software livre, 1., 2019. Anais [...]*. Universidade Federal de Minas Gerais.

Troiani, L., Sehnem, S., & Carvalho, L. (2022). Moda sustentável: uma análise sob a perspectiva do ensino de boas práticas de sustentabilidade e economia circular. *Cadernos Ebape.Br, 20*(1), 62–76. https://doi.org/10.1590/1679-395120200214

van der Velden, N. M. (2016). *Making fashion sustainable: The role of designers*. Delft University of Technology. Fashion & Textile. Disponível em: http://repository.tudelft.nl/islandora/object/uuid%3A8c66ca0a-605e-4f22-a4f1-f59b7e9ac874. Acesso em: 10 jan. 2017

Williams, D. (2016). Transition to transformation in fashion education for sustainability. In *Engaging stakeholders in education for sustainable development at university level* (pp. 217–232). Springer.

Study of Sustainable Thickeners for the Development of the Marbling Technique in Textiles

Catarina Costa and Joana Cunha

Introduction

Currently, the textile industry is considered one of the most polluting industries globally, and it is urgent to find solutions to minimize this impact. The environmental impact of the textile industry is felt in several areas throughout the product life cycle (Fletcher & Grose, 2012). In addition to waste in production and retail, particularly with the increase in post-consumer textile discards, many of the methods used for product development also have a considerable impact on the environment. In the product development phase, the consumption and pollution of water during the textile dyeing and colouring processes is particularly important, and it is estimated that about 20% of all water pollution is caused by these processes (Nayak et al., 2020). Given this problem, the present work focuses on a sustainable design approach to textile surface design through means of obtaining designs printed on the fabric using unconventional procedures, in this case the Marbling technique.

The Marbling technique is a craft-based technique that has its origins in the ancestral techniques of Suminagashi—a Japanese technique that means "floating ink" with origins in the twelfth century[1]—and Ebru—a Turkish technique applied in the production of painted papers with origins in the thirteenth century[2] and which was widely disseminated in Europe from the seventeenth century, having started to

[1] In: https://suminagashi.com/overview/

[2] In: *Ebru:* Turkish Art of Marbling, (https://aregem.ktb.gov.tr/TR-132387/ebru-turkish-art-of-marbling.html)

C. Costa (✉) · J. Cunha
Universidade do Minho, Centro de Ciência e Tecnologia Têxtil – 2C2T, Braga, Portugal
e-mail: jcunha@det.uminho.pt

A. Cruchinho, J. C. Neves (eds.), *Sustainability in Fashion: Creativity, Education, and Innovation*, https://doi.org/10.1007/978-3-031-80991-0_23

be applied in textiles more commonly with the advent of acrylic paints in the second half of the twentieth century (Carels, 2018).

The technique consists on creating drawings over a thick wet bath, by applying two or more colours that can be manipulated to create marbled effects, similarly to the veins of natural marble. The way of handling the paints on the surface of the bath gives rise to different effects, many of which were "catalogued" by Halfer (1893), thus allowing the creation of unique patterns.

For the inks to float an aqueous solution is used, which is obtained from substances capable of increasing the surface tension called thickeners. The most common recipes for the Marbling bath are based on dissolving Carageen (a natural thickener made from red seaweeds such as *Chondrus crispus*) or Methylcellulose (a thickener derived from cellulose) in water (Dedic, 1997). However, other substances exist that can be applied as thickeners in the Marbling bath and that can be more economical or even more sustainable. For example, Carageen has a high cost and the bath obtained does not remain stable for a long time.

The Marbling technique, although being craft-based, presents itself as an interesting technique from the point of view of sustainable design. Mainly due to the following reasons: Firstly, because when done well, it allows an almost zero waste of ink. Since the ink, when remaining completely on the surface, is transferred in its entirety to the material to be printed. Secondly, since the baths can be reused, depending on the thickeners used, the technique becomes even more interesting from the point of view of the environmental impact of the process. And finally, because it can be applied using large bath containers, thus allowing the creation of unique fabric pieces. However, this is a field that has not yet been fully explored.

Thus, as part of a more in-depth study on the feasibility of applying the technique in the creation of sustainable textile surfaces for use in clothing and home textiles, the chapter presents the first phase of this work that focuses on the study of thickeners that can be used in the Marbling bath and their influence on the visual effect obtained using acrylic fabric inks and textile pigment. Based on the results of this study, it is intended to select the thickeners with the best results for application with different textile dyes and different fabrics, studying the visual effects obtained and their colour fastness to washing and laundering.

Materials and Methods

The implementation of the Marbling technique is carried out through the synergy between the bath, the inks and the textile support. The present study aims to explore the bath component, to promote a more effective application of the technique on textiles suitable for clothing or home textiles.

One of the key points of this work focuses on the sustainability factor, in this way it is intended to obtain a sustainable practice in all aspects of the project itself. Therefore, it was decided above all to use substances that are not aggressive to the environment and particularly of natural origin. Additionally, it is intended to make the most of the thickeners, being for this important that the thickeners provide an adequate surface tension both to sustain the inks on the surface and to prevent of any type of contamination from occurring in the baths, in order to enable the reuse of the completed bath.

Another crucial factor to consider focuses on the scope of the intended final application, clothing and interior textiles. Bearing this in mind, whatever the thickeners and other products used, it is not at all advisable that they contain components that are harmful to health, to guarantee their use free of any type of danger.

At this stage of the experiment, different thickeners will be studied that demonstrate their ability to create a certain surface tension, allowing the inks to float on the bath, selected based on research in the literature: corn starch; cassava flour; agar-agar; alginate-based printing paste thickener (Matalgin MR-40); biodegradable wallpaper glue and Guar gum. Within these first experiments, different concentration values will be tested for each thickener, in order to understand which concentration is most suitable for each one, taking into account the technique to be explored.

For this study, a plain weave structure 100% cotton white fabric was used, which was pre-treated with potassium alum to allow the inks to adhere to the fabric without staining and thus obtain a result of greater colour fastness. For the experimental tests were used fabric acrylic inks, from five different brands that will be described as M1, M2, M3 M4 and M5, and textile pigment that will be represented as P.

The methodology adopted for each study comprised the following phases: (i) Bath preparation; (ii) Preparation of fabric samples; (iii) Placing the bath in the tub; rest; measure pH; (iv) Placing the inks/pigment on the bath; (v) Handling of inks for drawing formation; (vi) Printing on the textile substrate; let it soak for 30 s.; rinse; air dry; (vii) Heat setting in a mini thermofixation oven at 150 °C for 3 min.

The preparation of the baths was always carried out with distilled water and without the addition of any substance other than the thickener. In the case of baths with corn starch, cassava flour and agar-agar, these were brought to a boil for 2 min with constant mechanical agitation. They were then allowed to cool to room temperature, with rest, before proceeding with their use. The remaining baths were prepared with mechanical agitation at room temperature and then left to rest in order to allow the release of all air bubbles formed during the process. Before proceeding with the tests with acrylic inks and/or pigments, the pH of each bath was measured.

Table 1 presents the inks used in each of the experiments, and in situations where the results were worse, it was understood that there was no benefit in testing the entire range of inks, since the purpose of the study focuses on the ability of the bath to suit to the technique itself, whatever the ink used and not to compare the performance of different brands.

The analysis of the suitability of the bath for the Marbling technique was carried out based on: (i) the observation and analysis of the behaviour of the inks on the bath and (ii) the observation and analysis of the process of transferring the design to the fabric.

In the analysis of the behaviour of the inks the parameters analysed were the following: their ability to suspend in the bath, their diffusion over the bath (dispersion), the type of stain they originate, the ability to manipulate to form drawings and the

Table 1 Tested materials

Thickener	Concentrations	Acrylic inks
Corn starch	10 g/l; 20 g/l; 25 g/l	M1; M2; M3; M5
Cassava flour	10 g/l	M1; M5
Agar-agar	0,5 g/l; 1 g/l; 2 g/l; 4 g/l	M2; M3; M5; P
Alginate-based printing paste thickener	10 g/l	M1; M2; M3, M4; M5
Biodegradable wallpaper glue	10 g/l; 15 g/l; 20 g/l; 25 g/l; 35 g/l; 40 g/l	M2; M3; M4; M5; P
Guar gum	4 g/l	M2; M3; M4; M5; P

Table 2 Criteria for analysing the behaviour of inks in the bath

Parameters analysed	Level 1 (good)	Level 2 (satisfactory)	Level 3 (inadequate)
Ink's ability to suspend in the bath	Inks suspend completely without causing deposition	Most paints suspend well, but there is a slight deposition	When paints in general are not suspended
Ink's ability to disperse on the surface of the bath	Inks disperse without difficulty even when the surface is already heavily coated with ink	Inks have a lower dispersibility, or a decrease in dispersibility when the surface starts to be covered	Inks do not disperse, only a drop remains
Uniformity of the ink stain	Stain is uniform and circular with sharp contours	Stain is circular but less uniform or slightly streaked	Fully streaked stain and merges with surface bath
Manipulation ability to form drawings	Allow the creation of drawings easily and maintain the created drawings	Ink can be manipulated, but drawing effects are not as sharp or there is partial mixing of colours	Colour mixing or ink deposit occurs in the bath
Intensity and opacity of the colours obtained	Intense and opaque colours	Shades with medium intensity and slight opacity	Little or no intense and watery tones

intensity and opacity of the colours obtained. For each of these parameters, three levels of performance were considered: 1—good, 2—satisfactory and 3—inadequate, which are described in Table 2.

Regarding the results after creating the design, the following were analysed: the transfer of the pattern to the fabric; the degree of contamination of the bath; the amount of paint released during rinsing and the result obtained after thermosetting. In a similar way to the one described for the analysis of the behaviour of the inks in the bath, three levels were used: 1—good, 2—satisfactory and 3—inadequate. The criteria used are described in Table 3.

Table 3 Criteria for analysing the design transfer process

Parameters analysed	Level 1 (good)	Level 2 (satisfactory)	Level 3 (inadequate)
Quality of pattern transfer to fabric	The design transfers well, with sharpness and good colour intensity	The design transfers well, but with less colour intensity	The design does not adhere to the fabric and/or the design is too faint
Bath contamination after transfer	The bath has no or only light surface contamination	The bath has some surface contamination and/or inks sink after transfer without mixing into the bath	The bath is completely contaminated The inks mix in the bath and/or the inks sink and mix in the bath
Rinse	Little ink release during rinsing and design stays sharp	Significant ink release during rinsing, but the design remains relatively sharp	Significant or complete loss of colour and the design is faint or almost non-existent
Result after thermofixation	The pattern obtained retains the colours, is sharp and defined	The pattern presents slight changes in colour but remains sharp and defined	Significant change in colours and/or fading of the design

Table 4 Corn starch

Parameters analysed	10 g/l (pH 6)	20 g/l (pH 7)	25 g/l (pH 7)
Ink's ability to suspend in the bath	M2—2	M1—1; M2—1; M3—1; M5—1	M1—1; M3—1
Ink's ability to disperse on the surface of the bath	M2—2	M1—2; M2—1; M3—1; M5—1	M1—3; M3—2
Uniformity of the ink stain	M2—3	M1—2; M2—3; M3—2; M5—2	M1—3; M3—3
Manipulation ability to form drawings	M2—3	M1—1; M2—2; M3—2; M5—2	M1—2; M3—2
Intensity and opacity of the colours obtained	M2—2	M1—1; M2—3; M3—3; M5—3	M1—1; M3—1
Quality of pattern transfer to fabric	M2—3	M1—1; M2—2; M3—3; M5—3	M1—2; M3—2
Bath contamination after transfer	M2—3	M1—3; M2—2; M3—2; M5—1	M1—3; M3—2

Results

Tables 4, 5, 6, 7, 8 and 9 present the results observed in each of the developed baths, organized by type of thickener.

Table 5 Cassava flour

Parameters analysed	10 g/l (pH 6)
Ink's ability to suspend in the bath	M2—1; M5—1
Ink's ability to disperse on the surface of the bath	M2—1; M5—1
Uniformity of the ink stain	M2—1; M5—2
Manipulation ability to form drawings	M2—2; M5—2
Intensity and opacity of the colours obtained	M2—2; M5—2
Quality of pattern transfer to fabric	M2—3; M5—3
Bath contamination after transfer	M2—3; M5—2
Rinse	M2—3; M5—2
Result after thermofixation	M2—1; M5—1

Corn Starch

Baths with three different concentrations of corn starch for food use were tested, namely, 10 g/l, 20 g/l and 25 g/l.

Cassava Flour

The bath carried out with cassava flour for food use, at a concentration of 10 g/l, presented an adequate thickness for the technique.

Agar-Agar

As in the previous cases, a product from the food industry was also used here, and four different concentrations were tested.

Thickener Based on Sodium Alginate

Here, only the concentration of 10 g of Matalgin MR-40 per 1000 ml of distilled water was used, which showed a pH of 6. The experiment in question used only M3 inks and pigment.

Table 6 Agar-agar

Parameters analysed	0.5 g/l (pH 7)	1 g/l (pH 6)	2 g/l (pH 6)	4 g/l (pH 6)
Ink's ability to suspend in the bath	M2—1; M3—1; M4—1	M2—1; M3—1; M5—1	M2—1; M3—1; M5—1	M2—1; M3—1; M5—1
Ink's ability to disperse on the surface of the bath	M2—2; M3—2; M4—2	M2—2; M3—2; M5—3	M2—3; M3—3; M5—3	M2—3; M3—3; M5—3
Uniformity of the ink stain	M2—2; M3—3; M4—3	M2—2; M3—2; M5—3	M2—3; M3—3; M5—3	M2—3; M3—3; M5—3
Manipulation ability to form drawings	M2—3; M3—3; M4—3	M2—3; M3—3; M5—3	M2—3; M3—3; M5—3	M2—3; M3—3; M5—3
Intensity and opacity of the colours obtained	M2—1; M3—3; M4—1	M2—1; M3—1; M5—1	M2—1; M3—1; M5—1	M2—1; M3—1; M5—1
Quality of pattern transfer to fabric	M2—[a]; M3—[a] M4—[a]	M2—3; M3—3; M5—3	M2—[a]; M3—[a]; M5—[a]	M2—3; M3—3; M5—3
Bath contamination after transfer	M2—[a]; M3—[a] M4—[a]	M2—3; M3—3; M5—3	M2—[a]; M3—[a]; M5—[a]	M2—3; M3—3; M5—3
Rinse	M2—[a]; M3—[a] M4—[a]	M2—2; M3—2; M5—2	M2—[a]; M3—[a]; M5—[a]	M2—2; M3—2; M5—2
Result after thermofixation	M2—[a]; M3—[a] M4—[a]	M2—3; M3—3; M5—3	M2—[a]; M3—[a]; M5—[a]	M2—2; M3—2; M5—2

[a]Due to poor results regarding the ability to manipulate the inks to create patterns, the transfer to the fabric step was not carried out

Wallpaper Glue

Since the glue for wallpaper, compared to the other examples, presents a faster preparation methodology, six baths were performed in total.

Guar Gum Powder

Still within the scope of food products, Guar gum powder was used, a thickener characterized by the diversity of applications.

Table 7 Alginate-based thickener

Parameters analysed	10 g/l (pH 6)
Ink's ability to suspend in the bath	M3—1; P—1
Ink's ability to disperse on the surface of the bath	M3—1; P—1
Uniformity of the ink stain	M3—1; P—1
Manipulation ability to form drawings	M3—1; P—1
Intensity and opacity of the colours obtained	M3—1; P—1
Quality of pattern transfer to fabric	M3—1; P—1
Bath contamination after transfer	M3—1; P—1
Rinse	M3—1; P—2
Result after thermofixation	M3—1; P—1

Analysis and Discussion

Corn Starch Corn starch is a natural thickener well known in the food industry from the grain produced by corn and is mainly indicated for thickening various mixtures, such as sauces. Since it is aimed at the mentioned industry, it must be properly prepared to create the necessary consistency taking into account the intended effect, as already mentioned initially.

In general, and after some adjustments to the concentrations experienced, corn starch showed effective results in certain parameters, namely, the suspension of inks on the surface of the bath, in the opacity presented and in the matter of handling. However, in the criterion of transferring the printed pattern to the fabric, it did not show promising results, demonstrating in turn a dull and not very intense tone. Given the example of Fig. 1, it is possible to observe the result mentioned by comparing the two figures.

Cassava Flour As a natural thickener, cassava flour was also used, which is a product also widely used in the food industry from the cassava plant. It can be used for different purposes, mainly to thicken different consistencies of sauces or even desserts.

Similar to corn starch, cassava flour showed unpromising results. In the experiment carried out, the inks show good results in handling and consistency in the bath, however, in one of the most crucial stages of Marbling, which focuses on the aesthetic value of the designs transferred to the fabric, the inks have a less intense tonality, resulting in a watercolour effect like shown in Fig. 2.

Agar-Agar As an alternative medium for the thickener function, agar-agar was also used, another product widely used in the food industry, which has a gelatinous formula derived from seaweeds. This is available in powder form.

Table 8 Wallpaper glue

Parameters analysed	10 g/l (pH 7)	15 g/l (pH 7)	20 g/l (pH 7)	25 g/l (pH 7)	35 g/l (pH 7)	40 g/l (pH 7)
Ink's ability to suspend in the bath	M2—3; M3—3; M5—3	M2—3; M3—2; M5—2	M2—3; M3—2; M5—2	M2—1; M3—1; M4—2; M5—1; P — 1	M2—2; M3—2; M5—2	M2—1; M3—1; M4—1; M5—1
Ink's ability to disperse on the surface of the bath	M2—[a]; M3—3; M5—3	M2—[a]; M3—2; M5—2	M2—[a]; M3—2; M5—2	M2—1; M3—2; M4—3; M5—2; P—1	M2—3; M3—3; M5—2	M2—3; M3—1; M4—3; M5—2
Uniformity of the ink stain	M2—[a]; M3—3; M5—3	M2—[a]; M3—2; M5—2	M2—[a]; M3—2; M5—2	M2—1; M3—2; M4—3; M5—2; P—1	M2—; M3—3; M5—2	M2—3; M3—1; M4—3; M5—2
Manipulation ability to form drawings	M2—[a]; M3—3; M5—3	M2—[a]; M3—3; M5—3	M2—[a]; M3—3; M5—3	M2—1; M3—2; M4—2; M5—2; P—2	M2—; M3—3; M5—1	M2—3; M3—1; M4—3; M5—1
Intensity and opacity of the colours obtained	M2—[a]; M3—3; M5—3	M2—[a]; M3—3; M5—3	M2—[a]; M3—3; M5—3	M2—1; M3—2; M4—1; M5—2; P—1	M2—3; M3—3; M5—1	M2—1; M3—1; M4—1; M5—1
Quality of pattern transfer to fabric	M2—[a]; M3—[a]; M5—[a]	M2—[a]; M3—[a]; M5—[a]	M2—[a]; M3—[a]; M5—[a]	M2—1; M3—1; M4—1; M5—1; P—1	M2—1; M3—1; M5—1	M2—2; M3—1; M4—1; M5—1
Bath contamination after transfer	M2—[a]; M3—[a]; M5—[a]	M2—[a]; M3—[a]; M5—[a]	M2—[a]; M3—[a]; M5—[a]	M2—1; M3—2; M4—2; M5—2; P—2	M2—3; M3—3; M5—3	M2—2; M3—1; M4—1; M5—1
Rinse	M2—[a]; M3—[a]; M5—[a]	M2—[a]; M3—[a]; M5—[a]	M2—[a]; M3—[a]; M5—[a]	M2—2; M3—2; M4—2; M5—2; P—2	M2—2; M3—2; M5—2	M2—3; M3—1; M4—3; M5—2
Result after thermofixation	M2—[a]; M3—[a]; M5—[a]	M2—[a]; M3—[a]; M5—[a]	M2—[a]; M3—[a]; M5—[a]	M2—1; M3—1; M4—1; M5—1; P—1	M2—1; M3—1; M5—1	M2—3; M3—1; M4—3; M5—2

[a]Due to poor results regarding the ability to manipulate inks to create patterns, the transfer to the fabric was not carried out

Table 9 Guar gum

Parameters analysed	4 g/l (pH 5)
Ink's ability to suspend in the bath	M2—3; M3—1; M4—1; M5—1; P—1
Ink's ability to disperse on the surface of the bath	M2—3; M3—1; M4—1; M5—2; P—1
Uniformity of the ink stain	M2—; M3—2; M4—1; M5—1; P—1
Manipulation ability to form drawings	M2—; M3—1; M4—1; M5—1; P—1
Intensity and opacity of the colours obtained	M2—; M3—1; M4—1; M5—1; P—1
Quality of pattern transfer to fabric	M2—; M3—1; M4—1; M5—1; P—1
Bath contamination after transfer	M2—; M3—1; M4—3; M5—1; P—2
Rinse	M2—; M3—2; M4—3; M5—2; P—2
Result after thermofixation	M2— ; M3—2; M4—1; M5—1; P—1

Fig. 1 Result of the inks in the corn starch bath (**a**); result obtained in the transfer to the fabric (**b**)

Regarding the use of agar-agar, it can be concluded that this thickener does not meet the necessary conditions for performing the Marbling technique. The agar-agar produced baths with a more granular consistency, visible on the surface, which had a negative influence on the way the inks dispersed on it. For this reason, the inks did not diffuse in a uniform circular way as intended, making the granulation of the bath stand out. This granulation was observed at all concentrations tested. Regarding the parameter of handling the inks on the surface of the bath, it was found that the inks spread and mix with each other, making it impossible to create drawings (Fig. 3).

Alginate-Based Thickener The thickener based on sodium alginate was one of those that showed the most interesting results for application in the Marbling technique, both in terms of surface tension for the placement of inks, as well as the level of consistency of the designs created and the transfer quality of the inks into the fabric. As soon as the inks come in contact with the bath, they expand in defined circles, allowing the addition of several drops of colour. They also show a very

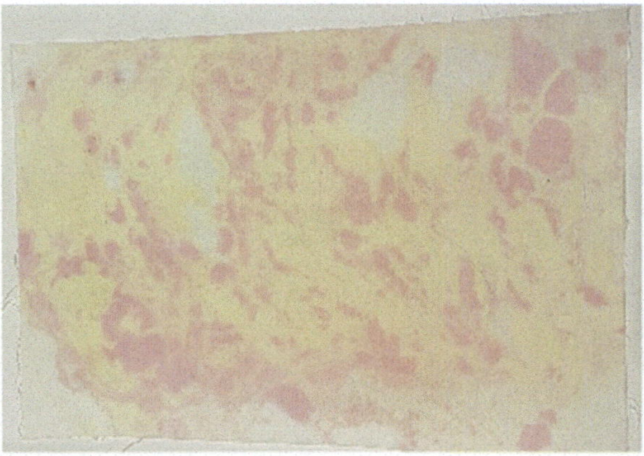

Fig. 2 Result obtained in the transfer to the fabric resulting from the cassava flour bath

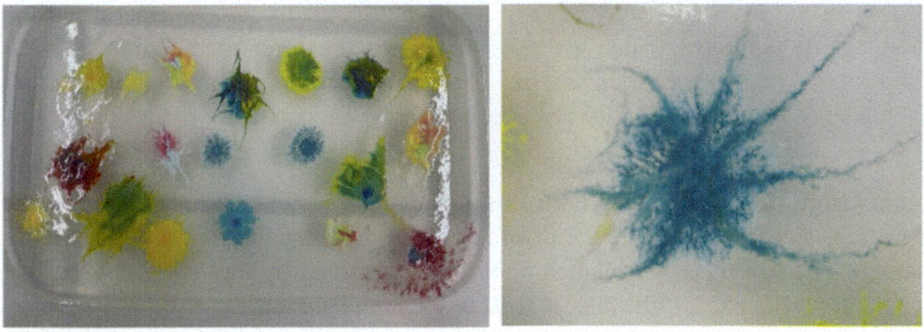

A B

Fig. 3 Result of the inks in the agar-agar bath (**a**); result of the reaction of the inks in the agar-agar bath through a close-up view (**b**)

strong and consistent tone. Manipulation proved to be easy and effective for both acrylic inks (Fig. 4) and textile pigments.

Wallpaper Glue As a thickener, the use of wallpaper glue was experienced, given that some sites on Marbling report that wallpaper glue has good characteristics to be used as a thickener, particularly glues based on Carboxyl-methylcellulose (CMC). In the search for wallpaper glues, some CMC-based glues were found that indicated they were toxic, so it was decided to test a glue that was non-toxic, biodegradable and of plant origin.

Regarding the use of wallpaper glue, it was possible to identify different reactions to the different parameters that need to be evaluated given the Marbling technique. For example, some of the acrylic inks (M2 and M4) did not show good results in certain concentrations of this thickener, while in others they showed more promising results. The M5 inks have a very light tone, almost imperceptible, and the M3 have a good consistency in the various concentrations. The manipulation of the inks

Fig. 4 Result of the paints in the bath with alginate-based thickener

proves to be favourable, with an easy manipulation for the proper creation of the drawings. When transferring the drawings made in the different experiments to the fabric, a good result is verified in terms of tone and definition of the drawings for all the inks used, even in situations where the dispersion or manipulation of the inks was only satisfactory (Fig. 5).

Guar Gum Still within the scope of food products, powdered Guar gum was used, a thickener characterized by the diversity of applications and for being above all natural. The way it is intended to be used is described by the manufacturer, and when properly mixed with water it creates a gelling solution.

Guar gum showed promising results considering the purpose of the technique itself. The way most inks react on the surface of the bath makes the results very promising in the first instance, due to the fact that it is possible to create consistent shapes. The concentration used to create the bath creates the viscosity necessary for the execution of the technique, allowing the inks to be properly manipulated, creating the desired motifs. Regarding the transfer of the designs to the fabric, they are presented in strong tones (Fig. 6), and not all the thickeners tested present this result.

Conclusions

The study carried out made it possible to analyse the suitability of different natural thickeners for use in the Marbling bath, as a technique for the sustainable textile surface design. The requirements for the choice of thickeners were being of natural origin, and not aggressive to the environment or to humans.

The parameters analysed were the ability of the bath to generate the surface tension necessary for the suspension of the inks, the ease of handling the inks for the formation of drawings, the non-formation of ink deposits in the bath, the non-contamination of the bath, the quality of the drawings obtained and transferred to

Fig. 5 Result obtained in the transfer to the fabric resulting from the bath with wallpaper glue

Fig. 6 Result obtained in the transfer to the fabric resulting from the powdered Guar gum bath

the fabric, and the maintenance of the design after rinsing and thermosetting. One of the most important requirements is related to the possibility of reusing the bath in order to reduce the consumption of water and chemicals.

It was found that, depending on the thickener used, certain designs, when transferred to the fabric, do not show the tone they had in the bath. That is, in the bath they show a very intense tone and in the fabric this tone ends up being lost, creating a faded or watercolour effect.

The study also showed that the technique, when carried out with a suitable bath, significantly reduces ink waste, since it is all transferred to the fabric and there is no significant loss in washing. And it also allows the reuse of the same bath to make new designs.

The results obtained allow us to conclude that the thickeners that present the best properties for the purpose in view are: thickener for printing paste based on Sodium Alginate at a concentration of 10 g/l; corn starch at concentrations of 20 g/l and

25 g/l; Guar gum at a concentration of 4 g/l; in the case of wallpaper glue, two baths with a good consistency were achieved, and in the case of this material there is a need to adjust the concentration according to the inks used—at a concentration of 25 g/l, the most suitable inks were M2 and pigments, and at a concentration of 40 g/l, the most suitable inks were M3 and M5.

As a future work, we intend to study the colour fastness to domestic washing and rubbing, using both textile pigments and acrylic inks with the three thickeners that emerge as the most interesting for this technique from the present study. In the sequence, we also intend to study the application of other textile dyes, in addition to pigments, with a view to applying the technique as a practice for a sustainable textile design for clothing and home textiles.

Acknowledgements This work is financed by Project UID/CTM/00264/2021 of 2C2T—Centro de Ciência e Tecnologia Têxtil, funded by National Founds through FCT/MCTES.

References

Carels, C. (2018). Pattern research project: An investigation of the pattern and printing process – Marble. *Pattern Research Project*. [online]. https://scholarscompass.vcu.edu/prp/21/. Accessed 26 Apr 2022.

Dedic, B. (1997). *Marbling magic*. [online] University of Kentucky. http://www2.ca.uky.edu/agcomm/pubs/fcs2/fcs2838/fcs2838.pdf. Accessed 7 Apr 2022.

Fletcher, K., & Grose, L. (2012). *Fashion & sustainability: Design for change*. [online] Google Books. https://books.google.pt/books?id=_PQgEAAAQBAJ&pg=PT325&dq=Fletcher. Accessed 25 Jul 2022.

Halfer, J. (1893). *The progress of the marbling art from technical scientific principles*. LH Kinder.

Nayak, R., Panwar, T., & Nguyen, L. V. T. (2020). Sustainability in fashion and textiles. In *Sustainable technologies for fashion and textiles* (pp. 3–30). https://doi.org/10.1016/b978-0-08-102867-4.00001-3.

Sustainability: Buzzword?

Sónia Seixas, Gianni Montagna, and Maria João Félix

Introduction

In Portugal, of the seven Portuguese bachelor's degree in Textile and Fashion Design, only three have specific curricular units (CUs) on sustainability issues: the bachelor's in *Fashion and Textile Design* of the Higher School of Applied Arts, Castelo Branco Polytechnic Institute (ESART); the bachelor's in *Fashion Design* of the University of Beira Interior (UBI) and the bachelor's in *Fashion and Textile Design and Production* of the Higher Institute Manuel Teixeira Gomes (ISMAT); which leads us to ask what knowledge future textile and fashion designers acquire in the field of sustainability.

The world is constantly changing, with new problems emerging on the cultural, social, economic, environmental, and ethical levels that are becoming increasingly complex. Higher education in textile and fashion design must meet these challenges and provide future industry professionals with a set of skills and ways of thinking (Manzini, 2009; Meyer & Norman, 2020; Swanson, 2020) to enable them to be active members of a circular economy that benefits the planet, society, and human beings.

Textile and fashion designers need to integrate sustainability concerns into their various pillars—environmental, economic, social, cultural, and ethical—during the creative process and inspire a new mindset and attitude throughout the value chain. The decisions made in the selection of materials and production processes significantly determine the climate and environmental impact of textile objects in their life cycle, and it is essential to consider their recycling, reuse, recovery, or disassembly (Ellen MacArthur Foundation, 2017, 2020; Duhoux et al., 2022).

S. Seixas (✉) · G. Montagna · M. J. Félix
CIAUD, Research Centre for Architecture, Urbanism and Design, Lisbon School of Architecture, Universidade de Lisboa, Lisbon, Portugal
e-mail: mfelix@ipca.pt

A. Cruchinho, J. C. Neves (eds.), *Sustainability in Fashion: Creativity, Education, and Innovation*, https://doi.org/10.1007/978-3-031-80991-0_24

The Portuguese textile and clothing industry (TCI) plays an important role in the social and economic Portuguese weave, is one of the oldest and remains one of the largest and most important sectors of the economy due to its wealth and employment creation (DGAE, 2018). In 2021, despite the pandemic COVID-19, its export volume increased by 16% compared to 2019, representing an export value of 5.4 million euros (ATP, 2019, 2022). Despite these positive results, TCI is criticized for the amount of waste produced and for the significant impact on environmental, energy, water, and resource consumption and is in a process of change (Ellen MacArthur Foundation, 2013, 2020; Boström & Micheletti, 2016; UNECE, 2018). The connection with nature is essential to the human being and integrating strategies that promote resource efficiency in a circular way is important in TCI (Berg et al., 2020). These include decarbonization, sustainability, circular economy, and digitalization practices (Ellen MacArthur Foundation, 2017, 2020; ATP, 2021) to restore and regenerate natural systems and production processes, taking into account the entire flow of the production chain, from fiber to disposal (Ellen MacArthur Foundation, 2017, 2020; ATP, 2021), with different players responsible for decision-making in the adoption of circular and sustainable practices. The linear model that drives social, cultural, ethical, and environmental degradation must be redesigned to a circular model that keeps textile products and materials in use while regenerating nature.

The main objective of this exploratory study was to understand how Portuguese higher education in textile and fashion design addresses the issue of sustainability in the preparation and training of undergraduate students. Through a non-interventional investigation and using literature review and data collection, it was possible to observe that the topic of sustainability is only addressed in an embryonic way. It is believed that the main contribution to design with this study was to understand that there are key competencies in the field of society, circular economy, and sustainability for which the textile and fashion designer should be prepared as a way to contribute to the construction of a future and promising textile ecosystem.

As a result, this study aims to draw attention to the need for academia update curricula to understand how issues related to sustainability, circular economy, and society can be pedagogically addressed in textile and fashion courses so that future textile and fashion designers can act as agents of change who, as social innovators (European Commission, 2020a), are equipped with tools that enable them to make informed decisions directed toward a more socially just and developed future (UNESCO, 2017, 2019).

Theoretical Considerations

The Textile and Clothing Industry and (in)Sustainability

In Portugal, TCI accounts for about 10% of national exports, 19% of employment, and 8% of manufacturing turnover (ATP, 2019), with exports increasing by 16% compared to 2019 in 2021, despite the pandemic COVID-19, which corresponds to 5.4 million euros, illustrating the degree of its importance in the social and economic Portuguese weave (ATP, 2022).

It is scientifically proven that the TCI is an industry that harms the environment, whether through the consumption of resources or the amount of waste produced (Boström & Micheletti, 2016; UNECE, 2018; Berg et al., 2020; Ellen MacArthur Foundation, 2020). But not everything is bad, these problems can also be turned into opportunities. The search for strategies that minimize the impact on the environment leads TCI to the need to restructure its business models and production processes, taking into account the spheres of decarbonization, circular economy, sustainability, and digitalization throughout the production chain, from fiber to disposal, with different players being responsible for their choices throughout the process in adopting more sustainable and circular practices (Fletcher & Grose, 2012; Gwilt, 2020; Seixas et al., 2021).

The European Commission (2021a, b) has considered the textile industry as one of the strategic sectors to boost the economy and technological innovation in the industrial strategy for the European Union (EU), and its role in digitalization and decarbonization (European Commission, 2021a) is also relevant. In view of the transition to a more digital, resilient, sustainable, and competitive economy at a global level, business models are sought that take into account their efficiency and the use of resources in a circular way: the development of sustainable production processes, recycled fibers, sustainable textiles, and where design can and should intervene. The role of the textile designer is fundamental to decision-making throughout the textile product development process, as 80% of the environmental impact of textile products is determined in the design phase (European Commission, 2021a, b). TCI continues to have a significant impact on the use of resources, water, energy, and the environment (Ellen MacArthur Foundation, 2013, 2020; UNECE, 2018; ATP, 2021; Duhoux et al., 2022). The linear model that drives social, cultural, ethical, and environmental degradation must be redesigned to a circular model that keeps textile products and materials in use while regenerating nature.

Higher Education in Textile and Fashion Design and (in) Sustainability

Higher education in design must adapt to the ever-changing world, especially in updating curricula to better enable future professionals to fulfill their mission. The designer must be knowledgeable in their field, but that alone is not enough; they must also be empathetic, have collaborative and communication skills, and be able to articulate their knowledge in design with other fields. New cultural, social, ethical, and environmental challenges are emerging, leading to increasingly complex problems. An interconnected complexity of social and human behavior where knowledge of behavioral science and political and social issues is also essential. Stimulating critical thinking in future designers to ensure that a set of social and ethical commitments are fulfilled (Findeli, 2001; Friedman, 2012; Meyer & Norman, 2020; Swanson, 2020).

Designing a textile object requires continuous thinking, it is a complex process in which not only the material but also the sociocultural aspects must be taken into account, placing the design at the center of the socio-ethical dimensions of sustainability (Papanek, 1995; Vezzoli & Manzini, 2008). Textile and fashion designers are located throughout the production chain and should bring a vision of sustainable thinking to their creative and critical thinking process. In this sense, higher education in textile and fashion design must incorporate sustainability in its different pillars—cultural, social, environmental, economic, and ethical—transversally and in different disciplines, in order to provide new learning experiences by acquiring a new vision of knowledge, a new way of thinking to respond more effectively to change. To enable future textile and fashion designers to actively shape change and contribute to significant changes in society (Manzini, 2011; Meyer & Norman, 2020), by acting as social innovators (European Commission, 2020a) and having tools that enable them to make informed decisions that are oriented toward a more socially just and better developed future (UNESCO, 2017, 2019).

It is important to align education systems with the changes in TCI to promote a sustainable textile system that regenerates and repairs natural systems by minimizing waste and pollution, which benefits society, the environment, and the economy (Ellen MacArthur Foundation, 2017, 2020; ATP, 2021; Seixas et al., 2022a, b), because only then can education be part of the solution. Go beyond the traditional approach in teaching and prepare future textile and fashion designers to apply theoretical and practical knowledge to real-world problems through formal and non-formal learning (European Commission, 2020b) in a closer relationship between academia and industry.

The design activity is a collaborative, multidisciplinary, interdisciplinary, and transdisciplinary activity. It is a complex process that involves multiple phases. It is anything but an isolated activity.

Methodology

This exploratory and non-interventionist study aims to understand how Portuguese higher education in textile and fashion design addresses the issue of sustainability in the preparation and training of undergraduate students. The data collection tool was carried out through searches on the official websites of the Directorate General for Higher Education (DGES) and higher education institutions (HEIs). The following keywords were defined: fashion design, textile design, sustainable design, and/or sustainability. Triangulation was used as the main method for the treatment and analysis of the data.

Data Collection and Analysis

Higher Education in Textile and Fashion Design in Portugal: Bachelor's Degree

To understand how sustainability is integrated into Portuguese undergraduate textile and fashion design, we first filtered the available programs by searching the DGES website for the keywords fashion design + textile design. This yielded seven programs, four from public schools and three from private schools (DGES, 2022) (Table 1).

In view of these results, it can be observed that higher education in this field is still quite young in Portugal and that, out of seven HEIs only two integrate the textile field, namely the bachelor's in *Fashion and Textile Design* of ESART from 1999 and the bachelor's in *Fashion and Textile Design and Production* of ISMAT from 2022.

The study plans were analyzed to understand the component that affects the predominant scientific area (CA) of the course, the design CA. It was found that UMINHO is the HEI that includes a higher percentage with 77.78% of the degree program that concerns the design CA; UBI follows with 58.33%; FAUL with 57.22%; ULHT and ISMAT with 53.33%; ESART with 44.44%; and finally ESAD with 25% (ESAD, 2020; ESART, 2020; FAUL, 2020; UMINHO, 2020; UBI, 2021; ISMAT, 2022; ULHT, 2022) (Fig. 1).

The Teaching of the Topic of Sustainability in Portuguese Bachelor's Degree in Textile and Fashion Design

To analyze the issues related to sustainability, we examined the study plans of the bachelor's degrees included in this analysis and found that only three of the seven existing programs have specific CUs related to this topic: CU *Sustainable Fashion*

Table 1 Bachelor's degree in textile and fashion design in Portugal

Acronym	HEIs	Estabelishment	Course	Year of implementation
FAUL	Lisbon School of Architecture, Universidade de Lisboa	Public university	Fashion design	1991
ESART	Higher School of Applied Arts, Castelo Branco polytechnic institute	Public polytechnic	Fashion and textile design	1999
UBI	University of Beira Interior	Public university	Fashion design	2000
ESAD	Escola superior de Artes e design	Private polytechnic	Design (with a specialization in fashion)	2004
UMINHO	Universidade do Minho	Public university	Fashion Design and marketing	2005
ULHT	Lusophone university of humanities and technologies	Private university	Fashion Design and production	2021
ISMAT	Higher institute Manuel Teixeira Gomes	Private university	Fashion and textile Design and production	2022

Author (2022)

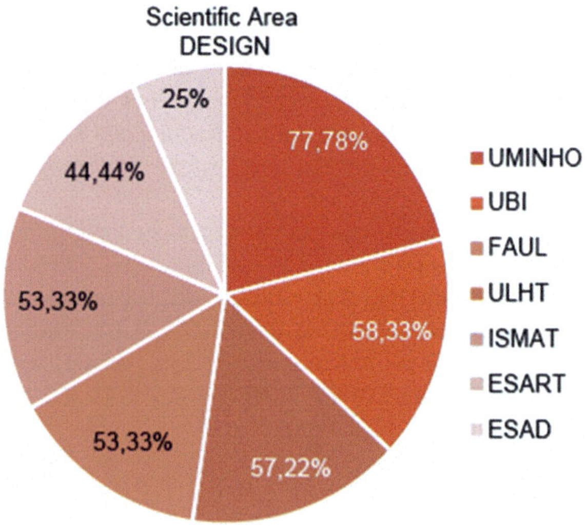

Fig. 1 Percentage of CA design in the study plans of Portuguese bachelor's degree in Textile and Fashion Design. (Author, 2022)

Design of the UBI bachelor's degree in Fashion Design, which includes five European Credit Transfer and Accumulation System (ECTS); *Sustainability in Fashion* of the ESART bachelor's degree in Fashion and Textile Design, which includes four ECTS; and *Sustainable and Circular Fashion* of the ISMAT bachelor's degree in Fashion and Textile Design and Production, which includes four ECTS (ESART, 2020; UBI, 2021; ISMAT, 2022) (Table 2).

Table 2 CUs in the context of the sustainability topic of Portuguese bachelor's degree in textile and fashion design

HEIs	Course	CU	ECTS
UBI	Fashion design	Sustainable fashion design	5
ESART	Fashion and textile designe	Sustainability in fashion	4
ISMAT	Fashion and textile design and production	Sustainable and circular fashion	4

Author (2022)

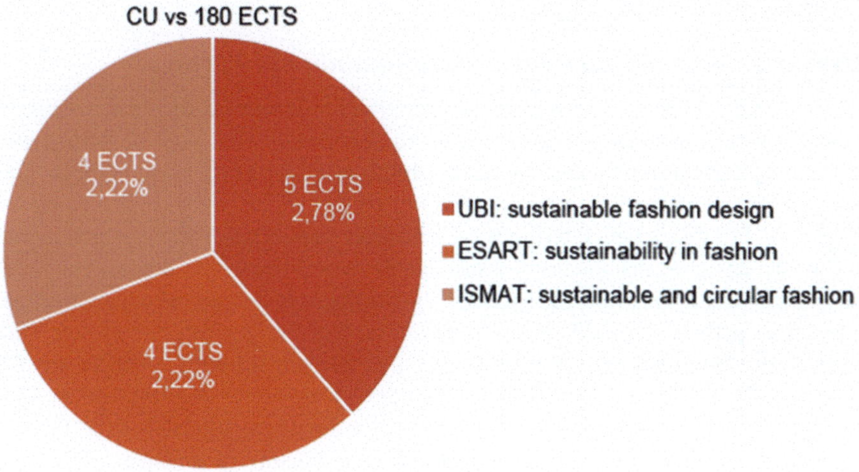

Fig. 2 Percentage relationship between ECTS of the bachelor's degree (180) and CUs in relation to the topic of sustainability at UBI, ESART, and ISMAT. (Author, 2022)

And, in a ratio relationship between the number of ECTS of the bachelor's (180 ECTS) and the weight of the CU concerns the issue of sustainability, the problem is even more frightening that students do not acquire even 3% of knowledge in this area in their academic career (Fig. 2). Is it not important to think and, above all, to act in the face of this scenario?

These findings are concerning because sustainability issues are a twenty-first century concern, and TCI is one of the largest polluters. It is heavily criticized for its less than positive impact on the environment and the amount of waste produced, which is in a process of change (Boström & Micheletti, 2016; UNECE, 2018; Ellen MacArthur Foundation, 2020). In order to protect the planet, society, and people, it is essential to restructure production processes and practices in the development of textile products, taking into account the entire production chain, from fiber to disposal (Fletcher & Grose, 2012; Seixas, 2021; Seixas et al., 2021) to protect the planet, society, and human beings. It is therefore urgent to update the curricula of bachelor's degree in Textile and Fashion Design to integrate sustainability issues in the different pillars—cultural, social, environmental, economic, and ethical. However, this does not mean that these issues will not be addressed in other bachelor's degree programs through optional CUs or teachers who are able to address these issues, but it will always be at the discretion of students and teachers.

Table 3 Academic degrees in design + sustainability that exist in Portugal

HEIs	Degree	Estabelishment	Course	Year of implementation
UTAD	Bachelor's	Public university	Sustainable Design	2022
FAUL FBAUL FCUL ICS ISEG	Master's	Public university	Design for Sustainability	2019

Author (2022)

Higher Education in Design + Sustainability in Portugal

In view of this analysis, it was important to understand what kind of Portuguese academic programs are focused on sustainability in the scientific field of design. For this purpose, a new filtering was carried out on the DGES website, indicating as keywords: For this, a new filtering was carried out on the DGES website, placing as keywords: design + sustainability + sustainable. At the bachelor's degree, out of 58 programs, there is only one, the bachelor's in *Sustainable Design,* which includes an alternative path in the field of *Sustainable Product Design; Interior Design and Sustainable Environments,* from the University of Trás-os-Montes e Alto Douro. At the master's degree, out of 74 programs, there is also only one, the master's in *Design for Sustainability,* lectured in association between five HEIs—the FAUL, the Faculty of Fine Arts of Lisbon (FBAUL), the Faculty of Sciences of Lisbon (FCUL), the Institute of Social Sciences of Lisbon (ICS), and the Higher Institute of Economics and Management of Lisbon (ISEG) (DGES, 2022) (Table 3).

These results show that Portugal is still at a very embryonic stage in terms of teaching design in combination with sustainability in higher education, which in turn raises questions: Is it not important to think and, above all, act in the face of this scenario?

Conclusion

This conclusion begins with the title: is it not, in fact, a *buzzword* sustainability, since, according to the collection and analysis of the data presented, this topic is still very little present in Portuguese higher education in the field of design? Is it not also the mission of academia to train professionals who feel prepared to face the current and future challenges of our planet? Especially in the field of design? The design activity also consists in anticipating problems and seeking solutions that take into account all dimensions of environmental, economic, social, cultural, and ethical sustainability.

The main objective of this chapter was to reflect and understand how Portuguese higher education in textile and fashion design addresses the issue of sustainability in the preparation and training of undergraduate students. From the triangulation of the applied methodological strategies, the literature review and the analysis of the national bachelor's in Textile and Fashion Design, it was possible to observe that the topic of sustainability is only addressed in a rudimentary way. Of the seven existing programs (DGES, 2022), only three integrate CUs that address sustainability: *Sustainable Fashion Design* of the UBI bachelor's in Fashion Design; *Sustainability in Fashion* of the bachelor's in Fashion and Textile Design of ESART; and *Sustainable and Circular Fashion* of the bachelor's in Fashion and Textile Design and Production of ISMAT (ESART, 2020; UBI, 2021; ISMAT, 2022). As for higher education in design + sustainability, there is only a bachelor's in *Sustainable Design* from the University of Trás-os-Montes e Alto Douro, with 58 programs, and a master's in *Design for Sustainability* lectured in association between five HEIs—FAUL, FBAUL, FCUL, ICS, and ISEG—with 74 programs (DGES, 2022), which leads us to consider the importance of thinking and, above all, acting in the face of this scenario.

TCI is in a process of change, looking for strategies that minimize environmental, social, economic, and cultural impacts, with a focus on decarbonization, sustainability, the circular economy, and digitalization (Ellen MacArthur Foundation, 2017, 2020; ATP, 2021). The linear model that drives social, cultural, ethical, and environmental degradation must be redesigned to a circular model. Restructure business models and production processes, including the development of textile products, taking into account the entire production chain, from fiber to disposal (Fletcher & Grose, 2012; Gwilt, 2020; Seixas et al., 2021) to protect the planet, society and human beings.

Higher education in design, in turn, must adapt to the world that is constantly changing, especially in updating curricula so that future professionals can better fulfil their mission. New cultural, social, ethical, and environmental challenges are urgent and lead to increasingly complex problems. Higher education in textile and fashion design should be more aligned with the changes in TCI and incorporate sustainability in its different pillars—cultural, social, environmental, economic, and ethical—in all areas and in different disciplines that contribute to building a promising textile ecosystem for the future. Provide new learning experiences by acquiring a new vision of knowledge, a new way of thinking to respond more effectively to change, and ensure that a range of social and ethical commitments are fulfilled (Findeli, 2001; Friedman, 2012; Meyer & Norman, 2020; Swanson, 2020) by acting as a social innovator with tools that enable them to make informed decisions that are oriented toward a more socially just and better developed future (UNESCO, 2017, 2019).

It is believed that the main contribution to design in this study was to alert to the need for higher education in design to update its curricula more rapidly to be in line with the changes of the planet, society, and human beings that are proving to be increasingly complex and rapid. Public policies should also be alert and enable these updates to be less bureaucratic and time-consuming. Only in this way can HEIs prepare students for the challenges of the present and the future, always keeping in mind the socio-environmental well-being of the human being and the planet.

As a future study, it is intended to continue this research to understand how issues related to sustainability, circular economy, and society can be more expressively considered in higher education in textile and fashion design, so that these future professionals can capitalize the knowledge and face the emerging challenges of the sector, society, and the planet.

Acknowledgments This work is financed by national funds through FCT—Fundação para a Ciência e a Tecnologia, I.P., under the Strategic Project with the reference UIDB/04008/2020 and UIDP/04008/2020, and through the individual research grant 2020.04735.BD.

References

ATP. (2019). *Fashion from Portugal 4.0. Diretório 2019*. ATP – Associação Têxtil e Vestuário de Portugal.
ATP. (2021). *Visão Prospetiva e Estratégias ITV 2030. Contributo para um Plano Estratégico para o Sector Têxtil e Vestuário Português até 2030*. ATP – Associação Têxtil e Vestuário de Portugal.
ATP. (2022). Press Release: exportações de têxteis e vestuário ultrapassarão os 5,3 mil milhoes de euros em 2021, 12 January. https://atp.pt/wp-content/uploads/2022/01/PressRelease-Exportacoes-da-ITV-Novembro21_10.1.22.pdf. Accessed 12 Jan 2022.
Berg, A., et al. (2020). *Fashion on climate: How the fashion industry can urgently act to reduce its greenhouse gas emissions*. McKinsey, Global Fashion Agenda.
Boström, M., & Micheletti, M. (2016). Introducing the sustainability challenge of textiles and clothing. *Journal of Consumer Policy, 39*(4), 367–375. https://doi.org/10.1007/s10603-016-9336-6
DGAE. (2018). *Indústria Têxtil e Vestuário (Sinopse)*. Direção Geral das Atividades Económicas.
DGES. (2022). *Direção Geral de Ensino Superior, Licenciaturas*. Portugal. https://www.dges.gov.pt/pt/pagina/licenciatura. Accessed 7 Aug 2021.
Duhoux, T., et al. (2022). *Textiles and the environment: The role of design in Europes circular economy*. ETC/CE Report 2/2022. European Environment Agency (EEA).
Ellen MacArthur Foundation. (2013). *Towards the circular economy. Economic and business rationale for an accelerated transition*.
Ellen MacArthur Foundation. (2017). *A new textiles economy: Redesigning fashion's future*.
Ellen MacArthur Foundation. (2020). *Vision of a circular economy for fashion*.
ESAD. (2020). *Escola Superior de Artes e Design*. Portugal. https://esad.pt/pt/cursos. Accessed 25 Feb 2020.
ESART. (2020). *Escola Superior de Artes Aplicadas*. Portugal. https://www.ipcb.pt/esart/escola-superior-de-artes-aplicadas. Accessed 25 Feb 2020.
European Commission. (2020a). *Press statement by President von der Leyen on the New European Bauhaus*. https://ec.europa.eu/commission/presscorner/detail/en/STATEMENT_20_1902
European Commission. (2020b). *Skills for Industry. Curriculum Guidelines 4.0: Future-proof education and training for manufacturing in Europe: final report, Office*. European Commission Executive Agency for Small and Medium-sized Enterprises. https://doi.org/10.2826/097323.
European Commission. (2021a). *Annual single market report 2021*. European Commission.
European Commission. (2021b) *Sustainable product policy*. https://ec.europa.eu/jrc/en/research-topic/sustainable-product-policy. Accessed 23 June 2021.
FAUL. (2020). *Faculdade de Arquitetura de Lisboa*. Portugal. https://www.fa.ulisboa.pt/index.php/pt/cursos. Accessed 25 Feb 2020.
Findeli, A. (2001). Rethinking design education for the 21st century: Theoretical, methodological, and ethical discussion. *Design Issues, 17*(1), 5–17. https://doi.org/10.1162/07479360152103796

Fletcher, K., & Grose, L. (2012). *Fashion & sustainability. Design for change*. Laurence King.
Friedman, K. (2012). Models of design: Envisioning a future design education. *Visibile Language, 46*(1–2), 132–153.
Gwilt, A. (2020). *[2014] A practical guide to sustainable fashion* (2nd ed.). Bloomsbury.
ISMAT. (2022). *Instituto Superior Manuel Teixeira Gomes*. https://www.ismat.pt/. Accessed 19 Jul 2022.
Manzini, E. (2009). Viewpoint. New design knowledge. *Design Studies, 30*(1), 4–12. https://doi.org/10.1016/j.destud.2008.10.001
Manzini, E. (2011). *Design schools as agents of (sustainable) change*. In *International symposium CUMULUS//DRS for design* (Vol. 1, pp. 9–16).
Meyer, M. W., & Norman, D. (2020). Changing design education for the 21st century. *She Ji, 6*(1), 13–49. https://doi.org/10.1016/j.sheji.2019.12.002
Papanek, V. (1995). *Green imperative: Ecology and ethics in design and architecture*. Thames & Hudson.
Seixas, S. (2021). The emergence of sustainability and the textile and fashion design education. In *PAD 20. Fashion and textile design reconstruction* (pp. 19–38). Aiap Edizioni.
Seixas, S., Montagna, G., & Félix, M. J. (2021). Materials matters in textile and fashion design education. In C. S. Shin et al. (Eds.), *Advances in industrial design. AHFE 2021. Lecture notes in networks and systems* (pp. 681–688). Springer. https://doi.org/10.1007/978-3-030-80829-7_84
Seixas, S., Montagna, G., & Félix, M. J. (2022a). Where is textile design in Portugal? Interlace the past to weave the future. In D. Raposo et al. (Eds.), *Advances in design, music and arts II. EIMAD 2022 (Springer series in design and innovation)* (pp. 220–237). Springer. https://doi.org/10.1007/978-3-031-09659-4_17
Seixas, S., Montagna, G., & João Felix, M. (2022b). The textile designer skills to face the emerging challenges of textile industry and the role of academia. In G. Montagna & C. Carvalho (Eds.), *Human factors for apparel and textile engineering. AHFE (2022) international conference* (pp. 23–30). AHFE Open Access. https://doi.org/10.54941/ahfe1001534
Swanson, G. (2020). Educating the designer of 2025. *She Ji, 6*(1), 101–105. https://doi.org/10.1016/j.sheji.2020.01.001
UBI. (2021). *Universidade da Beira Interior*. Portugal. http://www.ubi.pt/. Accessed 7 Aug 2021.
ULHT. (2022). *Universidade Lusófona de Humanidades e Tecnologias*. https://www.ulusofona.pt/. Accessed 19 Jul 2022.
UMINHO. (2020). *Universidade do Minho*. Portugal. https://www.uminho.pt/PT/ensino/oferta-educativa. Accessed 25 Feb 2020.
UNECE. (2018). *Traceability of Sustainable Value Chains. Enhancing transparency in the garment and footwear sector for informed and responsible choices*. https://unece.org/trade/traceability-sustainable-garment-and-footwear. Accessed: 5 June 2021.
UNESCO. (2017). *Education for sustainable development goals learning objectives*. United Nations Educational, Scientific and Cultural Organization.
UNESCO (ed.). (2019). *UNESCO, education for sustainable development*. https://en.unesco.org/themes/education-sustainable-development. Accessed 27 Oct 2020.
Vezzoli, C., & Manzini, E. (2008). *Design for environmental sustainability*. Springer.

Environmental and Social Sustainability: Unite Textile and Fashion Design Education with a Social Cause

Alexandra Cruchinho, Liliana Pina, and Paulo Martins

Introduction

Upcycling is the concept that marks the starting point of this study for the development of an exploratory research applied in a classroom context. The main objectives of this study are, in a first phase, to allow students a contact with different textile materials and to know how to identify their typologies. In a next phase, and being this the main objective, to transform used into new Fashion Design Products.

There is also a solidarity aspect, as the students are encouraged to create pieces, blankets/quilts that will be delivered to humanitarian institutions. In fact, it can be observed that the Portuguese textile and clothing industry continues to increase its production levels and to place on the market a very high number of textile products, sometimes with a reduced life cycle. The question is, however, what becomes of these garments or fabrics when they reach the end of their life cycle. Can they be recycled, destroyed, dumped, sent to other countries or can they simply be reused to create new products?

The research methodology is a mixed methodology, non-interventionist and interventionist, and went through a first bibliographic collection of frameworks of the concepts of upcycling and sustainability. The development of the project in a classroom context with a class group marks the interventionist phase of the methodology. Throughout the development of the project, the students' learning was reinforced in terms of knowledge of materials, techniques and, more related to the main theme of the project, social sustainability and circularity.

A. Cruchinho (✉)
Universidade Lusófona – Centro Universitário de Lisboa – CICANT, Lisbon, Portugal
e-mail: alexandra.cruchinho@ulusofona.pt

L. Pina · P. Martins
ISMAT – Instituto Superior Manuel Teixeira Gomes, Portimão, Portugal
e-mail: liliana.pina@ismat.pt; paulo.martins@ismat.pt

A. Cruchinho, J. C. Neves (eds.), *Sustainability in Fashion: Creativity, Education, and Innovation*, https://doi.org/10.1007/978-3-031-80991-0_25

Sustainability Theoretical Framework

The word "sustainability" has a broad meaning, and it is a general concept that can be attached to several areas and fields of study. According to Saricam and Okur (2019, p. 4), "it can be created only if it is adopted and improved by all the people, the companies and the institutes serving the people, the countries, the governments and the globe as a whole". The Sustainable Development Report mentioned 17 different sustainable goals that United European countries must implement or must improve. Independently of the countries different traditions and cultures, governmental and private companies, associations, citizens and different communities must be committed to change and make efforts in the fields of Quality Education, Sustainable Cities and Communities, Responsible Consumption and Product and finally Climate Action. Even though Portugal it is in the 26th-rank position with a score of 76.4%, challenges remain in the expected goals mentioned above (e.g. Quality Education and Sustainable Cities and Communities). Specifically on Responsible Consumption and Product and Climate Action Goals, the report concluded that in 2019 significant challenges were not yet properly focused (Sachs et al., 2019).

The truth it is that sustainable development is just possible with the production and the usage of sustainable products by balancing the ecological, economic and social dimensions (Saricam & Okur, 2019). This matter is quite important once that global fashion industry produces about 53 million tonnes of fibre every year (Amed et al., 2020). The term "sustainable fashion" stands for clothing that is designed for an extended lifetime use, produced in ethical production systems using resources and processes that are not dangerous to environment and the employees (Saricam & Okur, 2019). However, in terms of textile and products waste, more than 70% of that ends up in landfills or gets incinerated. To intensify the issue, less than one percent is reused to make new clothes (Amed et al., 2020).

Nowadays, sustainable practices in the fashion field considers the application eco-labelled or recycled materials employed in different fashion products (Saricam & Okur, 2019). But the purpose of sustainability philosophy goes beyond its practical application. Is about create "a strong and nurturing relationship between consumer and producer [...] It is about designing confidence and capability inducing pieces that encourage versatility, inventiveness, personalization and individual participation" (Fletcher, 2008, p. 125). In the last decade, these practices have been cherished by the fashion industry specialists, fashion designers, brand marketeers and also within the most important pawn in this business: the consumers (Saricam & Okur, 2019). By implementing sustainable marketing policies, fashion companies and brands stand up for environmental and social good causes. By reducing the consumption of natural resources and pursuing sustainable development and standards, fashion brands sustain and improve the relationship with their consumer (Pero et al., 2020). In fact, ethical practices such as offering recycling service and recyclable products in stores are suggested to increase fashion consumers' perceived effectiveness of environmental protection (Saricam & Okur, 2019).

Furthermore, it was found that understanding the ethical dimensions of the product enables the consumers to purchase the product and even increase the willingness to pay more for the product (Saricam & Okur, 2019).

In the scope of the textile and fashion products, Fletcher (2008) explains that to reduce waste is possible to establish waste management strategies, namely reduce, recycle and reuse. Upcycling methods are based into cradle-to-cradle, a philosophy developed by Braungart and McDonough (Aakko & Koskennurmi-Sivonen, 2013). Regarding its beginnings, upcycling is understood as the act of repurposing items and until the nineteenth century most items were used till they no longer could (Singh et al., 2019). The concept as upcycling is original from the time of the Second World War, when scarcity of materials was patent (Mantyi-ncube & Juba, 2020). The growing trends in upcycling clothes are becoming an interesting niche regarding counter-culture activities providing an alternative pathway for used clothes. (Ericsson & Brooks, 2015, p. 92). It reflects an environmental concern with the over-consumption of new clothing.

Even though people may change clothes out of necessity, for fun or as store service, the fashion industry still struggles with how to keep old clothes out of landfills. So, upcycling initiatives offer a simple way to reuse products by paying attention to the life cycle of materials and offering at least one near-term solution. As an alternative of throwing products away, their material is reused and its lifetime expanded (Fletcher, 2008; Aakko & Koskennurmi-Sivonen, 2013; Fletcher & Tham, 2014; Becker et al., 2015; Ericsson & Brooks, 2015; Amed et al., 2020). For Teli et al. (2015) and Vadicherla et al.(2017), upcycling is a process in which new products or materials are created with resource to waste or products that no longer serve a purpose. Upcycling can also be understood as repurposed, re-paired, upgraded or remanufactured products (Singh et al., 2019; Sung, 2020). These are also at least of equal quality or better than the original ones. Flowers, et al. (2018) also agree by referring that the products can come from an original creation or in alternative from the modification of the original product in order to create something with higher quality and value (Kamble & Behera, 2021). Upcycling as said also helps increasing the value of the products contrarily to recycling that typically reduces the value (Koch, 2019; Singh et al., 2019). If the process is done in a successful way, textiles or garments can be upcycled indefinitely (Aoife, 2020).

Upcycling can also help to reduce the manufacture of new products and consequently the environmental problems coming from this (Singh et al., 2019; Aoife, 2020) contradicting the notion that a disposed product no longer as value (Wegener, 2016), but for that hard work is needed as long with creativity and innovation (Vadicherla et al., 2017) and as importantly, the design of the upcycled product (Shuhailo et al., 2022) and the craft techniques applied (Mantyi-ncube & Juba, 2020). Lucietti et al. (2018) also agree that processes such as upcycling, downcycling or recycling help diminish the environmental impacts and are able to increase profits with materials that otherwise would be discarded. And upcycling also as a substantial difference in energy consumption being in this way one of the most sustainable solutions to achieve circularity in the textile supply chain (Shuhailo et al., 2022) and also to achieve social responsibility and ethical buying from companies and consumers (Kamble & Behera, 2021).

Keeping the quality of the original material(s) that can come from waste before or after consumption or be a combination of both is also relevant, although "upcycling in design and production is mainly applied to unique products or for minimal series by tailors or small companies" (Shuhailo et al., 2022, p. 8).

It is important to understand that the growth of the fashion industry brought many economic, social and environmental consequences, being one of the most polluting industries (Mantyi-ncube & Juba, 2020) and that nowadays most products are design to last only for some time (Teli et al., 2015). More, being able to identify and follow the whole chain till the disposal of the textile products is very complex (Mantyi-ncube & Juba, 2020) and how many garments are reused or recycled in unknown. Poor textile waste management also contributes for this (Kamble & Behera, 2021). Also, a large percentage of the textiles nowadays are made from a mixture of fibres, making the recycling process harder and the removal of all the trims can also be time consuming (Kamble & Behera, 2021) as it is the process to market these products that can be sometimes unique garments (Singh et al., 2019). It is known that the maintenance or reparation of products has been fading in the western countries (Singh et al., 2019) but in the past few years the upcycling trend has been revived by concerns with the environment (Singh et al., 2019). Upcycling can also be understood as a social movement (Koch, 2019) and can help create green jobs and consequently growth (Teli et al., 2015), but according to Singh et al. (2019), most consumers still prefer mass-produced goods, opposed to upcycled products that is only preferred by a small percentage. And this lack of attention is also referred by Sung (2020) as being happening on universities, regarding the use of upcycling for teaching and learning activities.

Additionally, some universities have been looking into upcycling activities. A Midwestern US university created a course called green prototyping and upcycling focusing on the product design and development, making students leave the consumer view and assume the designer and manufacturer view or the University of Wolverhampton that made students upcycle shirts into new garments, already back in 2016 (Flowers et al., 2018). Lee and DeLong (2018) resourced to a fashion company to give Fashion Design students stocks from various apparel brands, so that they could redesign, refine and remanufacture them. Also Park (2014) analysed a project between a non-profit organization and the Colorado State University where was made the reuse of military parachutes and wool. This project also had an additional social purpose as it was supposed to develop entrepreneurial solutions military veterans that were disabled or disadvantaged. Sung (2020) also refers that upcycling in education is important if a transition towards sustainability is wanted and also because entrepreneurship many times is faced with lack of resources. Aoife (2020) also states that there are opportunities in the textile sector for training in upcycling. According to Shuhailo et al. (2022), there is also demand for professionals capable of working within circular economy models.

On a business level, some brands implemented programs to encourage consumers to participate in the textile waste management, like Patagonia, Zara and H&M

introduced the collection of used garments (Kamble & Behera, 2021). The ReCrafted line was in this sense conceived by the brand Patagonia. The sustainable brand struggled with how to gather, organize and transform thousands of irreparable old clothes into desirable items while turning gain to the brand. So, to minimize the brand impact in the environment, in November 2019, Patagonia produced around 10,000 upcycled jackets, sweaters and bags (Amed et al., 2020). Still, the method of upcycling can be more than just a way to pursuit profits. It can be a way of schools and universities be more socially active whereas being sustainably active. Upcycling skills and techniques such as "restyling, reshaping, embellishing, and over-printing to give discarded, torn and stained fabrics added value, a new life and divert (or delay) waste from landfill" are being used by a small but significant group of designers and producers in the fashion field (Fletcher, 2008, p. 103).

Similarly to the work of charitable organizations such as TRAID or Junky Styling (Fletcher, 2008), first year students of Fashion Design degrees can also redesign second-hand garments by transforming them not only into one-off fashionable pieces but rather home textiles with a solidarity-based approach. Plus, the students could also improve their skills on the sewing and creative side. Additionally, by proposing an upcycling initiative, it will be possible, within the academic community (e.g. professors, technicians and students) to begin a discussion, invoke a deep sense of meaning and motivate the critical sense. Therefore, it is crucial to implement such project in the academia.

Project Description

Taking as a starting point the environmental concern and the search to reduce the amount of textile waste in such high values as has been seen in recent times, the (educational institution) developed a proposal to challenge students in Fashion Design Courses to think about the environment in which they seek to integrate after completing their training.

The challenge has several important focuses to highlight for the formation of young students. On the one hand, students entering their first year of training who need to be confronted with the problem in order to find their place and position in the area, at the same time as they come into contact with different materials, suitable for different types of pieces, and start to become familiar with their characteristics. On the other hand, students more advanced in their Fashion training or even in a second cycle of studies who live the experience already with a greater awareness and define their position towards this problem that worries specialists, designers and researchers so much. The educational institution has courses in the area of Fashion Design functioning in two of its institutions; thus, the work proposal was presented in the HEI (a) to three distinct groups of students: 55 first year students, 38 second year students and 8 master students. In HEI (b), the work proposal was addressed to a group of 8 first year students. All groups are quite heterogeneous and made up of people of both genders; however, the male gender is clearly the smaller group compared to the female gender.

The main objectives of this project are, in the first phase, to allow students to have contact with different textile materials and to know how to identify their typologies and in the next phase, being the main objective, to transform used products into new Fashion Design Products. There is also a solidarity aspect, as the students are encouraged to create pieces, blankets/quilts that will be delivered to humanitarian institutions. It is, therefore, an upcycling project that aims to raise awareness and responsibility of these new generations of professionals for the current issue of sustainability and circularity. Social responsibility is also involved, as new products will be created to be delivered to charitable institutions.

The initiative to collect textile materials and clothes began in September and will continue throughout the school year. During about 3 weeks, the groups will go through moments of explanation and contextualization of the project's themes, as well as the methodological framework of the creative process. The techniques will be explored as needs arise, and the group will always be accompanied by teachers who will ensure all the necessary technical and creative advice. At the end of 3 weeks, the groups will present their proposals for new products, and these will be selected to integrate one of the exhibitions of the (Conference). During the school year, they will use the textile materials to create blankets for a patchwork exercise that will be delivered to charitable institutions. This last exercise will be opened to all schools in the region and surrounding area of the University, with arts education, in order to disseminate this environmental concern involving groups of high school students, future students of higher education in the area of Fashion.

To finalize seems now suitable to propose the design of the product that is expected to be produced by the students: a patchworked duvet. Fig. 1 demonstrates the proposal of the design and technical specification of the duvet assembly.

Future Path and Improvements

The next path to follow is to put in practice this project and allow the students to understand the importance of social and environmental activities in the field of textile and fashion design. In the future may be interesting to consider involving communities around the geographical location of the universities where the production of the duvets will happen. After all, it is important to disseminate the project not only through the scientific community but also incite people to gather life ending clothing for future upcycling projects under a social cause. For the improvements and methods applied in a first round of products, it is also crucial to gather the feedback of the students and allowing the professors and technicians involved to understand possible project weaknesses and propose new strengths, solutions and new goals.

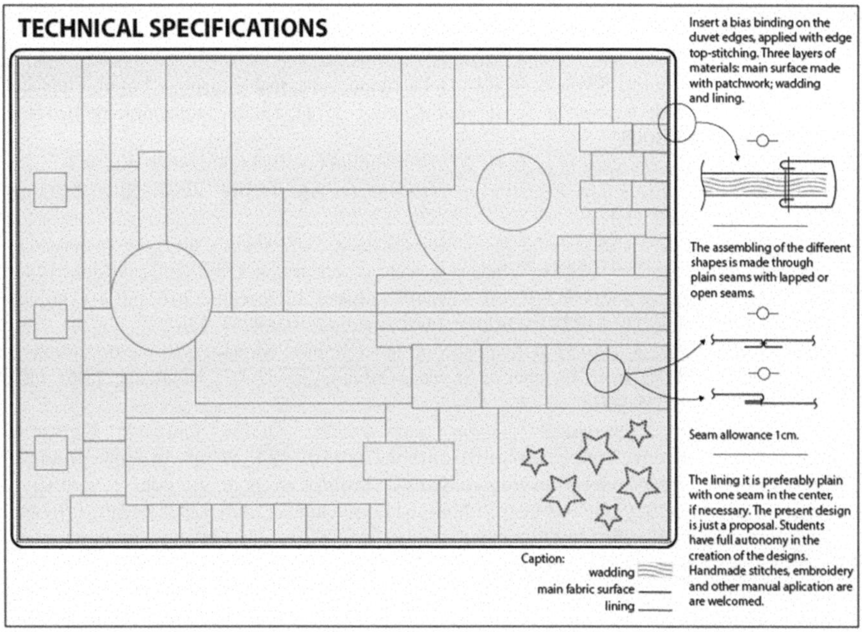

Fig. 1 Duvet design proposal and technical specifications

Conclusions

We highlight the importance of initiatives that focus on sustainability in the area of Fashion Design. We all have in mind that the fashion industry is the second most polluting and that we can do a little more to trigger a greater awareness of people. More than thinking about creating products from sustainable raw materials or using sustainable production technologies, it is urgent to rethink the use of textile products and their waste.

The study that begins with this small project and that is intended to deepen seeks to question the best way to reduce textile waste and the use of this waste in the creation of new products.

It is believed that this proposal will contribute, in a significant way, to triggering a greater awareness of the younger classes to the problem of reducing consumption. It is also important to highlight the preponderant role that educational institutions may have in this awareness-raising, which may involve challenges similar to the one presented here.

References

Aakko, M., & Koskennurmi-Sivonen, R. (2013). Designing sustainable fashion: Possibilities and challenges. *Research Journal of Textile and Apparel, 17*(1), 13–22. https://doi.org/10.1108/RJTA-17-01-2013-B002

Amed, I., et al. (2020). *The state of fashion, 2020.* https://doi.org/10.4324/9780203519943.

Aoife, R. (2020). *A case study of textile upcycling in an educational setting.* Sligo National University of Ireland, Galway.

Becker, F. G., et al. (2015). *Sustainable fashion and textiles, Syria studies.* https://www.researchgate.net/publication/269107473_What_is_governance/link/548173090cf22525dcb61443/download%0Ahttp://www.econ.upf.edu/~reynal/Civilwars_12December2010.pdf%0Ahttps://think-asia.org/handle/11540/8282%0Ahttps://www.jstor.org/stable/41857625.

Ericsson, A., & Brooks, A. (2015). African second-hand clothes: Mima-te and the development of sustainable fashion. In K. Fletcher & M. Tham (Eds.), (pp. 91–99). Routledge. https://doi.org/10.4324/9780203519943

Fletcher, K. (2008). *Sustainable fashion and textiles: Design journeys.* Earthscan. https://www.researchgate.net/publication/269107473_What_is_governance/link/548173090cf22525dcb61443/download%0Ahttp://www.econ.upf.edu/~reynal/Civilwars_12December2010.pdf%0Ahttps://think-asia.org/handle/11540/8282%0Ahttps://www.jstor.org/stable/41857625.

Fletcher, K., & Tham, M. (2014). *Routledge handbook of sustainability and fashion* (Ed. K. Fletcher & M. Tham). Routledge. https://doi.org/10.4324/9780203519943.

Flowers, J., Rauch, C., & Wierzbicki, A. (2018). Teaching upcycling to impact environmental attitudes. *Journal of Technology Education, 30*(1), 30–45. https://doi.org/10.21061/jte.v30i1.a.2

Kamble, Z., & Behera, B. K. (2021). Upcycling textile wastes: Challenges and innovations. *Textile Progress*, 65–122. https://doi.org/10.1080/00405167.2021.1986965

Koch, K. (2019). Clothing upcycling, textile waste and the ethics of the global fashion industry. *Zonemoda journal (Online), 9*(2), 173–184. https://zmj.unibo.it/article/view/10053/10166

Lee, Y. K., & DeLong, M. (2018). Rebirth product development for sustainable apparel design practice in a design studio class. *Fashion Practice, 10*(1), 34–52. https://doi.org/10.1080/17569370.2017.1413797

Lucietti, T. J., et al. (2018). *O Upcycling Como Alternativa para uma Moda Sustentável.* In 7th Academic international workshop advances in cleaner production. Barranquilla.

Mantyi-ncube, B., & Juba, S. (2020). Environmental sustainability through textile and apparel upcycling. *RAIS Journal for Social Sciences, 4*(2), 39–46. https://doi.org/10.5281/zenodo.4286649

Park, J. (2014). Upcycled parachutes project at Colorado State University. *Fashion Practice, 6*(1), 119–124. https://doi.org/10.2752/175693814x13916967095037

Pero, M., Arrigo, E., & Fionda-Douglas, A. (2020). Sustainability in fashion brands. *Sustainability, 12*(14), 10–13. https://doi.org/10.3390/su12145843

Sachs, J. et al. (2019). *Sustainable development report: Transformations to achieve the Sustainable Development Goals.* https://s3.amazonaws.com/sustainabledevelopment.report/2019/2019_sustainable_development_report.pdf

Saricam, C., & Okur, N. (2019). Analysing the consumer behavior regarding sustainable fashion using theory of planned behavior. In S. Mthu (Ed.), *Consumer behaviour and sustainable fashion consumption*. Springer. https://doi.org/10.1007/978-981-13-1265-6_2

Shuhailo, Y. V., et al. (2022). Incorporation of upcycling techniques into technology education. *Journal of Physics: Conference Series, 2288*. https://doi.org/10.1088/1742-6596/2288/1/012041

Singh, J., et al. (2019). Challenges and opportunities for scaling up upcycling businesses – The case of textile and wood upcycling businesses in the UK. *Resources, Conservation and Recycling, 150*. https://doi.org/10.1016/j.resconrec.2019.104439

Sung, K. (2020). Upcycling for teaching and learning in higher education: Literature review. In *World sustainability series* (pp. 371–382). Springer. https://doi.org/10.1007/978-3-030-15604-6_23

Teli, M. D., et al. (2015). *Upcycling of textile materials*. In Global textile Congress (pp. 164–167).

Vadicherla, T., et al. (2017). *Fashion renovation via upcycling*. https://doi.org/10.1007/978-981-10-2146-6_1.

Wegener, C. (2016). Upcycling. In *Creativity – A new vocabulary* (pp. 181–188). https://doi.org/10.4324/9781315270326-194

Reconstruction and Upcycling to Create a Fashion Installation

Sara Vaz, Alexandra Cruchinho, Ana Sofia André Bentes Marcelo, and Cristina Patrício

Introduction

More and more clothes are cheaper, yet human and environmental costs have grown dramatically. Deconstruction/construction/reconstruction will meet the representation of garments that were the protagonists of a project that aims to enhance the value of the piece that reaches the end of its life cycle and that may or may not have affective meaning. The whole project is supported by theoretical research based on themes such as fast fashion, slow fashion, and upcycling, being born after watching the documentary "The True Coast" by Andrew Morgan, which portrays precisely the impact that the textile industry is having on the world and society.

Is the world keeping its eyes open to this impact?

The project contemplated the deconstruction of a whole set of discarded garments, giving them a new language and approach, joining probably with improbable, extrapolating beyond the apparent horizon... trying to emphasize creativity, and finding solutions for the garments we no longer use?

S. Vaz (✉) · A. Cruchinho
Universidade Lusófona-Centro Universitário de Lisboa – CICANT, Lisbon, Portugal
e-mail: sara.vaz@ulusofona.pt; alexandra.cruchinho@ulusofona.pt

A. S. A. B. Marcelo · C. Patrício
ESART – Instituto Politécnico de Castelo Branco, Castelo Branco, Portugal
e-mail: ana.marcelo@ipcb.pt

A. Cruchinho, J. C. Neves (eds.), *Sustainability in Fashion: Creativity, Education, and Innovation*, https://doi.org/10.1007/978-3-031-80991-0_26

Fast Fashion

Fast fashion focuses on a pattern of production and consumption in which products are manufactured, consumed, and discarded very quickly, thus depending on the supply and production in terms of costs and time of marketing the products in the market, the essence to guide and meet the demand of unbridled consumption for new styles at low costs.

According to Lipovetsky (2007), consumers are inserted in the "civilization of desire," in which easy access to goods and services, resulting from the abundance of offers coming from the industrial revolution and new production processes, generates an incessant stimulation to demand.

Appearing in 1900 with the massification of the sewing machine, fast fashion from then on has been growing and there has been a great evolution until today. With the Second World War, there was a greater number of women entering the labor market, emerging with a new purchasing power that was impossible for them before. While before they were able to take back the clothes they had at home, now they can begin to have some freedom with regard to their purchasing power and decision-making. It was then that celebrities appeared who caused fashion to skyrocket at great speed, characterized by the 1950s, the model and actress Twiggy and the rock band The Beatles, who were great references for a gigantic mass of potential consumers at the time.

The first synthetic materials appeared, such as PVC and Polyester, which contributed to paradigm shifts, thus providing new uses of materials in garments, a revolution in a fashion that changed forever the way people dressed. In the 1960s, the textile industry took off and gained strength. When in 1970 the production prices increased, many companies chose to move their factories to Asia, however, the quantities could be increased and, on the contrary, the wages were lower, thus making clothing cheaper, on the other hand, advertising more expensive as an essential investment for sales.

The 1990s were marked by the global colonization of fashion, in the sense that international brands entered local markets, thus destroying existing local shops. It was no longer possible to take 4 months from design to sale, and everything was changed. The garments are thus made by women or even children living in developing countries (Fig. 1) who work in inhumane conditions, even working up to 16 h a day.

According to Cietta (2010), the great advantage of the fast fashion model lies in the ability to offer in very short terms what the market offers, supplying the shops constantly with articles designed on top of consumer trends. This way, this system manages better the risk inherent to the launch of a collection, reducing the chances of the articles not being sold.

Aiming to conquer a diversified audience, fast fashion chains also invest in exclusive collections, in addition to the basic lines aimed at mass consumption, initially presenting collections every 15 days and currently new pieces every week, catching the consumer's eye, making them want to buy new clothes, which are

Fig. 1 Sweatshops poor working conditions. (Source: https://www.dailymail.co.uk/)

constantly being advertised on social networks, often attracting the attention of the final consumer.

Advertising messages that sell happiness translates into the idea that one's life problems can be solved through consumption, placing increasing importance on the possession of ephemeral goods, and the constant search for something new that will finally satisfy us. One of the biggest examples of this is the disposable fashion industry or fast fashion. We are used to looking at the final, finished product, we need to step back to consider its origin and the impact that unbridled, unconscious consumption has on the living conditions of workers in low-cost economies and on the environment.

Slow Fashion

Slow Fashion/Kate Fletcher—Sustainability Design Fashion

"Fast fashion is about greed." Kate Fletcher (2007) says it is time to slow down and consider the true cost of choosing quantity over quality.

Fashion speed is a defining characteristic of the textile and clothing industry of today. Industrial production is extremely fast, turning a sample sketch or design into a finished product in just 12 days, and fast consumption. Fast fashion is not really about speed, but greed: selling more, making more money. Time is just one factor of production, along with labor, capital, and natural resources that are manipulated and squeezed in the pursuit of maximum profits Fletcher (2007). But fast is not free and free of great costs. Short delivery times and cheap clothes are only possible by exploiting labor and natural resources. Slow fashion is about designing, producing, consuming, and living better. Slow fashion is not based on time, but on quality (which has some components of time). Slow is not the opposite of fast—there is no dualism—but a different approach in which designers, buyers, retailers, and consumers are more aware of the impacts of products on workers, communities, and ecosystems Fletcher (2007).

The concept of slow fashion was born in the 80s/90s and is completely influenced by the slow food movement which aimed to reduce the consumption of fast food as much as possible. Therefore, slow fashion is linked to the slowing down of fashion, and the reuse of pieces that will serve from season to season. The products have more quality, coming from a more durable fashion. The time factor does not weigh so much since the planning is done in the long term and does not require workers to work overtime. In this case, the cost of the pieces will be higher compared to fast fashion, as it requires other care/time for the product to be 100% finished. In short, slow fashion is a preview of a different and more sustainable future of the fashion system and a business opportunity that respects the worker, the environment, and the consumer (Minney, 2016).

Kate's pioneering work ranges from developing slow fashion ideas and practices to directional sustainability projects, including Fashion Ecologies and Local Wisdom, which has engaged thousands of people around the world with the art of "wearable craft" and "post-growth." She is the founder of design for sustainability consultancy Slow Fashion where she works with businesses, educational establishments, and non-governmental organizations to promote the shift towards sustainability ("Kate Fletcher—Routledge & CRC Press Author Profile"). She is also recognized personality as an inspirational speaker.

Being sustainability a theme of the twenty-first century, Kate Fletcher reports and examines how sustainability has the potential to transform the fashion system and those who work in the area. The book fashion & sustainability—design for change (Fletcher, 2007) is divided into several parts, being the first phase analyzes the transformation of fashion products, defining ways in which the impacts of garments can be reduced and their development throughout the life cycle of the garment, never forgetting innovation with regard to materials, manufacturing, distribution, use, and reuse. The second part presents us with ideas that link the transformation of fashion systems at the root, including emerging business models that find commercial opportunities in the reduction of material flow. Finally, it makes reference to practical ways to transform the practice of fashion design and explores new opportunities for designers who want to go beyond the traditional maker design, to include work as a way of delivering a strong message thus being an activist to promote systemic change.

Sustainable Fashion vs. Circular Economy

The post-modern world faces several dilemmas in problematic situations concerning the environment and the economy, experiencing crises of all kinds and social instability. To overcome the problems and ensure the survival of businesses and the planet, it is necessary to find sustainable means of producing, trading, consuming, and disposing of products, seeking to reduce environmental and social impacts. The fashion industry, in the production of textiles, is among the most polluting activities of the twentieth century, still producing many impacts due to the use of chemicals in its production processes, causing the contamination of water and air, among other problems (Berlin, 2012).

The documentary *The Cotton Film: Dirty White Gold* tells us the real consequences of the use of highly toxic pesticides in traditionally cultivated cotton plantations with excess water, in textile processes, shortages in the treatment of effluents, energy consumption, and gas emissions in the production and distribution of products for analysis.

The circular economy is the opposite of the linear economy, proposing that the resources that are extracted are kept in circulation. In this way, the final destination of a material is no longer a pile of waste, but the process of developing products and systems, thus eliminating the concept of waste. The circular economy aims to keep products, components, and materials at their highest level of usefulness and value at all times. It consists of a continuous positive development cycle that preserves and enhances natural capital, thus optimizing the production of risk-minimized resources.

The mountains of rubbish that accumulate in landfills are considered a great waste for the Cradle-to-Cradle theory—the title of a book/manifesto published in 2002 by the American architect William McDonough and the German chemical engineer Michael Braungart, which has become one of the most influential works of ecological thinking worldwide. Cradle-to-Cradle thinking arises in opposition to the idea that the life of a product should be considered "from cradle to grave"—an expression used in life cycle analysis to describe the linear process of extraction, production, and disposal. For a Cradle-to-Cradle, William McDonough (2002), industry, the central idea is that resources are managed in a circular logic of creation and reuse, where each cycle passage becomes a new cradle for a given material. As such, the linear model is replaced by cyclical systems, allowing resources to be reused indefinitely and circulate in safe and healthy flows—for humans and for nature, where many of these nutrients are wasted, contaminated, and even lost. To then eliminate the concept of waste, design products from the beginning with the understanding that waste does not exist, after all the planet is cyclical, nature is not linear, it is holistic intuitive, and nurturing. The circular economy is good for the planet and also makes it possible to save money. It is founded on three principles, each of which is geared towards the pertinent resource challenges facing the industrial economy.

Upcycling

The term upcycling was first used in 1994 by German entrepreneur and environmentalist Reine Pilz. She released a book in 2002 *Cradle to Cradle: Remaking the Way We Make Things*, which had quite an impact on the public and began to raise awareness of the great topic of upcycling. William McDonough and Michael Braungart (William McDonough, 2002) state that the aim of upcycling is to avoid the waste of potentially useful materials, reducing the consumption of new raw materials during the creation of new products and energy consumption, air and water pollution, greenhouse gas emissions, resulting from the industrial processes of recycling. This makes the practice even more positive, from an ecological point of view, than recycling itself.

Upcycling is still the "famous customization of the already much talked about pieces." The difference is that upcycling is associated with the concept of sustainability, with the aim of developing eco-conscious projects that increase the life cycle of clothing, thus preventing more textile waste to be dumped into nature.

We can also define upcycling as a creative way of renewing garments.

However, there are always doubts, and when it comes to upcycling, there is a question that arises regarding purchasing; if we buy less that is good for the planet, but for all the millions of people who depend on manufacturing industries it is not so good, because they have to remain active in the market. So how can we maintain economic growth without it being devastating? We can thus come to the conclusion that even if we take advantage of the parts there are always details/parts of the piece that have to be rebuilt, this often leads us to buy from certain parts, that is, we are thus contributing so that these industries remain active, but at the same time, the level of pollution is lower.

We can also mention some examples of projects that were, and are being developed in the upcycling aspect, namely the partnership of the Levi's® (Figs. 2, 3, and 4) brand with the fashion designer Joana Duarte, representative of the Béhen brand, who created garments by recycling stopped stocks and old Levi's® pieces.

> In the process of production and dyeing of the pieces in FRESS' workshops in Lisbon, the use of environmentally friendly techniques and materials was privileged. For the creation of all the bottoms, surplus stocks of Levi's® High Loose and Levi's® Stay Loose (the first sustainable jeans of the brand and made with Softened Hemp) that still existed in the showroom were used. For the creation of all the tops, old denim pieces were used and that would be stalled stocks and/or "waste/waste." (Levi's®, 2021)

Fashion Installation

Installation is conventionally used in contemporary art to define a form that uses or occupies space. For Adam Geczy (2019), at Fashion Installation, installation asks that these very qualities are essential to the way the object is understood. As such, installation is an approach that is material and immaterial, in the sense that it rejects

Fig. 2 Levi's® + BÉHEN _1. (Source: Vogue Portugal)

any possibility that a work of art can be viewed objectively, or in isolation of space, time, culture, or history. According to this definition, "installation," as it has come to be known in art, is aptly suited to the ways in which fashion is presented and displayed, and increasingly, over the last 20 years, when designers have availed themselves of more and more sophisticated systems by which to convey the fashion object, which exists not in isolation but across a network of representations and narrative connotations.

Redesconstruction Project

Methodology

Projetual Methodology Based on the Designer Bruno Munari

The methodology used by Bruno Munari met the project's initial idea, starting from a sudden idea, a sudden inspiration or even improvization without method, seeking to emphasize the importance of the word creativity.

Fig. 3 Levi's® + BÉHEN _2. (Source: Vogue Portugal)

> The problem cannot be solved by itself; however, it already contains all the elements for its solution. It is necessary to know them and use them in the solution project. (...) in the field of design one must not design without a method, think in an artistic way, looking for the solution straight away, without having done research to document what has already been done similar to what one wants to design; without knowing which materials to use for the construction, without having well specified their exact function. (Munari, 2004).

Research

The first part of the project development went through the research demonstrated in the first chapter presented in this chapter. Articles were searched, and information in books about fast fashion and sustainable fashion, and subsequently, it was discovered areas that were entirely linked to the theme of upcycling, circular economy, etc. and that everything would give to sustainability and the way people can make a difference with a small gesture.

Silhouettes

The silhouette created (Fig. 5) was a very structured silhouette, with volumetry and somewhat exaggerated. Exaggeration is made to really understand the message of the concept in the best way. The intention was always to create pieces of clothing

Fig. 4 Levi's® + BÉHEN _3. (Source: *Vogue Portugal)*

Fig. 5 Silhouettes panel. (Source: Author)

Fig. 6 Panel designed for session. (Source: Author)

(jackets) where several parts of different pieces would be placed, which would be conjugated with each other, thus making the pieces with an exaggerated volumetry.

Collection of Items and Materials

For this project, we collected many clothing items, obtaining pieces for the continuation of the project (Fig. 6). During the collection, many messages were heard, namely: "ah you can take it, I don't need it at all!" or "what do you need the clothes for? I want to see the result! I really liked the concept!", this led me to think even more on this issue of how people see their garments, what message the garments send to each of us individually.

After this moment of reflection on the author, we thought about doing a photo shoot whose goal was to convey to the viewer a strong message, how day after day we consume more pieces of clothing.

The session (Figs. 7 and 8) developed was thought to emphasize both the piece of clothing displayed there and the process that the piece takes to be built, all the stages it goes through and most importantly, how long the piece will be made.

The pieces thus appear suspended on crosses, with a backdrop of knotted ties interweaving into pieces. A background image emerges that is supposed to be shocking! Shocking to look at the image and wonder why that scar, and the number of trousers next to it, possibly made in a day, or maybe in a few hours by each child.

Fig. 7 Panel parts supplied—1. (Source: Author)

Sketches/Inspirations

After receiving the pieces, the creative process is important (Fig. 9). To achieve what was intended, we had to play with pieces that already existed on the internet and "play" always trying to get as close as possible to what was intended. The goal was always to create pieces that would meet the concept, the mixture of fabrics, use fronts on the back and back on the front, and put details. So, five sketches were developed so that, from there, the final products were developed (Figs. 10, 11, 12, and 13).

Fig. 8 Panel parts supplied—2. (Source: Author)

Fig. 9 Line up sketches. (Source: Author)

Fig. 10 Coat illustration—1. (Source: Author)

Illustrations Developed Parts

Development of the Piece

Technical Sheet/Flat Design

In this project, drawings identified the parts' plans (Figs. 14 and 15) that probably will not be executed again. Still, they were identified, even so, the stitching, cuts, and the different parts that constituted the piece.

As the final presentation of this project, it was decided to do directly in final pieces, because the project worked directly with the disassembly of the piece, cutting, and disassembling, to see where would be better the shell and the collar. As the pieces were being put in place, I combined what looked best with what; I made changes in the stitching... When I "realized", I already had all the pieces on the table because I could only see what kind of pieces I had and how to use them. It was a creative process that was a pleasure to develop, which worked almost like draping,

Fig. 11 Coat illustration—2. (Source: Author)

because the piece was patternmaking on the bust, and only then could you see the result and the aesthetics.

The first piece, the shorter jacket, was the first to be developed. Until I got something to come out, I had to undo and end up "wasting" some parts of pieces just because they either didn't connect aesthetically or they were marked.

Editorial

The editorial of the final pieces (Fig. 16) was thought to highlight the details and the project's primary goal at a civilizational level to alert consumers to reduce the consumption of clothing to combat something present in our daily lives, *consumption*. Before the development of the garments, it was thought in an installation that resorts to images merely descriptive and illustrative of what would be the day-to-day of a child who works mainly for big brands that produce what is called fast fashion.

Fig. 12 Development panel/coat detail 1. (Source: Author)

Fig. 13 Development panel/coat detail 2. (Source: Author)

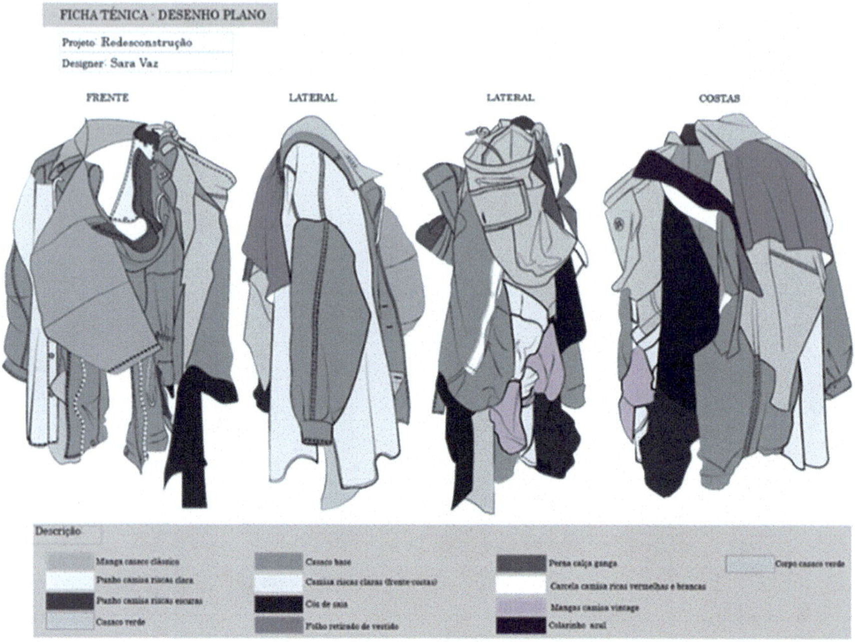

Fig. 14 Technical drawing 1. (Source: Author)

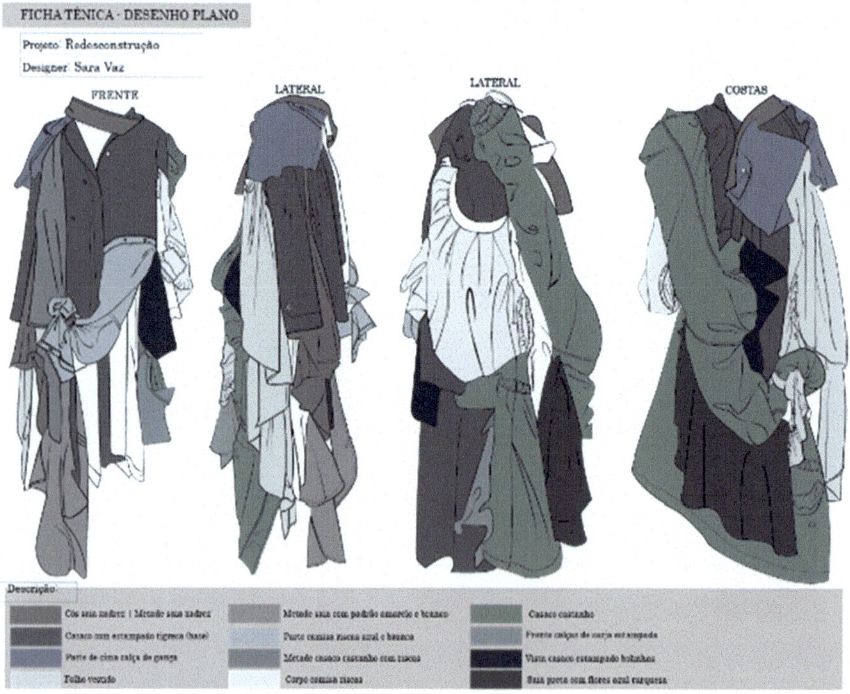

Fig. 15 Technical drawing 2. (Source: Author)

Fig. 16 Editorial panel. (Source: Author)

Conclusion

The growing environmental and social impacts of consumerism in a fashion characterize the concern for the planet's future. It is crucial to consider the end of the earth regarding the development of conscious consumption alternatives and meeting the use, reuse, and recycling of parts called upcycling.The focus on developing the upcycling technique in companies is essential for the fashion industry because it can transform and replace fast fashion with slow fashion, valuing quality more than quantity. The project aimed, thus, the reuse something that would be discarded, transforming it into a new product, with the same or a new function, without transformations and use of chemical processes.In the case developed, the fashion installation was to mobilize what we call "fashion," building an identity that aimed to characterize what could be the daily life of a garment that will be discarded. The transmitted image will be seen as transitory and contextual, putting the viewer to question what he is observing; however, he feels part of the work, fitting into the scenario.

References

Adam Geczy, V. K. (2019). *Fashion installation—Body, space, and performance* (1st ed.). Bloomsbury Visual Arts.
Cietta, E. (2010). *A relução do fast fashion. Estratégias modelos organizativos para competir nas indústrias híbridas*. Estação das Letras e Cores.
Fletcher, K. (2007). *Ecologist informed by nature.* [Online]. https://theecologist.org/2007/jun/01/slow-fashion. Accessed 14 maio 2019.
Lipovetsky, G. (2007). *Afelicidade paradoxal: Ensaio sobre a sociedade de hiperconsumo*. Companhia das letras.
Minney, S. (2016). *Slow fashion: Aesthetics meets ethics*. New Internationalist.
Munari, B. (2004). *Das Coisas Nascem Coisas*. Edições 70.
William McDonough, M. B. (2002). *Cradle to cradle: Remaking the way we make things* (1st ed.). North Point Press.

Fashion, Sustainability and Artistic Experimentations

Carolina Bianque Galito and Maria Silvia Barros de Held

Introduction

Two decades into the twenty-first century, the issue of sustainability appears increasingly urgent and necessary in contemporary life. However, discussing this subject has still been a somewhat arduous task, since sustainability to date has not, in fact, reached people and transformed their realities as a proposal.

As consumption establishes itself as one of the main tools of modern capitalism, Bauman (2008) understands the existence of a consumerist culture with the character of a syndrome, since it degrades duration and increases ephemerality, raising the value of novelty in the face of the value of the stay. To this end, Lipovetsky (1989, p. 160) reiterates this dialectic by stating that:

> Economic logic has truly swept away any ideal of permanence, it is the rule of the ephemeral that governs the production and consumption of objects. From now on, the short temporality of fashion has engulfed the universe of merchandise, metamorphosed, since the Second World War, by a process of renewal and 'programmed' obsolescence, conducive to ever reinvigorating consumption.

Although many people still live under this logic, many others vehemently question its existence, as Krenak (2020, p. 26) asks: "How many Earths do these people need to consume before they understand that they are on the wrong path?"

In order to briefly analyze sustainability from economic, environmental, social, and cultural points of view, this chapter aims to discuss some points that can serve as reflections on the topic, in order to add investigative possibilities in areas that express creativity, such as art and fashion.

C. B. Galito · M. S. B. de Held (✉)
University of São Paulo-USP, Sao Paulo, Brazil
e-mail: carolbianque@usp.br; silviaheld@usp.br

A. Cruchinho, J. C. Neves (eds.), *Sustainability in Fashion: Creativity, Education, and Innovation*, https://doi.org/10.1007/978-3-031-80991-0_27

Some Sustainability and Fashion Issues

It is known that some concepts related to sustainability, according to authors in specific areas, vary and are still widely debated in science. In order to corroborate the study, some of them were presented and contrasted in a constructive way so that the reflective debate is based.

For sustainability, Bursztyn and Bursztyn (2012) understand that there are three spheres that constitute its basis: biosphere (environment), sociosphere (society), and econosphere (economy). Thus, the debate that touches on the biosphere and the econosphere is known as sustainable development. This development is one of the most controversial in contemporary times, since the search for society's well-being directly affects the natural environment and, therefore, thinking about sustainable development becomes somewhat paradoxical.

> since, to be development, it cannot be sustainable. (Bursztyn & Bursztyn, 2012, p. 49)

In turn, Veiga (2010) considers that the term sustainability is not yet precise, since its consolidation is recent. Its emergence took place in 1948 in the Universal Declaration of Human Rights by the United Nations, and since then it has been debated and analyzed from different angles. The author reiterates that:

> As the pessimist Schopenhauer said, every truth goes through three states: first it is ridiculed, then violently fought, and finally accepted as self-evident. With sustainability, a similar cycle was completed in three decades (Veiga, 2010, p. 13).

Sachs (2009, p. 71) understands that "social sustainability comes first, as it stands out as the very purpose of development, without taking into account the probability that a social collapse will occur before the environmental catastrophe." The author (Sachs, 2009) even considers the economy to be a necessity, but reinforces that it cannot be the focus, as it brings with it social inequality and, consequently, obstructs environmental sustainability.

In this sense, Gadotti (2000, p. 35) argues that sustainability encompasses the broad meaning of existence, "as beings of meaning and givers of meaning to everything that surrounds us," seen more from an ethical angle than from economic interests, in a transversal and transdisciplinary way.

Most of the authors presented here understand that the term sustainability is broad and encompasses not only the environment but also human issues from the perspective of social demands. And the big challenge is to consider the economy as an integral and important part of the development process, both in its positive and negative sides.

In fashion, research and authors advance the so-called green consumption. Berlim (2016) states that on the one hand there are sustainable behaviors that actually add value to the brand with well-known terms such as eco-chique, green-glamour, and ecofriend, but on the other hand they only function as mere consumerist marketing.

In this sense, fashion and sustainability may seem, at first, to be antagonistic elements since, as the author states (Berlim, 2016), the logic of fashion is to be consumed in an exaggerated way in contemporary times activated by fast fashion—a pattern of rapid production and consumption, in which products are manufactured, consumed, and then discarded.

However, there is nothing more justified than linking sustainable thinking to fashion, as it becomes an attempt to minimize the damage caused by it. Thus, Berlim (2016, p. 13) reiterates the need to recognize "the transgressive, reflective and expressive aspects that fashion points out in today's society [...] which we need to reflect on, then choose (if we can) and, finally, discuss."

Therefore, Cunha (2021) understands that the most obvious and immediate solution is to slow down clothing production. For the surplus, there are better alternatives such as sales in outlet stores with merchandise sold at lower prices, thrift stores, donations to charities, and even being reused through customization or recycled into new pieces.

However, some thoughts aimed at changing behavior are already being formed as described. Several issues in this gear—fashion and sustainability—still need to be resolved and require broad attention: such as environmental governance in the fashion industry and the breaking of consumerism paradigms, for example. Therefore, to broaden the reflection, a new chapter opens below.

Sustainable Consciousness, Paths, and Possibilities

With the significant increase in waste production since the post-war period, there is a growing concern about what to do with it, especially from the twenty-first century onward. One of the ideal solutions for preserving the environment is recycling, which in this chapter will be defined in accordance with the Ministry of the Environment (n.d.) as being the reuse of discarded materials using different techniques so that they can be reintroduced into the cycle productive.

Therefore, in addition to reducing the volume of waste, it reduces the consumption of natural raw material resources, as well as water and energy. Furthermore, Fonseca (n.d.) reiterates that recycling generates an income opportunity for many families in large Brazilian urban centers, through paper and aluminum collector cooperatives, for example. It also points out the materials that are most frequently recycled: glass (food jars, bottles, jars, and various containers); paper (newspapers, magazines, leaflets, cardboard boxes, and packaging); metal (aluminum cans, steel cans, nails, lids, copper, and aluminum in general); plastic (pots, PET bottles, bags, and various packaging).

However, it is known that waste disposal in general cannot always be allocated to cooperatives and recycling companies. What's left is dumped in unsuitable places, commonly known as dumps and landfills. According to Streit (2015), from an environmental point of view, dumps and controlled landfills differ little. What differs between them is the final treatment that each one receives.

> At the landfill, there is no preparation to receive the waste, it decomposes in the open, which causes major environmental impacts (mainly to the soil and submerged water resources) and also social problems related to the transmission of diseases and work in precarious conditions and unhealthy. The controlled landfill is the intermediary, as the waste receives coverage (generally clay, earth and grass) and already has the leachate and methane gas channeled. (Streit, 2015, p. 1).

Another non-viable method as it is not suitable for all types of waste, but which continues to be one of the current alternatives, are incineration plants. As stated by the VG Waste portal (2019), incineration consists of burning waste in dedicated ovens at high temperatures in order to reduce the large volume generated. This practice has advantages and disadvantages, as the portal points out.

Among the advantages are that of transforming combustion into thermal energy and then into electrical energy. From a disadvantageous point of view, it refers to the high cost of investment, and maintenance of equipment, which in a way, is one of the problems that every solution considered sustainable faces.

Having seen some possibilities and challenges in solving waste, sustainable consciousness also encompasses human concerns for a more balanced life, which leads to the concept, understood by Gadotti (2000), as planetary consciousness.

Under this logic, for the concept "sustainability" to move toward new pedagogical proposals, it is important to understand it under two main axes, as suggested by Gadotti (2012, p. 67):

1. Ecological, environmental, and demographic sustainability (natural resources and ecosystems): refers to the physical basis of the development process and nature's capacity to support human action, with a view to its reproduction and the limits of population growth rates
2. Cultural, social, and political sustainability: refers to the maintenance of diversity and identities, directly related to people's quality of life, distributive justice, and the process of building citizenship and people's participation in the development process

As sustainable possibilities, several possible paths that have not yet been explored are considered, which includes the selection and reuse of discarded materials for new functions and new languages. In this case, artistic creation is considered relevant to new actions and new proposals within environmental awareness and awareness.

To this end, Gadotti (2012, p. 62) reaffirms the importance that "the preservation of the environment depends on ecological awareness and the formation of awareness depends on education" and, with this, highlights the relevance of ecopedagogy in which it promotes learning between human relationships with the environment. The author (Gadotti, 2012, p. 63) calls this awareness "ecotraining," which would be related "to the studies of the imaginary, the transversality approach, transdisciplinarity and interculturality, constructivism and the pedagogy of alternation."

However, some directions and possibilities aimed at solving the problem are already being suggested, as described. Although doubts remain in these proposals, awareness needs to be worked harder. To broaden the reflection on this topic, a new chapter opens.

Artistic and Sustainable Perspectives

It is understood that sensitivity and creative capacity are inherent to human beings. In this sense, it is believed that no other animal on this planet can create shapes, objects, and things in a playful way in order to raise awareness through language. Costa (2004) reports that until 2020, no researcher was able to observe emotions in animals related to observing landscapes or sounds as in humans.

In order to ennoble sustainable perspectives, art is supported by enabling action in different ways in society, whether through political-social positions or through language itself.

However, it is worth remembering that the discussions that permeate the concepts of art and its actions are dense and complex. Realizing that art awakens pleasure in people is realizing its importance as a global material and immaterial culture.

In this way, in 2005, the Rio de Janeiro street block entitled "Vagalume o Verde" proposed completely sustainable ideas for carnival. A pioneer on the subject, it gained visibility and put thirty to forty thousand people on the streets. In addition to being a carnival group, they consider themselves a sociocultural and environmental organization, which aims to:

> Contribute to the strengthening of street carnival and encourage the practice of sustainable solutions in the event and its production chain, considering economically viable, socially fair and environmentally correct alternatives. (Portal Bemglo, 2019, n.d.).

As Novo (2015) states, VOV (Vagalume o Verde) first focused on holding its parade using reused materials through artistic workshops offered by the group to make costumes and props, which in a certain way contribute to the environmental awareness of the community in question. Thus, revelers are encouraged to collect their own trash, which is then sent to recycling cooperatives.

Another sustainable partnership between the street block considered satisfactory was with the "Casa da Empada" pie chain, which donated all marketing material, such as paper fans, for example, to be reused in the block's costumes, floats, and props. As Sérgio Murilo (Per Vivere Bene, 2019), director of the company, states, the decision to partner with the VOV block is that the waste would have a valuable destination by being transformed into art, as the block, in addition to providing joy to revelers, also raises awareness about environmental issues.

In this way, Ana Cristina Damasceno (Portal Per Vivere Bene, 2019), coordinator of Galpão das Artes Urbanas where the pieces are manufactured, also believes that by using urban waste in general to create and make art, environmental and ecological awareness is promoted, a practice that has become increasingly common in Rio de Janeiro carnivals (Fig. 1).

Fig. 1 Ana Cristina making costumes with waste materials. (Source: https://www.perviverebene.com.br/post/bloco-sustent%C3%A1vel-em-prol-da-folia-e-meio-ambiente)

> Transforming or recycling waste is part of conscious consumption, as, in addition to reducing the accumulation of waste, it contributes to respecting and preserving the environment and, consequently, improving the population's quality of life. Some people's trash is other people's luxury Conta Ana. (Portal Per Vivere Bene, 2019, n.d.).

From this perspective, in 2019, the research author proposed something similar during the Visual Arts workshop entitled "Artistic Fashion" under her teaching. In order to promote sensory, plastic, aesthetic and creative experiments to students through the investigation of materials considered, by the traditional teaching method, as unusual, sustainability appeared as a transversal axis in the workshop's Political-Pedagogical Project (Fig. 2).

This fact is justified by stimulating numerous developments when using unusual discarded material in the production of pieces and encouraging students to raise awareness of inputs through reuse.

Based on this premise, the purpose of combining clothing with sustainability was due to the fact that the teacher had previous experience in the area of fashion and costume and thus believed that the tangibility of clothing provides a unique sensory experience toward reflection and environmental awareness, which is achieved by enabling students to wear their own works.

Fig. 2 Revelers from the VOV block wearing sustainable costumes. (Source: http://vamosuniromundo.blogspot.com/2013/02/bloco-vagalume-o-verde-faz-o-carnaval.html)

In total, five artistic garments were selected to be presented in a parade format during the end-of-year presentation at the Núcleo de Arte Professor Albert Einstein, a public municipal school institution where the workshop was held (Fig. 3). The motivation for broadcasting the works in the form of a fashion show was to encourage and value the students' artistic production by highlighting and attracting the attention of the public present.

The selection of works was based on completion and the final appearance of the finish. Each artistic outfit was described according to its materials: the first of them is the bodice made from a milk carton and a skirt made from a garbage bag with a detail of milk carton arabesques; the second has two customized pieces, both a blouse and shorts, which were painted and decorated with metal caps and a milk carton, with a headpiece made from transparent plastic egg packaging; the third work is made up of three pieces—a milk carton bodice lined with fabric scraps, a transparent skirt made from a broken umbrella, and a wing made from reused fabric; the fourth work has a painted t-shirt decorated with plastic caps, a belt with garbage bag straps, a head and back adornment with plastic bottles, as well as a bag; and the last work features a bodice structured with a painted milk carton, a skirt with strips of repurposed fabric, and an arm ornament made from garbage bags.

Fig. 3 Making the garment with plastic caps. (Source: Personal collection)

Fig. 4 Photoshoot taken after the parade in the school gardens. (Source: Personal collection)

However, the direct contribution that the reuse of waste material in artistic clothing has on students was observed, as it was noted the change in the perspective that the actors involved in the research had on the materials, which went from trash to investigative experiments (Fig. 4).

As an example of this statement, the skirt produced in the third work with the disused umbrellas was brought by the student herself, who visualized this clothing possibility for herself. Therefore, it is worth emphasizing the concrete outcome that this research obtained so that it can serve as a methodological basis for future research of this kind.

Conclusion

In order to consider all the reflections discussed here, since some concepts regarding the environmental topic are still under wide discussion, it is believed that the transformative potential lies in knowledge and, above all, in education.

The experience of theoretical-practical work motivated an essay that goes beyond the walls of the classroom and, therefore, reinforces and establishes exchanges between students and society. The analyzes and conclusions presented during the work process through unique creation, artistic making, and sustainable production show how the mediation of the educator within cultural projects raises social awareness in order to obtain awareness regarding the issue of sustainability.

Thus, as proposed, this research sought to contribute to studies related to sustainability, fashion, and some examples of its artistic developments as viable and practicable solutions in relation to the reuse of waste resulting from excess consumption in a sensitive and reflective way.

It is expected, therefore, that approaches to sustainability will appear more and more frequently from different perspectives, in order to enrich proposals and solutions in future decision-making.

References

Bauman, Z. (2008). *Vida para Consumo: a transformação das pessoas em mercadoria*. Trad. Carlos Alberto Medeiros. Zahar.

Bemglo. (2019). *Vagalume O Verde*. 8 fev. 2019. Disponível em: https://blog.bemglo.com/vagalume-o-verde/. Acesso em: 06 out. 2020.

Berlim, L. (2016). *Moda e Sustentabilidade: Uma reflexão necessária*. Estação das Letras e Cores.

Bursztyn, M., & Bursztyn, M. A. (2012). *Fundamentos de Política e Gestão ambiental – Caminhos para a sustentabilidade*. Garamond.

Costa, C. (2004). *Questões de Arte; o belo, a percepção estética e o fazer artístico* (2nd ed.). Moderna.

Cunha, R. (2021). *Marcas geram desperdício destruindo toneladas de roupas novas não vendidas*. 15 jan. 2021. Disponível em: https://www.stylourbano.com.br/marcas-geram-desperdicio-destruindo-toneladas-de-roupas-novas-nao-vendidas/?fbclid=IwAR0_xmnSSsPUS68Kuk0I-Aj3eajZ_bCnZBHW6fdPUrQfeNVGp-AMa7KqIW2Y#.YAHBkfYXcCM.facebook. Acesso em 06. jun. 2022.

Fonseca, L. H. A. (n.d.). *Reciclagem: O primeiro passo para a preservação ambiental*. Centro Universitário Barra Mansa, s.d. Disponível em: https://semanaacademica.org.br/system/files/artigos/reciclagem.pdf. Acesso em: 21 out. 2020.

Gadotti, M. (2000). *Pedagogia da Terra*. Petrópolis.

Gadotti, M. (2012). *Educar para a Sustentabilidade: uma contribuição à década da educação para o desenvolvimento sustentável* (2nd ed.). Editora e Livraria Instituto Paulo Freire.

Krenak, A. (2020). *A vida não é útil* (1st ed.). Companhia das Letras.
Lipovetsky, G. (1989). *O Império do Efêmero: a moda e seu destino nas sociedades modernas*. Trad. Maria Lucia Machado. Companhia das Letras.
Ministério Do Meio Ambiente. *Reciclagem*. Disponível em: https://www.mma.gov.br/informma/item/7656-reciclagem. Acesso em: 21 out. 2020.
Novo, C. L. D. C. (2015). *O Lugar do Vagalume: uma análise sob as perspectivas da nova geografia cultural e humanística de um bloco carnavalesco do Rio de Janeiro no início do século XXI*. 192 f. Dissertação de Mestrado em Geografia – Instituto de Geografia. Universidade do Estado do Rio de Janeiro, Rio de Janeiro, 2015. Disponível em: https://www.academia.edu/26727252/O_Lugar_do_Vagalume_uma_an%C3%A1lise_sob_as_perspectivas_da_nova_geografia_cultural_e_human%C3%ADstica_de_um_bloco_carnavalesco_do_Rio_de_Janeiro_no_in%C3%ADcio_do_s%C3%A9culo_XXI. Acesso em: 05 out. 2020.
Per Vivere Bene. (2019). *Bloco sustentável em prol da folia e meio ambiente*. Disponível em: https://www.perviverebene.com.br/post/bloco-sustent%C3%A1vel-em-prol-da-folia-e-meio-ambiente. Acesso em: 12 out. 2020.
Sachs, I. (2009). *Caminhos para o desenvolvimento sustentável*. Org. Paula Yone Stroh. Garamond
Streit, J. A. C. (2015). *Política Nacional de Resíduos Sólidos – a Lei 12.305/10 e as Normas de Gestão Ambiental NBR ISSO 14001: uma análise comparativa*. Universidade Federal de São Carlos, 2015. Disponível em: http://www.ppgsga.ufscar.br/alunos/banco-de-dissertacoes/2015/jorge-streit.pdf. Acesso em: 19 out. 2020.
Veiga, J. E. D. V. (2010). *Sustentabilidade: a legitimação de um novo valor*. Editora Senac São Paulo.
VG Resíduos. *Quais são as vantagens e desvantagens da incineração do lixo*. 24 abr. 2019. Disponível em: https://www.vgresiduos.com.br/blog/quais-sao-as-vantagens-e-as-desvantagens-da-incineracao-do-lixo/. Acesso em: 19 out. 2020.

Portuguese Sustainable Fashion Micro Brands: Strategies in the Covid-19 Pandemic Market Context

Graça Guedes and Juliane Biz das Neves

Introduction

Brands compete for market share and space in consumers' minds. Posner (2015) states that a brand is a composite shaped internally by company strategy and externally by consumer perception. It combines tangible and intangible elements, including garments, retail environment, packaging and advertising, and meanings, values and associations attributed by consumers.

According to Carvalhal (2015), for fashion consumers, buying and relating to brands can be an exercise of searching for oneself, in tune with the world, other people and society. In this sense, a brand is idealised with a focus on its consumers but is also constantly rethought to follow their behaviours changes. Therefore, the trajectory of a brand follows the changes in the behaviour and values of its target consumers.

The Covid-19 pandemic affected billions of individuals, and the measures to fight it have made this event a global catalyst that forced fast and profound societal changes. This event transformed society, shaped new consumer attitudes and forced entire industries to adapt to a new reality that requires flexibility, resilience and creativity (Bell, 2020).

In this scenario, micro and small fashion brands stand out. Due to the size of their structure, they had more flexibility; they could make decisions quickly and assume the imperative of change and adaptation as necessary to survive the new market conditions. The initial survey of the Portuguese sustainable micro fashion brands showed these micro brands entered the market during the last decade. They also were created as sustainable brands, mainly by younger managers with well-defined

G. Guedes (✉) · J. B. das Neves
Universidade do Minho, Centro de Ciência e Tecnologia Têxtil – 2C2T, Braga, Portugal
e-mail: mgg@det.uminho.pt

A. Cruchinho, J. C. Neves (eds.), *Sustainability in Fashion: Creativity, Education, and Innovation*, https://doi.org/10.1007/978-3-031-80991-0_28

purposes and filling gaps left by big industry brands. Those micro brands seemed aligned with their market target values and open to exploring new concepts.

The research aimed to establish the strategies adopted by sustainable fashion micro brands during the Covid-19 pandemic and understand how this pandemic context influenced them. The research is exploratory and uses diverse data collection and analysis procedures. The theoretic framework focused on the pandemic scenario study and the reports published by the National and International Health organisations. The research technic was in-depth interviews with the managers of six Portuguese micro fashion brands selected from the survey carried out and having as selection criteria the accessibility and availability of managers to participate in the research.

Development

The interviewees were the directors of the selected six Portuguese fashion micro brands. The interview script specifically focused on the brand's strategy to face the pandemic context, as the goal was to distinguish these strategies and those that brands had formulated in the period that preceded it. After the content analysis and the interpretation of the interviews, the researchers synthesised the information provided by the brand managers as presented.

The studied micro brands were Cuscuz Design, SIZ, Wayz, Nüwa, Frambooesas and Zouri.

Cuscuz Design

Cuscuz Design is a fashion and design accessories brand that develops handmade products with reused materials. It was created in 2015 by Ana Mendes and became a father-daughter brand since Ana's father, Amâncio Neves, initiated his collaboration with the brand.

The brand (Fig. 1) focused on craftsmanship and became best known for its sunglasses, made from discarded wood. Its team is composed only of a father and daughter, responsible for all functions, including product manufacture.

SIZ Brand

SIZ is a womenswear brand created in 2017 by Raquel and Sofia Rodrigues. It presents an ethical, organic and sustainable proposal in which production is made in small quantities or on demand (Fig. 2). The two sisters manage the brand, and a local sewing atelier is in charge of the production. Each SIZ collection exposes a

Fig. 1 Cuscuz fashion campaign photos

Fig. 2 SIZ fashion campaign photos

theme that is a critical issue to know and solve, such as global warming, excessive plastic consumption or the thawing of glaciers (Marcas Portuguesas, 2021).

Wayz

Wayz (Fig. 3) is a Portuguese brand launched in 2018 by Pedro Maçana and Daniel Gonçalves. It develops sneakers in an ethical proposal, with timeless design and made to last. The brand uses only ecological materials for its production, made locally. Its partners are family businesses with decades of experience and knowledge in footwear manufacturing, where artisans are paid fairly in an environment with good working conditions (Marcas Portuguesas, 2021).

Fig. 3 Wayz fashion campaign and packaging photos

Fig. 4 Nüwa fashion campaign photos

Nüwa

Catarina Lopes founded Nüwa in 2019. Her family owned a textile factory, Confecções Calvi (founded in 1977), and this production company supported the creation of the new brand. Nüwa (Fig. 4) focuses on offering the market sportswear designed and produced under ecological and social values. Confecções Calvi has around 130 employees. Catarina is the Nüwa brand's manager and shares resources and employees with the leading company. The entire manufacturing process is carried out in-house, from the yarn (which they buy) to the making and finishing. The brand wants to make it easier to build an ethical wardrobe by offering only conscious and sustainable options. One per cent of its profits reverted to the AMURT association (Ananda Marga Universal Relief Team).

Framboesas

Launched in 2015 by Teresa Mendes and Marta Tavares, Framboesas is a womenswear brand that wants to bring happiness to its consumers (Fig. 5). Both founders are brand managers, performing all functions from design and photography, customer support and logistics to product design. The brand has four product lines: swimwear, loungewear, general clothing and accessories.

Zouri

Launched in 2018, Zouri is an eco-vegan brand that develops shoes using plastic waste collected from the Portuguese coast and other ecological materials. Adriana Mano and António Barros created the brand since both shared a strong desire to contribute to cleaning the ocean (Fig. 6). They established an environmental partnership with the city government to collect beach garbage. They receive plastic waste from Portuguese beaches that are processed and incorporated into the soles of

Fig. 5 Framboesas fashion campaign photos and scrunchies packaging

Fig. 6 Zouri fashion campaign pictures

the brand's shoes. The shoe structure is wholly produced with organic materials. Zouri is a social and environmental impact project awarded in 2019 by the European Investment Bank second place in Europe's Best Circular economy Project category.

Strategic Response from the Brands to the Pandemic

History shows that in times of crisis, creativity flourishes. For an uncertain scenario, brands need new plans. The Covid-19 pandemic triggered an era of innovation where companies rethink their marketing discourses, way of selling, business plans, product lines and even their purposes (Bell, 2020). So, the first question of the interview was intended to verify the changes the brands introduced.

Cuscuz mentioned they had more time to experiment and introduce new products and started producing decorative objects. SIZ also started to offer some decorative products developed with the macramé technique. Zouri mentioned that they tested various products, from furniture to ceramics and a series of applications to diversify the product line. Soon the brand will present some products other than footwear, consistently applying the brand's concept, marine litter and material reuse. Framboоesas, in addition to its usual line of bikinis and womenswear, launched a collection of elegant pyjamas that was very successful.

This information confirms the movement of fashion brands towards lifestyle products related to the house. This strategy aligns with the research from trends agencies, like WGSN, which shows that fashion brands were entering the decoration market to create ecosystems with various products that expand their universe (WGSN Forecast Team, 2020). It makes sense to think that these practices were driven by people being more at home and valuing this space. When they felt obliged to remain in isolation, they began to pay more attention to the house, and the home became a focus, seen as a true sanctuary.

Still on new products, Nüwa, to encourage positive thoughts, started to produce protective masks against Covid-19 in more cheerful colours. Also, they produced a line of t-shirts with the slogan "everything will be fine". These attitudes demonstrate a concern for getting involved with consumers' well-being, encouraging the feeling of hope, cited by WGSN (Bell et al., 2021) as one of the feelings that should shape consumer behaviour and post-pandemic consumption.

Nüwa also mentioned that they were encouraged by the pandemic to sell comfortable clothes and sportswear, products that people were looking for, because they were more at home. This statement is in line with the casualisation trend, a trend already developing before the pandemic and that during 2020 was highlighted, standing out in comparison to other movements (Amed et al., 2021). The pandemic social context reinforced the demand for casual yet sophisticated clothes, reinforcing the opportunity for success of the elegant pyjamas created by Framboоesas. The forced confinement and the generalisation of homework make comfort an essential differentiator for brands since consumers have become much more focused on physical well-being and mental health. Homes became multifunctional spaces for individuals and families (Bell, 2020).

According to Saunter (2020), the pandemic has made many brands take the lead by developing strategies to help, reassure and connect with the public. The next question of the interview intended to verify if the brands were taking any other action in response to the pandemic, such as differentiated communication actions with themes related to the fight against the virus, well-being or the participation or creation of online events, actions to promote small brands, among others.

Cuscuz cites an action it carried out in collaboration with other independent brands and local artists, where they gave their Instagram profiles to these artists to publicise their work. This action shows an intention to unite the community, a reinforcement of collective thinking for good and a movement encouraged by the pandemic. Brands are finding innovative ways to connect and make a difference as the coronavirus outbreak paralyses the global economy, fueling fear and anxiety (Saunter, 2020).

Also, in response to this question, SIZ, Nüwa and Frambooesas mentioned that they made actions on Instagram involving the well-being component. According to WGSN (Saunter, 2020), brands have offered resources and initiatives to help people care for their mental health in this time of uncertainty. Research shows that the prioritisation of well-being is one of the main trends encouraged by pandemics. Health concerns are one of the global drivers that should influence consumers and the new emotions arising in this context.

Nüwa mentioned the theme of positivity that influenced some actions the brand took in this period. Positivity is related to hope, and a new focus was implemented on this feeling, which recognises individual responsibility to make changes (Bell et al., 2021). Brands that show empathy and carry out philanthropic actions remain in consumers' minds will gain more strength (Saunter, 2020).

The interviewees were then asked what changes they noticed in consumers' behaviour during the quarantine periods and pandemics in general. The brand's managers of Cuscuz, Wayz and Nüwa highlighted the prioritisation of e-commerce during quarantines, which is aligned with what is stated by McKinsey & Company (Lund et al., 2021) that many people have discovered the convenience of e-commerce and other online facilities during the pandemic. Wayz notes that the pandemic has accelerated the power of online channels and brought more awareness to people more predisposed to seek information about products. Cuscuz mentioned that consumers are more interested in discovering small brands. SIZ believes that the more robust presence in the online environment has made the information sharing more intense and, therefore, the appreciation of sustainable brands.

McKinsey & Company claims that regarding strengthening digital culture, companies are now three times more likely than they were before the crisis to conduct at least 80% of their customer interactions digitally (Singhal & Sneader, 2021).

The next question focused on the digital channels and their importance to the brands. A central topic considered if the brand's digital presence after the pandemic confinement kept as vital as it was during it, both in terms of consumer behaviour and brand strategy.

All brands cited Instagram and websites as the most important digital channels. SIZ also cited newsletters and marketplaces, and Zouri referred to social networks but claimed that Facebook is more important in sales. As all brands have an online store, the website is of fundamental importance for sales, and Instagram is cited as a communication channel.

The answers confirm that these channels have become more critical for several reasons. Websites with e-shops were essential, when the physical stores closed. According to Wayz's manager, the pandemic accelerated everything associated with digital, and even people who lived close to the brand studio often prefer to buy online through the website.

Cuscuz explains that they could no longer receive people in their ateliers, so they used Instagram to communicate processes and brand history. Nüwa comments on the importance of the social network in sharing content and interacting with consumers. Frambooesas puts Instagram ahead of everything and realises that communication on the platform reflects sales on the website. Zouri, on the other hand, claims that it is through social networks that it connects with various audiences, of various nationalities, at a lower cost than standard advertising solutions. They argue that digital is more interesting because it can reach more people and mention that a digital communication strategy supports the brand.

In the digital environment, people found information, facilities, entertainment, fun and other experiences, when leisure activities, malls and physical stores were no longer an option (Nexo Jornal, 2020). Research indicates that the impact of digital technology in general and social media in particular on consumer behaviour is enormous and has accelerated dramatically due to pandemic effects. All companies (and everyone else) noticed this change, and small brands profited from it as much as possible.

The following question tested the finding of WGSN Trend Team (2020) that identified the search for truth as a driver influencing consumer behaviour and feelings in the pandemic scenario. The interviewees were asked if they had observed the appreciation of transparency in communication among their clients and how they had responded to this requirement.

All brands agree that there has been a greater appreciation of transparency in recent years. Cuscuz reveals that before, they thought they had to hide the fact of having a small structure and being a family business. However, when they started to show the reality, people gave more attention and valued the brand. In addition to communicating processes and showing the atelier and production in images and videos, the brand has also carried out actions such as the Live Atelier at Moda Lisboa, in which Amâncio demonstrated processes of working with wood to develop glasses in the brand's stand during the event.

Wayz and SIZ mentioned that they show the product's price breakdown, and Wayz shares information about each material used in the footwear components and the name of the partners with whom they work for production.

The research highlights that society is experiencing a period in which disinformation has much power, and the crisis has reinforced the importance of the truth. In this sense, consumers tend to seek more information and transparency in brand communication to develop a relationship of trust with each other and increase the loyalty between the brand and consumer (WGSN Trend Team, 2020).

Sustainability is unquestionably fundamental to the development of fashion brands. Kulsum (2020) writes that sustainable fashion is essential to rebuild the fashion industry after the pandemic. The author believes that the experience of the global pandemic creates awareness about the importance of sustainable fashion. The impacts he anticipated include various possibilities, such as the union of producers, brands, young designers, retailers and consumers to create sustainable fashion solutions for production, distribution and consumption and promote social and environmental changes.

The interviewees were asked if they perceived a greater demand or debate about sustainability since the pandemic's beginning among their clients and population in general.

Cuscuz, Wayz, Nüwa, Frambooesas and Zouri agree that people are more aware of sustainability in fashion. Cuscuz comments that the intensive sharing of local and sustainable projects has brought more visibility to the subject. With the more significant entrance of digital, people have found it easier to find information. Nüwa believes that the news helped spread the word and that pandemics acted as an accelerator, contributing to organisations already working to educate people, such as Fashion Revolution.

The following question focused on the social actions implemented or not by the brands.

SIZ and Cuscuz answered that they are not yet involved in social action but are analysing the implementation of activities with this objective. For the other studied brands, it appears that social actions have been a concern since their creation and are already involved in several actions. This perspective is essential because trends indicate that mindless consumption is no longer an accepted behaviour. Social actions stimulate brands to do more to encourage consumers to buy sustainable products through strategies that create purpose in consumption (Marian, 2020). Consumers tend to regard these attitudes that promote positive changes in a society highly.

Then, respondents were asked about people's interest in local brands to verify their perception about strengthening community sense.

The brands, except for SIZ, agree that people looked more at local commerce after the pandemic, even with an incentive from the Portuguese government. They mentioned that people realised they needed to support local businesses so they could overcome the most difficult periods. This behaviour confirms the reinforcement of a collective sense, concluding that people feel closer to their local communities and buy to support small, independent, local businesses.

The last question referred to brands' vision for the next 2 years and whether they intended to make any significant changes.

Zouri answered that one of its main plans is the diversification of product offers. In this way, they aspire to be a reference project in terms of sustainability. Sustainability is also a concern for Frambooesas, which is working on changing materials to develop more sustainable products. In addition, they mentioned what they have been doing at a social level: shortly, they intend to form an association with an organisation related to mental health. Nüwa cites introducing new products and differentiated lines to add to its offer, which is currently sportswear.

Núwa and Wayz mentioned the expansion to other countries, especially for the European market. Wayz intends to create a second-hand store with its products to show that sneakers can have a second life and encourage circularity. SIZ is focused on sustainability and exploring the issue of noise pollution in the oceans to create awareness about it. Cuscuz wants to be closer to people and its consumers, and it also wants to improve the product's exposure on digital platforms and continue to invest in its storytelling.

Therefore, the diversification of the product offer, contemplating the idea of exploring the lifestyle mentioned above, is one of the actions three of the six brands intend to carry out. Market expansion, sustainability and transparency in communication are strategic movements the brands intend to implement. They also consider it essential to develop their products by communicating the artisanal process behind each one or by reinforcing the concept of timelessness and durability. The improvement of the digital platforms is a development target.

A central concern of all studied brands was to remain faithful to themselves, to their concept and essence. They consider their values as trustworthy guides for their performance in the market.

Conclusion

During periods of significant change or times of crisis such as the Covid-19 pandemic, cohesive brand management becomes even more critical, as a well-built and managed brand can adapt to society, presenting coherence and keeping its purpose.

Small brands align with current times and issues such as sustainability, transparency, ethics, presence in the digital environment and a vital purpose. The pandemic context boosted interest in those values and, in this way, encouraged the consumer to pay more attention to small brands, mainly through the digital environment, which made it necessary to make a more significant effort to communicate and act in digital virtual and social media.

The research demonstrated that micro-sustainable fashion brands could quickly adopt changes to evolve with society by promoting well-being, guaranteeing acceptance, promoting circularity, valuing timeless products and acting with transparency, ethics and inclusion. Furthermore, they must, above all, take advantage of the moment to strengthen themselves through the collective, using the digital environment to create humanised digital connections.

Acknowledgements The authors thank all those involved in the organisation of CIMODE and to all the contributors that will enrichen this event. This work is financed by Project UID/CTM/00264/2019 of 2C2T—Centro de Ciência e Tecnologia Têxtil, funded by National Funds through FCT/MCTES.

References

Amed, I., et al. (2021). The state of fashion 2021, McKinsey & Company. Retrieved from https://www.mckinsey.com/~/media/McKinsey/Industries/Retail/Our Insights/State of fashion/2021/The-State-of-Fashion-2021-vF.pdf. Accessed 6 Mar 2021.

Bell, A. (2020). *Pós-coronavírus: proteja o futuro da sua marca, WGSN*. Retrieved from https://www.wgsn.com/fashion/article/86779. Accessed: 15 Apr 2021.

Bell, A., et al. (2021). *Future consumer 2023, WGSN*. Retrieved from https://createtomorrowwgsn.com/2513751-wp-future-consumer-2023/?_fsi=IsbKGdTp. Accessed: 15 Apr 2021.

Carvalhal, A. (2015). *A moda imita a vida: como construir uma marca de moda* (1st ed.). Senac Rio.

Kulsum, U. (2020). Sustainable fashion as the early awakening of the clothing industry post corona pandemic. *International Journal of Social Science and Business*, *4*(3), 422–429. Retrieved from https://ejournal.undiksha.ac.id/index.php/IJSSB/index%0ALOGO. Accessed 22 Feb 2021.

Lund, S., et al. (2021). O futuro do trabalho pós-COVID-19. McKinsey. Retrieved from https://www.mckinsey.com/featured-insights/future-of-work/the-future-of-work-after-covid-19/pt-br. Accessed 4 Mar 2021.

Marcas Portuguesas. (2021). Retrieved from http://www.marcasportuguesas.pt. Accessed 15 Dec 2021.

Marian, P. (2020). *Coronavírus: aceleradores globais de mudança, WGSN*. Retrieved from https://www.wgsn.com/fashion/article/86569. Accessed 14 Apr 2021.

Nexo Jornal. (2020). 20 características de nosso tempo que o ano de 2020 escancarou. *Nexo Jornal*. Retrieved from https://www.nexojornal.com.br/especial/2020/12/30/20-características-de-nosso-tempo-que-o-ano-de-2020-escancarou. Accessed 22 Feb 2021.

Posner, H. (2015). *Marketing fashion: Strategy, branding and promotion* (2nd ed.). Laurence King.

Saunter, L. (2020). *Coronavírus: estratégias das marcas – WGSN Fashion, WGSN*. Retrieved from https://www.wgsn.com/fashion/article/86633. Accessed 15 Apr 2021.

Singhal, S., & Sneader, K. (2021). *O novo normal chegou: tendências que definirão 2021 – e o futuro, McKinsey&Company*. Retrieved from: www.mckinsey.com/featured-insights/leadership/the-next-normal-arrives-trends-that-will-define-2021-and-beyond/pt-br. Accessed 6 Jan 2021.

WGSN Forecast Team. (2020). *Futuras Inovações 2023, WGSN*. Retrieved from: https://www.wgsn.com/fashion/article/89181. Accessed: 15 Apr 2021.

Fashion Design and Sustainability: From Concept to Production

Rui Barreiros Duarte, Ana Duarte, and Ana Paula Pinheiro

Introduction

> Sustainability (…) is a balanced use of natural, social and economic capital for the continued health of the planet and future generations (Sherin, 2008).

In the current market, competition is aggressive and consumers don't always appreciate the price difference that comes from opting for sustainability. Environmental values are not a prevailing option, as it can be perceived in the way plastics and other polluting materials are used.

Individual awareness of the problem is not enough to change behavior. There needs to be a social movement that acts as a value system. This promotes actions in environmental and animal protection areas with positive results. One of the next steps must involve clothing production from the ground up.

This is a current and important issue, but it is essential to promote communication.

Thus, C.L.A.S.S. (Creativity Lifestyle, and Sustainable Synergy) was created, an international Eco Hub that has been advocating a new generation of fashion where the fusion of design, innovation, communication, and responsibility shape an informed and competitive business, both economically and socially.

The media is starting to report the impact of clothing waste. It is in this area that material must be selectively reused in order to reintroduce it into the production

R. B. Duarte (✉)
CIAUD, Research Centre for Architecture, Urbanism and Design, Lisbon School of Architecture, Universidade de Lisboa, Lisbon, Portugal

A. Duarte
Duarte, Lisbon, Portugal
e-mail: duarte@duartehajime.com

A. P. Pinheiro
RBD.APP – Arquitectos, Lisbon, Portugal

A. Cruchinho, J. C. Neves (eds.), *Sustainability in Fashion: Creativity, Education, and Innovation*, https://doi.org/10.1007/978-3-031-80991-0_29

industry. We are facing new desirable situations: the reuse of fibers that would not have been used in the past, industrial waste, and research into new environmentally friendly materials.

The Duarte brand searches for surplus material and industrial waste to obtain working materials.

However, it must be ensured that the materials have the least possible impact on the environment and sustainability of the planet. Some of these materials, despite being recycled and contributing to zero waste, have a negative impact on the environment. One example is indigo denim, which needs a lot of washing to achieve the desired color, consuming a lot of water.

The behavior of materials in terms of resistance and suitability for each other is another matter that design must guarantee, in a process that has multiple variables.

Sustainability, and New Materials

At this stage of production development, there is a market problem that needs to be overcome: the sustainable materials used in fashion design are more expensive because they interfere with the industrialized production system. The need to adapt to new processes cannot be extended to all products. However, there is research on alternative environmentally friendly materials and exploring new sources of supplies which boost industrial innovation.

This is the case of New Generation Materials (*Amni Soul Eco, Roica*); Bio-Based Materials (*Agraloop, Appleskin, beLEAF, Crabyon, Desserto, Ingeo, Piñatex, Veg*ea); Man Made Materials (*Eastman Naia, Lenzing Ecovero*); or Bio-Fabricated Materials (*Zoa, Mylo, Fine Micellium*).

These are materials researched by the industry and disseminated by C.L.A.S.S., created through non-polluting alternative production sources, and which safeguard animal life.

In an article by Emily Chan, Vogue International UK also mentions 12 environmentally friendly materials that are being produced to create six alternatives to leather (three mentioned in C.L.A.S.S. plus *Reishi, Mirum,* and *VitroLabs*), recycled materials (*Circulose* and *NuCycl*), bio-based materials (*Kintra, Clarus*), and those with a negative carbon impact (*AirCarbon, LanzaTech*).

These are materials that use alternative sources and address the problems mentioned. Thus, in the context of recycled materials, we can mention the denim Levis has been using since the beginning of 2022, in its 501 *Circulose* jeans. It's a fabric made from 100% discarded clothes. In the manufacturing process, the cotton from the clothes is separated and dissolved, adding wood pulp afterward, in order to produce a type of viscose (Chan, 2022).

NuCycl also uses 100% recycled clothing. It is a product created from cotton waste in the post-consumer phase, used since 2019 by Stella McCartney when she launched the *Infinite Hoodie* with the Adidas brand.

We are dealing with two matters: recycled materials and new materialities, which intersect in this area. However, in the other areas mentioned—alternatives to leather, bio-based fabrics, and carbon-negative materials—there are used products that highly differ from traditional sources, which are now used to obtain new materials or equivalent products.

Thus, the alternatives to leather include the use of mycelium (fungal roots) used in *Mylo*, although it is not completely plastic-free.

Reichi is identical to leather, with the same smell and appearance, and is tanned without the use of chrome or synthetic products. Due to its characteristics, it can be finished in various ways and is biodegradable (Holzmeister, 2020).

Mirum has no plastic, and the material can be completely circular, being made from plants and minerals.

Piñatex is made from pineapple waste, although it contains PLA bio-based plastic. Its coating uses PU (polyurethane leather) which gives it greater durability.

Vegea is made from grape waste from the wine industry. This material still contains 45% PU.

VitroLab refers to a research that replicates the feel of leather using some animal cells.

C.L.A.S.S. also mentions *Appleskin*, which is a leather that comes from apple processing, or *Desserto*, a leather that comes from a process of processing cactus treated with non-toxic products.

Fine *Micellyum* makes it possible to obtain leather from *Micellyum* roots, a fine network of threads that forms mushrooms.

As for bio-based materials, we can mention *Kintra* (100% organic), which aims to be an alternative to polyester and nylon. It uses sugar derived from corn and wheat.

Clarcus transforms natural materials—cotton, hemp, and wool—into high-quality products with the same performance as synthetic materials.

Hemp is five times more resistant than cotton. Taking advantage of its qualities, *HEROTEX* (92% hemp, 4% lycra, and 4% polyester) is an elastic hemp patented in 2019 in Italy.

As for carbon-negative materials, *AirCarbon* absorbs CO_2 from nature by recycling natural processes. Thus, the marine organisms used convert methane and carbon dioxide into a molecule that can be melted.

LanzaTech converts CO_2 emissions from steel mills into ethanol and at the end of the process produces polyester.

Leading brands on the international market and large investments support this research in order to innovate in ecological terms, but the other side of the coin must always be taken into account. Replacing a base material or a process involves assessing what is gained and what is put at risk in the whole process. For example, using materials that conflict with the world's food supply is a problem, as is the case with corn and wheat. This is what happened in Mexico when they wanted to replace oil with an alcohol derived from corn, which increased the price of tortillas. In the creation of bioplastics, corn was also used (first generation), but in the next two generations, the bases are nonfood products, in order to avoid the negative impact of side effects.

What we are referring to is the production system, its sustainable viability, and its circularity, in addition to the type of material produced industrially in technologically developed countries. There were mentioned a few market dynamics that allow us to clarify the state of the art and position ourselves in terms of conquering the future.

However, consumers don't always value price differences. Their needs are driven by the economic value of the purchase. In terms of recycling, companies would need to introduce new tools and strategies into the production chain to create new sustainable value to improve business models in the future. It is a complementary alternative to production that exploits waste, as mentioned above. It's the same principle used in other industries such as stone, where some companies produce reconstituted stone from waste and stone dust.

In addition to sustainability, in ethical terms, health and justice are now priorities in the fashion industry.

The social distancing caused by COVID-19 has highlighted the importance of digital channels.

The legislation that has been produced in recent years is part of these concerns, involving NGO, governments, suppliers, and buyers. Everyone is more aware, more selective and demanding, more informed about the product cycle (origin, material characteristics and impact on health, production and transportation conditions), and the possibility of fabrics carrying a label is being considered, informing the public and institutions about the environmental impact they generate.

Another aspect that would boost the implementation of this system would be the creation of incentives such as reducing taxes and speeding up the dissemination of this type of product. It's a social aspect that involves politics, culture, and communication working together toward a social and environmental goal.

Traceability

Traceability is "the ability to identify and trace the history, distribution, location, and application of products, parts, and materials, to ensure the reliability of sustainability claims, in the areas of human rights, labor (including health and safety), the environment and anti-corruption." (Househam et al., 2014).

Traceability is a sustainability tool that guarantees the best conditions throughout the production process by linking goods and products, guaranteeing good production practices, and respecting people and the environment. It allows control of the entire life cycle by monitoring certifications.

The development of innovative technologies that respond to ecological needs and the growing demand for organic products has led to the optimization of the sustainability framework according to a process that works as a whole (Fig. 1).

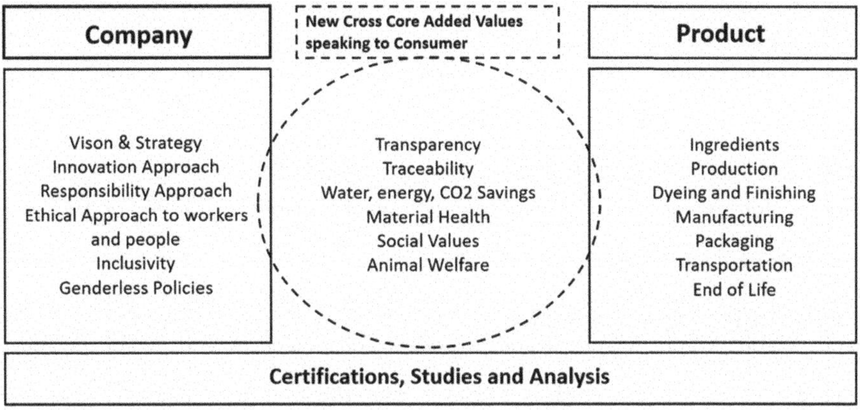

Fig. 1 Responsibility: sustainability model proposed by Bettoni, during the C.L.A.S.S. Workshop for MODALISBOA, November 5, 2021

Cradle to Cradle

The integrated cradle to cradle philosophy considers the world to be an integrated global system that functions as a network of interactive ecosystems where the misuse of resources jeopardizes these systems and the future.

We can outline three concepts of development toward sustainability, progressively more sustainable: *eco-efficiency; eco-effectiveness; eco-effectivity*.

Eco-efficiency refers to the reduction of environmental impacts in the supply of goods and services, but does not alter the already established system of production.

The inadequacy of this concept has led to the operationalization of *eco-effectiveness*, which is more appropriate as it indicates the direction to be followed in order to solve the sustainability problem. Adopted by Manzini and Vezzoli, this concept separates sustainable technical and socio-cultural innovations from non-sustainable ones.

The concept of *eco-effectivity*, developed in "cradle to cradle" by McDonough and Braungart, refers to a perpetual industrial production that works in a closed cycle capable of feeding back on itself. Using only its materials, it does not generate any emissions or waste. This system can be a beneficial form of assessment in ecological terms, as it considers a change in the production paradigm without reducing it, but by structurally changing the way we produce. It emphasizes the need to change the means and modes of production by focusing on innovation.

Scientific research is moving beyond reuse, which is just one step in the process, to focus on *eco-effectivity*, which emphasizes trade and the common good in which it is rooted: ECOLOGY | EQUITY | ECONOMY (Braungart & McDonough, 2014).

Eco-effectivity systematizes the processes of changing products and systems into five stages:

Step 1: Abstain from the use of hazardous substances such as mercury, cadmium, lead, PVC.
Step 2: Select products that can be returned to industrial metabolism; respect the consumer and the environment; give priority to celebration and enjoyment.
Step 3: Create a list of damage caused by manufacture or use.
Step 4: Design using only substances that are known to be benign.
Step 5: Create products that actually improve the environment (Pinheiro, 2017).

Case Study: Duarte Brand

Duarte brand, created in 2015, is guided by the principles of sustainability, aiming to raise awareness of environmental and social problems. With pieces produced in Portugal with sustainability in mind, the brand monitors production, ensuring that Fair Trade standards are followed. The logistics, from making the garments to sending them to the consumer, are all done within a radius of less than 300 km, which contributes to a reduction in the carbon footprint.

Ana Duarte (the brand's founder) designs unique prints for each collection and creates different types of textures using natural and technological knits and fabrics.

Having practiced judo since the age of four, the designer is particularly sensitive to ergonomics and movement, creating comfortable pieces of excellent quality.

The garments are generally unisex, to cover different body types, using recycled, organic or surplus materials, always with the goal of zero waste.

Besides entering the market using raw materials that don't harm the planet, sustainability can be talked about through storytelling. Some of the themes of the collections produced by the Duarte focus on environmental and social issues in the field of streetwear.

In this sense, four collections have already been presented with concepts relating to environmental concerns: AW20 Third Pole; AW21 Reef; SS22 World Keeper; AW22 Season2.

AW20 Third Pole: Crossing environmental concerns with travel, the "Third Pole" theme draws attention to the Himalayas, which holds more fresh water than any other region in the world (apart from the North and South Poles). The first expedition to Everest in the 1950s radicalized the exploratory adventure in adverse environments, which today involves other regions in danger due to climate change.

In partnership with the street artist *Edis One*, who is developing the "Extinction" Project, the collection draws attention to five types of endangered animals in the Himalayas: the Snow Leopard, the Golden Takin (goat-antelope), the Red Panda, the Red-necked Calau, and the Black-necked Heron (Fig. 2). Based on these references, the project used recycled cotton and warm wools in shades of blue and green.

AW21 Reef: The theme of the great coral reef draws attention to the rapid destruction of the ecosystem and how we can reverse the situation (Fig. 3). On the imaginary journey you discover underwater cities, where life has been replaced by increasing stillness. The colorful patterns express the living systems and the white

Fig. 2 Duarte. AW20: Ana Duarte's drawing based on *Edis One*'s drawings

Fig. 3 Duarte. AW21Reef: Happening to draw attention to the destruction of the coral reef, Moda Lisboa. (Photograph: Ugo Camera)

Fig. 4 Duarte. (SS22 World Keeper: Photographs: Ugo Camera (left and right). Ruben Branches (center))

tones, the dead corals. The cool sustainable style uses recycled organic cotton, upcycled marine plastic from the *Sequal Initiative*, technical fabrics such as neoprene in different shades of coral, produced in small quantities just for the collection.

At the end of the fashion show, Duarte created a performance in which the models undid a mesh (created with leftovers from previous collections), which represented the act of algae coming off the coral—known as coral bleaching. Underneath, white T-shirts symbolized the lifeless corals, which read "*Coral Bleaching*" and "*Reef in Danger.*"

SS22 World Keeper: Using simple and appealing communication in defense of sustainability, the project stars the dog Tadao, "the guardian of the world," who fights against the protagonists of adversity: Smog Man (air pollution), Fire Man (climate change), Deforestation Man (destruction of forests), and Wave Man (excessive water consumption) (Fig. 4).

The garments, mostly unisex, make production profitable and use environmentally friendly materials, investing in sustainability and traceability (Fig. 5).

The dark tones of gray, cream, blue, and black are combined with colors that indicate hope in an urban culture that transforms the negative aspects of the past.

Fig. 5 Duarte. AW22 Season 2. (Photograph: Ruben Branches)

C.L.A.S.S. Icon Award 2021

C.L.A.S.S. (Creativity Lifestyle, and Sustainable Synergy) is a driving force behind sustainability. Since 2007, it has been promoting the development of a new generation of sustainable businesses, linking the author and the market.

It has a four-pronged approach: Design, Sustainability, Innovation, and Communication. Design focuses on circular economy strategies; Sustainability refers to the company, production, and well-being; Innovation covers circular business models, processes and ingredients; Communication is part of storytelling values and should combine story-making values.

According to Giusy Bettoni (CEO of C.L.A.S.S.), 86% of people think that transparency in business is more important than ever. Many consumers see durability as a good investment.

Ana Duarte was the winner of the *C.L.A.S.S. Icon 2021* competition.

This competition, promoted by C.L.A.S.S., aims to alert contemporary consumers to the values of sustainable fashion. Each year, it rewards designers who combine sustainable design with innovation and responsible communication.

> How can we activate a new generation of values that appeal to contemporary consumers?
>
> With a perspective focused on design, innovation, responsibility, and communication, in the four-dimensional approach of the C.L.A.S.S platform, we went deeper and discovered how the Duarte brand is putting all this into practice (Designers, 2022).

Among other advantages, this Award supported Brand Communication and allowed free attendance for two seasons (September 2021 and February 2022) to the *White Milano Fair* in Italy.

Conclusion

We are in front of an innovative industrial production dynamic that focuses on the circularity of the production system, avoiding the use of polluting materials that degrade nature and interfere with animal life. The case of leather is a clear example of how alternative sources can be used without losing quality. We are talking about a product with a lifespan of around 10 years that is compatible with the changes introduced by fashion and innovation. Other cross-cutting aspects of the industry are meta-materials that introduce characteristics that don't exist in nature. Although this "leap forward" is important, the principle of "*cradle to cradle*" cannot be ignored. Systems must be closed; the entire product cycle must be taken into account in order to avoid waste. We can say that attitudes and status are changing. This symbolic value of positioning social success has gone from minks and mink fur, which are still sold on a large scale in Denmark today; to all kinds of products that resulted from the consumerist society of the 1960s with the "use-it-and-throw-it-away" spirit, generating pollution that has indestructible plastics as its corollary; to the philosophy in which luxury and elegance derive from design using sustainable products.

We are undergoing a paradigm shift and sustainability, from concept to production, and this is a dynamic that also interacts and needs to be further developed. An ethical, responsible sense with a great aesthetic and environmental effect is successfully replacing the old status icons. Progress changes the sense of the future that must be grasped in order to exist. Duarte has explored sustainability, while looking to the next step, always according to "*cradle to cradle.*"

References

Braungart, M., & McDonough, W. (2014 [2002]). *Cradle to Cradle, create and recycle limitlessly*. Translated from English by F. Bonaldo (pp. 9–11, 165–184). Editora G. Gili.

Chan E. (2022). *12 more sustainable materials that could change the future of fashion*. Vogue International, [online]. https://vogue.globo.com/um-so-planeta/noticia/2022/07/12-materiais-sustentaveis-que-podem-mudar-o-futuro-da-moda.html. Accessed 16 Sept 2022.

Designers, A. (2022). *Fast talks*. [online] Modalisboa.pt. https://www.modalisboa.pt/pt/fasttalks/modalisboaandnowwhat. Accessed 11 Sept 2022.

Holzmeister, S. (2020). *Discoveries of tissues from plants grow* [online]. https://harpersbazaar.uol.com.br/bazaar-green/crescem-as-descobertas-de-tecidos-vindos-das-plantas/. Accessed 15 Sept 2022.

Househam, A., Bombis, E., & Liew, D. (2014). *A guide to traceability. A practical approach to advance sustainability in global supply chains* (p. 6). United Nations Global Compact Office. [online]. https://www.bsr.org/reports/BSR_UNGC_Guide_to_Traceability.pdf. Accessed 10 Sept 2022.

Pinheiro, A. P. (2017). *Architectural rehabilitation, sustainability and design*. PhD. Lisbon School of Architecture, Universidaden de Lisboa. [pdf] online. https://www.repository.utl.pt/handle/10400.5/14115. Accessed 10 Sept 2022.

Sherin, A. (2008). *SustainAble: A handbook of materials and applications for graphic designers and their clients* (p. 12). Rockport.

Art, Crafts and Design: Interrelations

Maria Silvia Barros de Held

Some Clarification on Art, Design and Craftsmanship

According to Juan Acha (1978:3), "art is very far from beginning and ending with production (work)". In the case of this study, this statement can also be extended to crafts, noting that production becomes only one of the starting points for analysing an object, whether it is considered art or craft, since it also includes other aspects, such as distribution and consumption.

Acha does not see distribution as the ultimate cause that determines production, but, in contrast to Acha, for Lauer (1983: op. cit.: 77), in the case of pre-capitalist plastics, "the pressure applies to distribution, which is the phase in which the profits that legitimise activity in the capitalist system are generated, as well as the phase from which production is influenced" (ibid.: 70).

To distribution concerns one of the aspects relevant to where sales occur. For example, small object production workshops have exhibitions, usually annual, as an almost traditional procedure. Most of the time, what you find there is what Chiti (1971: 8) calls "entertainment with a pseudo-aesthetic cocktail, an elementary level of exhibitionism, or the cheap narcissism of the middle class".

We must also consider the imitative aspect of reproducing objects. In this sense, Chiti (op. cit.: 7,8), although commenting on ceramic production, believes it to be pertinent and relevant in the case of this study, both for art and crafts:

> ...glue vans have always occurred in art, but never as much as today (...) work based on imitating the exterior of each style, but ignoring the philosophical connotations behind it (...). You don't understand its cultural meaning, you ignore almost everything, and so you adopt handouts or external and superficial symbols of the thing to achieve an only apparent update. And then there are quite ridiculous incongruities where, to cite one case, a certain

M. S. B. de Held (✉)
São Paulo University, São Paulo, Brazil
e-mail: silviaheld@usp.br

A. Cruchinho, J. C. Neves (eds.), *Sustainability in Fashion: Creativity, Education, and Innovation*, https://doi.org/10.1007/978-3-031-80991-0_30

person with a super-bourgeois, conservative and reactionary mentality finds himself making and exhibiting 'poor art'. (...) Gaffes of this kind are committed when one works without a clear ideological and philosophical sense of what one is doing and simply repeats what others do without fully understanding it.

Still on the subject of reproduction of copying, Mário de Andrade, in 1938, already made the following comments:

Another distressing issue concerns the repetition of forms or creations, seen as novelty or revelation. Commitments to maintaining models or incorporating new themes to build objects go beyond mastering techniques or individual discoveries. The model exists as a mark of the identity of that moment, which the group may want to continue, but with the autonomy to partially transform the model or even replace it with another. There is also a fantasy of what is a type when others see the culture as perpetuating formal aspects that fit in with the desired characteristic or typical content. It is also necessary to be aware of the implications of commerce, tourism, the state, fashion and the intervention of intellectuals in the discoverers of the typical, sometimes as artisans and sometimes as objects. The repetition of the model lies in the use of a technique for an accepted product and creation; what there is of it emerges in the rebellion of this model as a transgressive form of repetition. First and foremost, handicrafts bear witness to the man/nature complex. It is through material culture that the mastery of technique and the type of object will be telling about the space in which it is made, either through physical aspects or through the ideology of the culture itself. (Andrade, 1938: http://www.eba.ufmg.br/alunos/kurtnavigator/arteartesanato/filos-03-artesao.html).

It would not be appropriate here to discuss the erudite and popular aspects. Still, we can observe Canclini's reflection (1980: op. cit.: 78) regarding the determination of the artistic in plastic production, in other words, to which variations the artistic aspect is subject, according to the time and, within them, to the relationships that men establish with objects, there being no constant properties either in the objectives of the producer or in work and even less in the perceptive habits of the receivers throughout history, in other words, everything is mutating. In this respect, fashion also fits in.

In this sense, Mário de Andrade also states that:

When we talk about people, we talk about identity; when we talk about social groups, regions, communities, and nations, we have a culture. The initial and basic image that guides what is handmade is born on the level of doing, mastering knowledge and technology, and executing with one's hands is the most representative of the prototype of being an artisan, making crafts, and characterising the handmade object.

What matters is that the support of the tools takes on the condition of an extension and projection of man's body, multiplying possibilities in transforming and revealing in the interventions that he, man, makes with nature as an individual and translator of his culture. In this way, the object becomes a testimony not only to technical knowledge but, above all, to a vision of the world, its revelation, man and society, engaging in dialogue in an attempt to tell who they are through what they do, signifying for themselves and their group the symbolic values of those who experience their cultural model (op. cit.: http://www.eba.ufmg.br/alunos/kurtnavigator/arteartesanato/filos-03-artesao.html).

As far as design is concerned, its origins have already emerged amid various controversies, made even worse by the aspect of its function, with discussions that at first involved both art and technique. These discussions seem to have evolved and, it seems, will never cease.

We agree with Mazza et al. (2007) when she states that, on the other hand, industrial design straddles two aspects: the practical nature of technique and the non-utilitarian nature of art. Still, a robust sign marks design for the same authors: its projetual aspect. Thus, they state:

> ...historically, however, the transition from one type of manufacturing, in which the same individual conceives and executes the artefact, to another, in which there is a clear separation between designing and manufacturing, is one of the fundamental milestones for characterising design. (Mazza et al. 2007: 3)

According to Ferreira Gullar (Salgado & Franciscatti, 2014:6), although craftwork has been seen as an inferior activity since antiquity, the profound difference between art and craft occurred from the Renaissance onwards, in the construction of churches at the end of the Middle Ages, when teams were divided on building sites between artists (who acted individually) and artisans.

Gullar also states that the artist:

> ...it differs from the craftsman who continues to produce objects of use and is tied to traditional forms. One of the characteristics of handicrafts, in contrast to the art that was born at the time, is that it is characterised by the search for new forms and styles, while handicrafts are conservative and repetitive. Experience is passed down from father to son, not as aesthetic knowledge or stylistic form, but as the object's shape, i.e. a cup is made like this, a tray is made like this, a chalice is made like this. (Salgado & Franciscatti, 2014: 6)

Although Ferreira Gullar's text was published in 1994 and Mário de Andrade's even further back, in 1938, they are believed to remain extremely relevant today.

At the time, Mário de Andrade stated that:

> Unfortunately, craftmanship is one part of art technique, the most despised, but art technique is not just about craftsmanship. Craftsmanship is a technique that can be taught, but there is a part of art technique that is the objectification, the concretisation of an inner truth of the artist. This part of the technique obeys secrets, imperative whims of the subjective being, in all that he is as an individual and a social being. This is not taught, and reproduction is imitation. This is what we call the technique of Rembrandt, Fra Angelico or Renoir, all three of whom differ profoundly not only in the conception of the painting, but consequently in the technique of making it. (Andrade, http://www.eba.ufmg.br/alunos/kurtnavigator/arteartesanato/filos-03-artesao.html)

The Artistic Object, the Craft Object and the Design Object

The discussion between art, handicrafts and design is intended to show that deep down, there are many more connections between the areas, almost complementary than many divergent aspects.

According to Alvarez and Barraca, a set of factors such as industrial production, the establishment of an art market and increased consumption led to a "revolution" in art, where new materials were introduced, as well as a change in thinking, such as the realisation of the possibility of attributing aesthetic value to mass-produced everyday objects, as well as experiments in the graphic-visual field. Art Nouveau was one of the artistic movements that emerged during this period, including in fashion, which shared much of this revolutionary new thinking in art.

These changes led to the creation of the German Bauhaus school, which aimed to unify art with crafts, engineering and fashion (Sonia Delaunay). In this context, the concept of design emerged. At the same time, an artist called Marcel Duchamp created sculptures with ordinary objects, which he called *ready-mades*.[1]

It is known that, for some, one of the characteristics of design is mass production, so, taking this approach, Duchamp's works could be considered an anticipation of design, not only because of the concept of mass production, but also because of the use of everyday objects, giving them a new perspective, which in design terms would be redesign, when objects are given a new meaning.

Charles Bezerra, in "The humble designer", goes a little further in his discussion of designers and says that anyone can be a designer as long as they create and/or adapt something to solve a problem. For him, design is present in all areas of life, including nature. This means a competitive advantage since there is evolutionary biology in nature, where design is evolution itself.

Some authors relate art to labour, such as Karl Marx. For him, art has its origins in work, since all variations of labour result from human needs, and in this case, art is man's aesthetic need.

Which leads to the reasoning that art and craft emerged as a single strand, and from the separation of the concepts of labour and art, there was also a need to separate art and craft. This is mainly due to the evolution of industry and the mechanisation of trades/work.

Regarding comparative literary research between Munari, Baxter, 2000, Bonsiepe (1983) and Löbach, 2001, Bruno Munari would be the introduction to the world of design methodology due to the dynamic presentation and understanding of the phases. This is followed by Baxter, where design methodology is analysed as a tool for implementing innovation, and then Gui Bonsiepe. Bernard Lobach presented his methodologies aimed at and applied to both social aspects and products. It is worth noting that none of the methodologies studied by the aforementioned theorists are proven to be definitive and exact in terms of their final results. Still, they are presented experimentally within the multiple literatures that have already been published.

This discussion of existing differences is recurrent in other authors, such as Richard Sennett in The Craftsman (2009:79). The excerpt below shows that this is not an easy discussion, as it involves society as a whole. The similarities between the two areas are more present in the art/craft relationship than in the craft/art relationship.

[1]*Ready-made* is the term coined by Marcel Duchamp himself to name some of his works. These works are produced with everyday objects chosen at random (without aesthetic concern) and made in series.

> Probably the most common question about crafts is how they differ from art. In numerical terms, it's an irrelevant question; professional artists comprise a tiny fraction of the population, whereas craftsmanship extends to all professions. In practice, there is no art without craft; the idea of a painting is not a painting. The dividing line between crafts and art may seem to separate technique from expression [...].

Marta Feghali (2010:2) also discusses this relationship between art and craft, emphasising that the art/craft relationship is more frequent than the craft/art relationship, as previously pointed out. To explain this phenomenon, she relies on four fundamental points: "the mastery of knowledge of the artist's work as a craftsman, thematic coherence, originality and commitment to contemporaneity". Mastery of technique is the point at which the segments come closer together since, for the work to be accomplished, there is a need for the "act of working", as Marta puts it:

> [...] the craftsman is uncompromised of the content of originality while the artist needs to be an artisan for the labour of the work. Still, he is entirely an artist when he starts from scratch, something genuinely new, his invention. That's why we usually find references to the "artist-craftsman" and not always the other way around. In what the craftsman proposes, in what he does or intends to do, originality is not a constant and repeated situation: it is merely occasional, so it is rarer to find a craftsman artist. (Feghali, 2010, p. 2)

This shows the relationship between art and craft, with the differences between the two areas. It can be seen that these differences are mainly present in the following points: originality and a commitment to contemporaneity, that is, the ability to generate new, innovative works that contain between their lines a portrayal of the historical moment, in the sense of contextualisation, since these factors are "demanded" more of artists than of craftspeople.

The semantic and symbolic aspects decided by the "designer-artist", an expression used by Mazza (op. cit.), will be projected into the aesthetic quality that the product must express.

Thus, interpretivism, based on the designer's creative process, is now considered, according to the same authors, to be one of the most important aspects in post-industrial societies, since the referent is no longer the object but the message. Thus, for Mazza (op. cit.), design also takes on an expressive form close to art.

As Cardoso (2004:15) says, "... design, art and crafts have a lot in common and today when the design has reached a certain institutional maturity, many designers are beginning to realise the value of rescuing the old relationships with handicrafts", in which case fashion is also included.

Although some authors believe that design does not relate to craft production, broader concepts recognise the use of craft as an instrument of design expression, highlighting the transformative potential of its intervention in traditional environments and its ability to add value to the object, including in fashion. In this sense, Maria Luiza Castro (2009) states:

> ...This theoretical discussion has important implications for the application of public policies and design methodologies in less industrialised regions since it accompanies emerging social trends and enables a coherent search for sustainable development, including not only environmental issues but also social and economic ones, as advocated by the World

Commission on Environment and Development. http://www.periodicos.uem.br/ojs/index.php/EspacoAcademico/article/viewFile/7356/4809

Conclusions

When faced with an unsolvable obstacle within a particular stage or phase of a design-related project, a retro solution is proposed based on Bonsiepe, the opposite direction of the same flow, to qualify the possible origin or gap found in the course of the design methodology. With the proposed visualisation of the method, focused on the creative and aesthetic spheres of design, the structure of a design method will be easier to understand and practise depending on the project's complexity.

As can be seen from this study, there is proof of the hypothesis that states: over time, there have been several intersections, in various aspects, between art, craft and design, and these have contributed to the conceptual evolution of each of the sectors and, at the same time, to the interaction between them. The cross-cutting link between them encompasses fashion and, nowadays, sustainability issues.

References

Acha, J. (1978). *Las Artes Plasticas como Sistema de Produccion Cultural*. Cuadernos Del Centro de Investigacion de Artes Visuales.

Baxter, M. (2000). *Projeto de Produto: guia prático para o desenvolvimento de novos produtos* (2nd ed., p. 262). Blucher. Trad. Itiro Iida.

Bonsiepe, G. (1983). *A Tecnologia da Tecnologia* (p. 196). Blucher.

Canclini, N. G. (1980). *A Socialização da Arte: teoria e prática na América Latina*. Editora Cultrix.

Cardoso, R. (2004). *Uma Introdução à História do Design*. Editora Edgar Blücher.

Chiti, J. F. (1971). *Curso Practico de Ceramica* (Vol. I). Ediciones Condorhuasi.

de Andrade, M. (1938). *O artista e o artesão* [online]. Aula inaugural dos cursos de Filosofia e História da Arte, do Instituto de Artes, da Universidade do Distrito Federal. Avaliável em: http://www.eba.ufmg.br/alunos/kurtnavigator/arteartesanato/filos-03artesao.html [Acesso em 07/02/2016].

de Castro, M. L. A. C. (2009). *Entre arte e indústria: o artesanato em suas articulações com o design*. Revista Espaço Acadêmico, no. 102. Avaliável em: http://www.periodicos.uem.br/ojs/index.php/EspacoAcademico/article/viewFile/7356/4809 [Acesso em: 10/01/2013].

Lauer, M. (1983). *Crítica do Artesanato*. Editora Nobel.

Löbach, B. (2001). *Design industrial: Bases para a configuração dos produtos industriais*. Trad. Freddy Van Camp. Editora Edgard Blücher Ltda. São Paulo.

Mazza, A. C. A., Ipiranga, A. S. R., & Freitas, A. A. F. D. (2007). O design, a arte e o artesanato deslocando o centro. *Cadernos Ebape. BR, 5*, 01–11.

Salgado, M., & Franciscatti, K. V. S. (2014). *Contraponto entre Arte, Artesanato e Trabalho: A falsa diferenciação e a atrofia da fantasia* [online]. Minas Gerais: Depto. de Psicologia (DPSIC) e Lab. de Pesquisa e Intervenção Psicossocial (LAPIP) da Universidade Federal de São João Del-Rei (UFSJ). Avaliável em http://www.ip.usp.br/laboratorios/lapa/versaoportugues/2c43a.pdf [Acesso em 20/10/2015].

An Interactive Artifact Promoting Cultural Sustainability: Case Study of the Sliiv Project

Selma Pereira and Rui Dias

Introduction

Sustainability in fashion is not limited to garments produced with organic or recycled fabrics. Sustainability in fashion is a process of change in ways of thinking and in the very practice of planning a fashion design collection, in the choice of raw materials, in the production, communication, dressing, and in the enjoyment of fashion. Aware of the interdependence of the individual, social, environmental, and cultural implications of the fashion industry, sustainable fashion comes to question the system that prioritizes profit over social and environmental concerns, valuing diversity, culture, traditions, prosperity, and the well-being of people and the environment.

The production of fashion products requires skilled labor, throughout its entire production process—from sourcing textile fibers to making and finishing fashion products. This cycle when well directed can generate livelihoods, support ancestral textile traditions, and inspire and influence creative communities around the world.

Fashion, besides being a form of creative expression, has an enormous social, cultural force, with the potential to foster communication between individuals, communities, and entire nations.

In addition to the functional and commercial side, fashion can be the subject of other approaches. In this chapter, we will focus on the art-fashion. Art-fashion is the result of contemporary art, influenced by the fashion system, in which the artistic

S. Pereira (✉)
ISMAT – Ensino Lusófona/CICANT, Lisbon, Portugal
e-mail: p7087@ismat.pt

R. Dias
Instituto Politécnico de Castelo Branco, Castelo Branco, Portugal
e-mail: ruidias@ipcb.pt

A. Cruchinho, J. C. Neves (eds.), *Sustainability in Fashion: Creativity, Education, and Innovation*, https://doi.org/10.1007/978-3-031-80991-0_31

merges with the commercial and fashion to create artistic objects embedded with characteristics of the fashion system (Carvalho et al. 2017).

The emergence of fashion-art in the contemporary art scene has flourished with post-digital aesthetics. Post-digital artifacts question and force us to rethink the way fashion and textiles are displayed in galleries/museums, their place as artistic objects, and the interaction between them and the public. In this sense, we interpret artifact as the fruit of a creative purpose, capable of addressing a specific aesthetic discourse, the result of the union of the notion of work of art with potentially utilitarian purposes. In many cases, they are objects of desire that become agents for inducing critical thinking. Post-digital art-fashion artifacts are artifacts with primarily artistic intentions and with less commercial and marketing focus (Pereira & Fernandes-Marcos, 2020).

The post-digital fashion artifacts have been exploring the textile materials in an imagetic perspective, assuming itself as an expression of itself inducing the appeal to the sensorial experience and to the immediate senses in the search for aesthetic pleasure. Fashion-art, more than the creation/exhibition of artifacts for passive appreciation of its visual aesthetic discourse, has also tended to the development of artifacts that appeal to the touch, smell, hearing through the handling of materials, providing a multisensory immersion of the work.

Theorical Background

Cultural Sustainability

Handmade textile work requires learning, in several cases time-consuming, manual dexterity, tacit knowledge, and access to suitable raw materials. However, handcrafted work is still one of the most exploited areas of fashion, because despite requiring a high level of skill, the high number of hours of manual production makes the process/products more expensive, making it impossible for them to compete with fast fashion products. It is considered that globalization has contributed to further devalue craft work, increasing global competition and allowing brands/companies from countries with more advanced economies to profit from low labor costs and insufficient labor legislation in disadvantaged countries (Condé Nast, 2022).

The knowledge and activities inherent in traditional, heritage handicraft practices are under the protection of the 2003 UNESCO Convention for the Safeguarding of the Intangible Cultural Heritage.

Traditional crafts are understood to be the skills and products derived from cultural traditions characteristic of a particular region and/or community, handed down from generation to generation as part of the local cultural territory. The 2003 Convention, mentioned above, warned about the risk that the processes of globalization and social transformation, along with the conditions they create for a renewed dialogue between communities, also bring with them, like the phenomena of intolerance, serious threats of degradation, disappearance and destruction of intangible cultural heritage, due in particular to the lack of means to safeguard it (UNESCO, 2003).

The study by The Heritage Crafts Association (2019) identified aging master craftsmen, lack of apprenticeship structures, and market competition with low-priced products as the main factors putting cultural crafts at risk worldwide.

In response to the impact of globalization, there is an increasing reliance on localism. Localism emphasizes the value of geographical proximity and the need to strengthen regional traditions in order to help local communities and natural systems endure and flourish in the face of profit-driven industries (O'Riordan, 2001). In the context of fashion, localism contrasts with the global fashion system, becoming a decentralized response to fast fashion, bringing designers, producers, and consumers closer together, and generating richer ways of understanding materials and the conditions of production (Fletcher & Grimstad Klepp, 2018).

Post-digital Aesthetics

The post-digital concept has been popularizing in the last decade, promoted through its own artistic production, and simultaneously it has been object of questioning, and there is no consensus among authors regarding the definition of the term or the relevance of its use. We agree with the definition of post-digital advocated by Florian Cramer (2013), Cramer and Jandrić (2021), Hans Obrist (2015), Jonathan Openshaw (2015), and Lucia Santaella (2018).

Post-digital initially appeared related to critical reflection about the concept of materiality of digital aesthetics, but it is not limited to the question of rematerialization. In post-digital, the boundaries between "old" and "new" media are blurred, and they are combined and re-signified under new critical looks. The digital logic and aesthetics are joined by touch and tangible materials that incorporate the digital continuum.

In post-digital artifacts, creativity overlapped technology, emerging a will to blur the boundaries between the physical and the digital, even though commonly maintaining a digital/computational aesthetic. The manual and the handcrafted are allied to logic and digital languages, adopting aesthetic discourses with their own characteristics that cannot be reduced to the epithet of hybrids (Pereira, 2018).

In fact, digital media have changed social interaction, but they do not replace, in their entirety, physical and immersive experiences. Now, post-digital rematerialization comes to allow the appeal to all senses. Beyond the exploration of sensations, the post-digital brings a field of creation and research oriented to the design of emotions. The material side of media objects and digital art, adopting (re)materialized perspectives, takes us back to Erkki Huhtamo's theories (2008) and to the capacity of these objects to arouse the desire of interaction and, essentially, to stimulate through the look the will to touch, but also to provoke desire in its three emotional dimensions: visceral, behavioral, and reflective (Norman, 2004).

In the art-fashion artifacts, and more broadly in textile art, the physical material tends to be the central element and the imagery created assumes an enormous potential precursor of development in areas of visual, digital, and interactive arts, and especially post-digital. These artifacts take on post-digital characteristics when they explore rematerialization aesthetics supported by digital technology (Pereira & Fernandes-Marcos 2020).

Project Description

The *Sliiv* is an ongoing project studying the use of recent electronics and e-textiles in the design of wearable interfaces for artistic expression and performance. The Sliiv comprises original hardware and software and is planned as a development platform that can be adapted to different configurations, shapes, and functions.

The main component is a wearable sleeve which can be used on the left arm and uses touch capacitive sensors as the main input for controlling a set of predefined parameters. The form factor of the Sliiv controller was decided for ergonomics and usability considerations, as well as for the versatility in the creation of different layouts and motives on the top layer. The sleeve wraps around the arm, closing with velcro on the lower end, and can be used directly on the arm or above other clothes.

The sleeve contains the sensors and other electronic components and communicates wirelessly over wi-fi to a computer running the custom software *Sliiv engine*, implemented in the MaxMSP programming environment (Fig. 1). The electronic

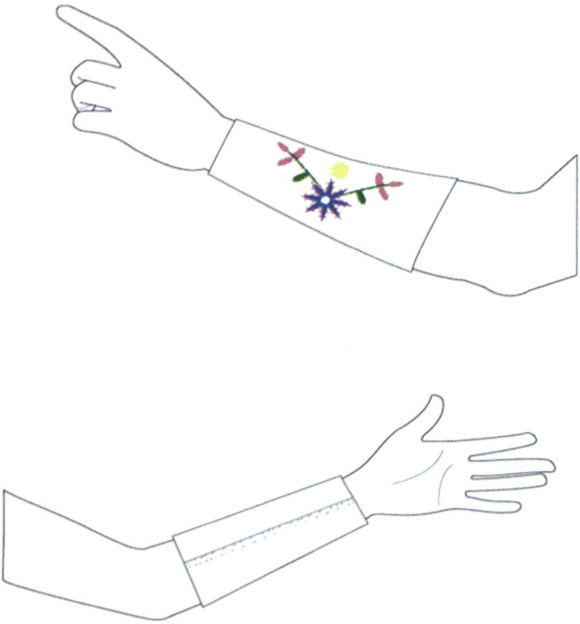

Fig. 1 Overall structure of the system

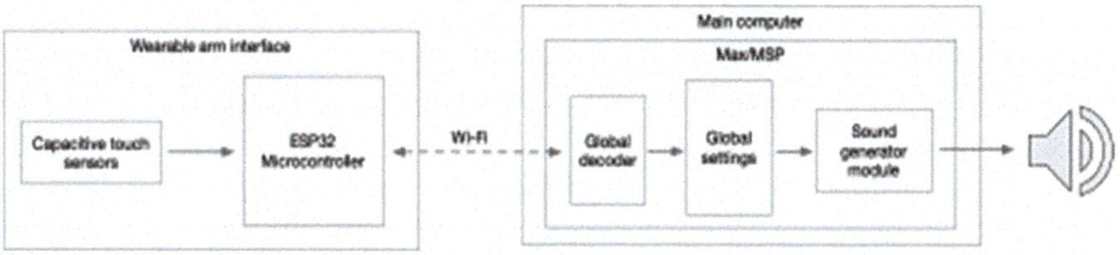

Fig. 2 The Sliiv arm controller

Fig. 3 Stylized motives of traditional embroidery of Monchique

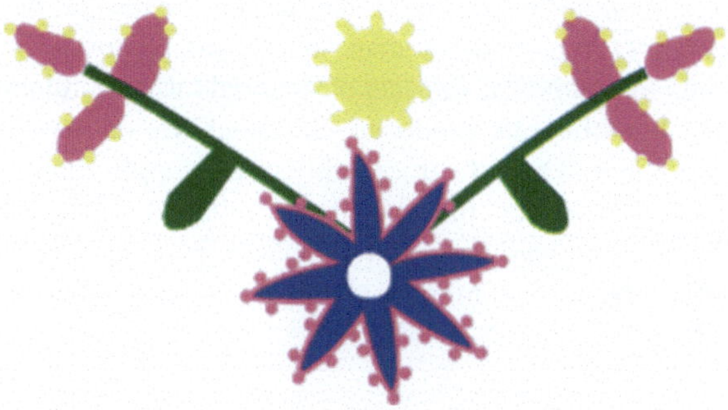

components comprise an ESP32 Espressif microcontroller development board and a battery, which are located in a small pouch in the lower side near the velcro, and the sensors, on the upper part, sewed to the decorative motifs on the top layer of the sleeve. The sensors are capacitive touch sensors made of conductive string, ribbons, and fabric, which are shaped to the motives on the top layer. The software receives and decodes the data coming from the Sliiv controller, mapping the data into parameters of the sound engine (Dias, 2022; Pereira & Dias, 2022).

The exterior of the prototype (see Fig. 2) is made of linen fabric, from the traditional weaving of Monchique. The embroidery (see Fig. 3) is done manually, with colored threads of industrial wool, with a wide stitch and outlined with a backstitch.

The interior is made of felt, to give stability to the sleeve and to support the cables and the microcontroller. The lining in contact with the arm is cotton fabric for comfort and easy care. The sleeve closure is velcroed to allow for easy attachment and adjustment to various arm widths. The hems are made with the spike stitch, made of wool in various colors, which in addition to highlighting the traditional language of the prototype make it more resistant.

Description of the Materials

Following Catherine Donnor's line of thought in *A Philosophy of Textile* (2020), the interpretation of the terms textile can be ambiguous, the textile can be seen as a material, a concept, a language, or a metaphor. Reflecting on the relationship between theory and artistic practice implies simultaneously considering the materiality and expressiveness of the textile as matter, but also the language elicited by the textile. Textiles are often associated with the feminine, with domestic and craft concerns, which often hides the richness and plasticity of the materials.

For the artist, it is essential to know the materials through handling, through doing, in order to perceive the potentialities of the fabric/fiber and know how to choose the most adequate to the project—this knowledge, besides being theoretical, is obtained through touch.

Natural materials fall into two broad categories: vegetable- or cellulose-based (e.g., cotton, hemp, linen) and animal- or protein-based (e.g., wool, silk, leather, feathers). While natural materials are often considered to be sustainable, the social and environmental impacts of materials vary in relation to where they come from and how they are produced. Sustainability of materials is complex and context-dependent, where the traceability of the supply chain, the selection of the appropriate material for the intended applicability in the manufacturing processes, distribution, washing and maintenance practices during use, but also at the end-of-life of the product must be taken into account (Fletcher & Tham, 2016).

Tradition Linen Weaving in Monchique, Algarve, Portugal

Traditional Algarve textiles do not currently have a high visibility. However, these textile traditions come from very ancient times and have their own characteristics that distinguish them from others: the cultivation and production of textile fibers, the technology employed, the techniques used, and the ways of decorating the pieces and their purposes.

Traditional weaving of linen, wool, and rags in the Algarve still follows the same processes and techniques that were used centuries ago in Al-Andalusia. Despite this, it is an area that is still little discussed and researched.

We understand traditional weaving as an activity with centuries-old traditions, carried out with raw materials grown or acquired in the region, using instruments and craft technologies characteristic of each municipality or district, manufactured, wholly or partially, in the region, and which can be purchased as loose pieces or as complementary accessories. We exclude from traditional weaving the use of technologies other than traditional ones, such as mechanical spinning wheels or mass-produced horizontal looms.

Until a few years ago, the linen cycle was done from North to South of Portugal. Currently, this process is falling out of use, the plantations have been lost, and those involved have aged and abandoned activities or sought other more lucrative professions.

Indigenous Portuguese linen comes in three varieties: Galician, Moorish, and national Riga. These varieties can be of spring/summer or autumn/winter culture.

Moorish linen is the variety grown in warmer climates. We find this variety south of the Tagus river and in the districts of Bragança, Santarém, and Castelo Branco.

This species has a long, dark, and coarse fiber. In the south of the country, we find the sub-varieties abertiço and serrano. Both linen and its sub-variants are autumn crops, sown after the first rains, in October and November, and harvested in May. They adapt to clayey, compact, or poor soils and do not have properties that are suitable for the production of fibers for industrial purposes (Oliveira, 1978).

In the Alentejo and Algarve, linen is grown on fresh, agricultural land that is already reasonably manured from previous crops. In October and November, the seed is sown, and the seed is thrown into the ground. This variety is the least demanding and does not require irrigation. Also in May and June, the linen is subject to tanning or watering. Traditionally, during the months of August to October, the flax is transformed into fiber through dulling, crimping, and staking. The fiber is transformed into yarn by hand spinning. In March, the fiber is spun, bleached, and folded. Between March and April, the linen is prepared for weaving, and still in April it is woven. Between April and May, the fabric is bleached.

Traditional textile production has managed to survive in southern Portugal by the stubbornness and persistence of some weavers. Currently, there have been initiatives and academic, municipal, and regional projects, praiseworthy, to publicize the regional textile heritage. However, it is necessary to educate a public to value, recognize, and be willing to pay the real value of handcrafted products, without belittling their labor.

The activities involved in the linen and wool cycle are disappearing, and why? Rural workers have aged, others have sought activities that would allow them to lead more fulfilling lives. Young people have become disconnected from the rural worlds, left the fields, studied, and do not apply these traditional techniques.

Traditional weaving worked, in this region, as a complement to the agropastoral activity, not occupying the artisans full-time, and in this way, this activity fit into the complex system of the rural economy (Pereira, 2016).

Decoration of the Algarve's Traditional Weavings

The most characteristic products produced by D. Maria Nunes—the last weaver in the municipality of Monchique still following the traditional weaving methods, and about whom we have written a long study, previously published—are the linen and rag weavings. Today, she sells mainly utilitarian objects, such as table cloths, kitchen cloths, and rugs. In sales the rag weavings stand out, carpets and rugs made of cotton strips under a cotton warp. Although these are the products that sell the most at the fairs, in the sales one feels the weight of competition with the industrial rag rugs, sold in large commercial stores produced in series and at lower prices, and that lead some consumers to be mistaken about the production and quality.

The linen weavings trimmed with handmade lace attract national and tourist lovers, but the price difference leads many to opt for the rag weavings. The typical "alforges" (saddlebags) and "alforginhos" maintain formal characteristics with traditional dimensions, colors, patterns, and decoration, however the fibers used in

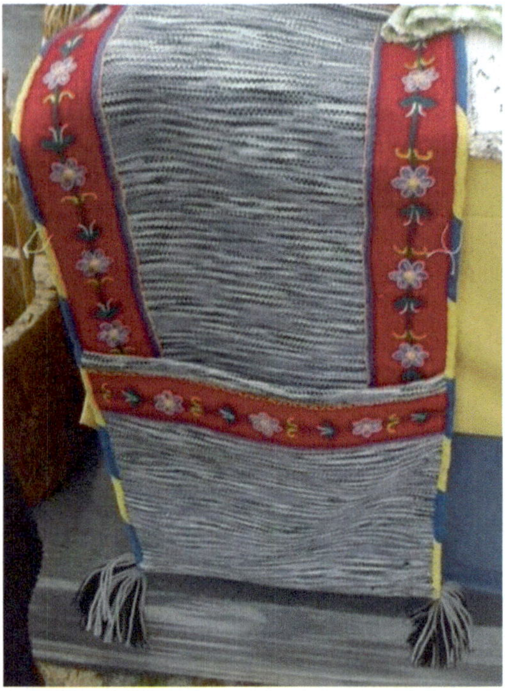

Fig. 4 Alforginho produced by the weaver D. Maria Nunes, Monchique, Algarve, Portugal. (Fonte: Pereira, 2012)

Monchique's "alforges" are industrially manufactured, sometimes with a mixture of synthetic fibers, purchased in local haberdashery. These pieces require a longer production time, greater investment, and knowledge of weaving, embroidery, and sewing techniques. It is important to emphasize that in the Algarve today, there are few people who know how to make the characteristic spiked hem, typical of the rustic garments of the Algarve region. The spiked hem is more durable and gives greater stability and security to the fabric, in addition to its decorative value (Fig. 4). The price and lack of functionality of the "alforges" and "alforginhos" nowadays means that they are rarely sold. D. Maria produces in limited numbers the saddlebags for exhibition in the stands where she sells, and only produces again after sale or by order (Pereira, 2012).

The "alforginhos" (see Fig. 5) are pieces used to transport products on horses—similar to "alforges" but smaller in size. Traditionally, they were made of woolen cloth or rags. It consists of a strip of cloth with the ends are folded and joined, with a spiked hem, so as to form two bags.

Analysis of the Project in Light of Post-digital Aesthetics

Like digital media artifacts, post-digital textile artifacts aim to provide an enriching experience of sensory, emotional, and intellectual pleasure while exploiting technological innovations to create new aesthetic discourses. Post-digital artifacts need not

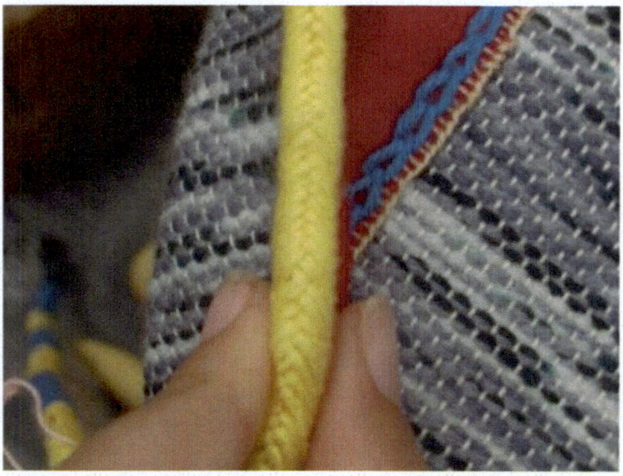

Fig. 5 Cob (cob hem): Traditional decorative hem for *alforges* and *alforginhos*. Made with lace thread forming V-shaped stitches. (Source: Pereira, 2012)

be produced exclusively by computational/digital materials. In fact, these post-digital textile art artifacts are mostly hybrids (tangible-offline and virtual-online) (re)materialized in various media thus adapting to exhibitions in tangible spaces and to exhibitions, increasingly in vogue, online (Fernandes-Marcos & Pereira, 2022).

If we analyze the artifact *Sliiv*, presented in this chapter, we verify the post-digital dimensions, such as sensory experience, ephemerality, performativity, immersion, spectator-participant-user, haptic ability, and appeal to memories and traditions.

In this artifact, traditional textiles are joined by digital technologies and physical computing, in a collaborative prototype for musical expression, promoting social interaction and the safeguarding of traditional knowledge.

Final Considerations

In the present form, the Sliiv is a music and sound controller, planned to be very simple and intuitive to use while providing a considerable level of fine control over the sound parameters. However, being a controller, the Sliiv can be used to control any type of media other than sound. Future implementations may explore the control over real-time graphics or video, whether simultaneously with sound, or collaboratively with other Sliiv interfaces.

In this artifact, we use traditional textiles, handmade with techniques and resources of artisan technologies that are being forgotten, highlighting the importance of cultural sustainability. In addition, the embroidery used in traditional pieces, currently without much practical use, gives them a new dimension through the integration of the interface of musical expression, which must be discovered through touch, bringing it closer to contemporary art through post-digital aesthetics.

References

Carvalho, A., Pereira, S., & Fernandes-Marcos, A. (2017). An expanded field in media digital art: Proposal of an aesthetics in the interconnection of media art, fashion and theatre. *ARTECH2017: Proceedings of the 8th international conference on digital arts*. Macau: Artech International, https://doi.org/10.1145/3106548.3106621.

Condé Nast. (2022). *The sustainable fashion glossary*. https://www.condenast.com/glossary/glossary-introduction

Cramer, F. (2013). *Anti-media: Ephemera on speculative arts*. NaiOlO Publishers and Institute of Network Cultures.

Cramer, F., & Jandrić, P. (2021). Postdigital: A term that sucks but is useful. *Postdigital Science and Education, 3*, 966–989. https://doi.org/10.1007/s42438-021-00225-9

Dias, R. (2022). Sliiv – Prototype design for a wearable musical interface. *ARTeFACTo Macao 2022, 3rd international conference on digital creation in arts*. Media and Technology Emerging Realities. 24–25 November 2022.

Fernandes-Marcos, A., & Pereira, S. (2022). Post-digital textile aesthetics: a critical reflection on the process of digital re-materialisation of artefacts. *ARTeFACTo Macao 2022, 3rd international conference on digital creation in arts*. Media and Technology Emerging Realities. 24–25 November 2022.

Fletcher, K., & Grimstad Klepp, I. (2018). A note from the editors of fashion practice. *Fashion Practice, 10*(2), 133–138.

Fletcher, K., & Tham, M. (Eds.). (2016). *Routledge handbook of sustainability and fashion*. Routledge.

Huhtamo, E. (2008). Tactile temptations: About contemporary art, exhibitions and tactility. In C. Sommerer, L. Mignonneau, & D. King (Eds.), *Interface cultures. Artistic aspects of interaction* (pp. 129–139). Transcript Publishers.

Norman, D. (2004). *Design emocional*. Rocco Edições.

O'Riordan, T. (2001). *Globalism, localism and identity: Fresh perspectives on the transition to sustainability*. Earthscan.

Obrist, H. U. (2015). Curation in the Postdigital age. In *Postdigital artisans. Craftmanship with a new aesthetic in fashion, art, design and architecture*. Frame Publishers.

Openshaw, J. (2015). *Postdigital artisans. Craftmanship with a new aesthetic in fashion, art, design and architecture*. Frame Publishers.

Pereira, S. (2012). *A última tecedeira da Serra de Monchique, Dissertação de mest., História do Algarve, Faculdade de Ciências Humanas e Sociais*. Univ. do Algarve.

Pereira, S. (2018). A moda na era do pós-digital. In *Tese de doutoramento*. Universidade Aberta.

Pereira, S., & Dias, R. (2022). An interactive artifact promoting cultural sustainability – Case study of the Sliiv project. *FL Conference – Fashion & Sustainability*, 3–5 November, 2022, Universidade Lusófona.

Pereira, S., & Fernandes-Marcos, A. (2020). O processo criativo na era pós-digital – uma reflexão crítica baseada na prática artística. *In Proceedings of the 2nd international conference on digital creation in arts and communication, ARTeFACTo 2020* (pp. 127–135). Bruno Mendes da Silva (Ed.). Nov. 26–27 2020, Faro, Portugal. http://hdl.handle.net/10400.2/10403

Santaella, L. (2018). Arte, ciência & tecnologia: um campo em expansão. Pablo Gabira, Percursos contemporâneos. Realidades da arte ciência e tecnologia, Belo Horizonte: UEMG.

Index

A. Cruchinho, J. C. Neves (eds.), *Sustainability in Fashion: Creativity, Education, and Innovation*, https://doi.org/10.1007/978-3-031-80991-0

The manufacturer's authorised representative in the EU is Springer Nature Customer Service Centre GmbH, Europaplatz 3, 69115 Heidelberg, Germany. If you have any concerns regarding our products, please contact ProductSafety@springernature.com

Printed and bound by CPI Group (UK) Ltd, Croydon, CR0 4YY
01/10/2026
02225962-0013